CRC SERIES IN ENZYME BIOLOGY

Editor-in-Chief
John R. Sabine, Ph.D.

3-HYDROXY-3-METHYLGLUTARYL COENZYME A REDUCTASE

Editor
John R. Sabine, Ph.D.
Department of Animal Sciences
Waite Agricultural Research Institute
University of Adelaide
Adelaide, Australia

PYRUVATE CARBOXYLASE

Editors
J. C. Wallace, Ph.D.
D. B. Keech, Ph.D.
Department of Biochemistry
University of Adelaide
Adelaide, Australia

CHOLESTEROL 7α-HYDROXYLASE (7α-MONOOXYGENASE)

Editors
Robin Fears, Ph.D.
Biosciences Research Center
Beecham Pharmaceuticals Research Division
Epsom, United Kingdom
John R. Sabine, Ph.D.
Department of Animal Sciences
Waite Agricultural Research Institute
University of Adelaide
Adelaide, Australia

Cholesterol 7α-Hydroxylase (7α-Monooxygenase)

Editors
Robin Fears, Ph.D.
Biochemist
Biosciences Research Center
Beecham Pharmaceuticals Research Division
Epsom, Surrey
United Kingdom

John R. Sabine, Ph.D.
Reader in Animal Physiology
Department of Animal Sciences
Waite Agricultural Research Institute
University of Adelaide
Adelaide, South Australia
Australia

CRC Series in Enzyme Biology

Series Editor-in-Chief
John R. Sabine, Ph.D.

CRC Press, Inc.
Boca Raton, Florida

Library of Congress Cataloging-in-Publication Data
Main entry under title:

Cholesterol 7 [alpha]-hydroxylase (7 [alpha]-
monooxygenase).

(CRC series in enzyme biology)
Includes bibliographies and index.
1. Cholesterol hydroxylase. 2. Bile acids--
Metabolism. 3. Cholesterol--Metabolism. I. Fears,
Robin. II. Sabine, John R. III. Title: Cholesterol
seven [alpha]-hydroxylase (seven [alpha]-monooxygenase)
IV. Series. [DNLM: 1. Steroid Hydroxylases--physiology.
QU 140 C547]
QP603.C45C47 1986 612'.01513 86-2298
ISBN 0-8493-6553-8

Direct all inquiries to CRC Press, Inc., 2000 Corporate Blvd., N.W., Boca Raton, Florida, 33431.

International Standard Book Number 0-8493-6553-8
Library of Congress Card Number 86-2298
Printed in the United States

SERIES PREFACE

The "CRC Series on Enzyme Biology" is a series of books, each one devoted to a single enzyme and each one endeavoring to draw together in a comprehensive and systematic manner all that is currently known about the chemistry, biochemistry, and physiology of that particular enzyme. Each volume, written or edited by one or more international enzyme specialists, draws together the latest information available on the occurrence, structure, function, role, and control of the biologically more important enzymes.

Each chapter or section of each volume is not written primarily for those specializing in that narrow area (innumerable specialist reviews do this), but rather to provide a coherent and integrated summary of that aspect for those working on other, quite different facets of the same enzyme. In this way several distinct classes of scientist, and student, will benefit from each volume, namely, those concerned with "type" situations of which that enzyme is an example, those working on the metabolic systems of which that enzyme is a key component, and those concerned with diseases in which the enzyme has an important role.

SERIES EDITOR

John R. Sabine, M. Agr. Sc., Ph.D. is a Reader in Animal Physiology at the Waite Agricultural Research Institute of the University of Adelaide, Adelaide, Australia. Dr. Sabine obtained his Bachelor's and Master's degree in Agricultural Science from the University of Melbourne and then his Ph.D. in Animal Nutrition in the laboratory of Dr. B. Connor Johnson at the University of Illinois, Urbana. After several research appointments — Monash University, Australian National University, and the University of California, Berkeley, he was appointed to the faculty of University of Adelaide in 1967. At various times since then he has held Visiting Professorships at Brandeis University (Graduate Department of Biochemistry), the University of Stockholm (Wenner-Gren Institute), the University of Oklahoma Health Sciences Center (Department of Biochemistry and Molecular Biology), and the University of Kuwait (Department of Biochemistry), as well as Visiting Scholar appointments at Oxford University (Department of Clinical Medicine) and Harvard University (Chemistry Department).

Dr. Sabine is a member of several scientific societies, including particularly the Australian Biochemical Society, the Nutrition Society of Australia, and the Australian Physiological and Pharmacological Society. He has presented his research findings at various national and international meetings, and has been chairman of a number of sessions at these meetings. He was co-convenor of, and leader of the Australian delegation to, the U.S./Japan/Australia Cancer Conference (Hawaii, 1975) and the convenor and chairman of the unique international symposium "Lipids in Cancer", which was held on board the Indian-Pacific Express as a satellite meeting to the 12th International Congress of Biochemistry, 1982. In 1979 he delivered the 12th Patricia Chomley Oration to the Australian College of Nursing.

Dr. Sabine has published some 50 research articles and 14 invited reviews. He is on the Editorial Board for *Nutrition and Cancer*, and his book *Cholesterol* was the first comprehensive coverage of this important field to appear in 20 years. His major research interests revolve around cholesterol physiology, with particular reference to the control of its synthesis and to its role in membrane structure and function, and with emphasis upon the role of cholesterol in the etiology of cancer. He has further research interests in such diverse fields as the role of earthworms in biological resource recovery, the physiology of goats for meat and fiber production and the interaction between scientists and society.

THE EDITOR

Robin Fears, B.Sc., Ph.D. is a biochemist at the Biosciences Research Centre of Beecham Pharmaceuticals Research Division, Epsom, Surrey, U.K. Dr. Fears obtained his bachelor's degree in Biochemistry from the University of Bristol in 1969 and his Ph.D. from Queen Elizabeth College, London, in 1973. The subject of his thesis was the nutritional and biochemical aspects of protein and amino acid metabolism during pregnancy. He joined Beecham Pharmaceuticals in 1973 and has worked on the identification and development of novel pharmacological agents in the Occlusive Vascular Disease and Thrombosis projects.

Dr. Fears is a member of several scientific societies and has presented his research findings at various national and international meetings as well as publishing some 60 scientific papers. His principal research interests include the biochemistry of fibrinolysis, the pathology of thrombosis, and the regulation of lipid and lipoprotein metabolism in health and disease with particular reference to atherosclerosis.

CONTRIBUTORS

Bo Angelin, M.D., Ph.D.
Associate Professor
Department of Medicine
Huddinge University Hospital
Karolinska Institutet
Huddinge, Sweden

Ashok K. Batta, Ph.D.
Associate Professor
Department of Medicine
University of Medicine and Dentistry of New Jersey
Newark, New Jersey

Kathleen M. Botham, Ph.D.
Lecturer
Department of Physiology, Biochemistry, and Pharmacology
The Royal Veterinary College
London, United Kingdom

Ingemar Björkhem, M.D., Ph.D.
Associate Professor
Department of Clinical Chemistry
Huddinge University Hospital
Karolinska Institutet
Huddinge, Sweden

Henry Danielsson, M.D., Ph.D.
Professor
Department of Pharmaceutical Biochemistry
University of Uppsala
Uppsala, Sweden

Kurt Einarsson, M.D., Ph.D.
Associate Professor
Department of Medicine
Huddinge University Hospital
Karolinska Institutet
Huddinge, Sweden

Emil Ginter, Ph.D., C.Sc.
Senior Scientist
Department of Biochemistry
Institute of Nutrition
Bratislava, Czechoslovakia

Ronnie Hansson, Ph.D.
Department of Pharmaceutical Biochemistry
University of Uppsala
Uppsala, Sweden

I. R. Hardie, M.D., MBBS(Syd), FRACS
Associate Professor of Surgery
University of Queensland
Princess Alexandra Hospital
Wooloongabba, Queensland
Australia

Douglas M. Heuman, M.D.
Assistant Professor of Medicine
Medical College of Virginia
Richmond, Virginia

Phillip B. Hylemon, Ph.D.
Professor of Microbiology and Medicine
Department of Microbiology and Immunology
Medical College of Virginia
Richmond, Virginia

William M. Kubaska
Department of Microbiology
Medical College of Virginia
Richmond, Virginia

C. T. Kwok, Ph.D.
Senior Research Officer
Department of Surgery
University of Queensland
Royal Brisbane Hospital
Brisbane, Queensland,
Australia

Christopher J. Packard, Ph.D.
Principal Biochemist
Department of Biochemistry
Royal Infirmary
Glasgow, United Kingdom

S. P. Pillay, M. B. Ch.B (Cape Town), FRCS (Ed)
Senior Lecturer in Surgery
Department of Surgery
University of Queensland
Royal Brisbane Hospital
Brisbane, Queensland, Australia

W. R. Rush, Ph.D.
Staff Researcher
Department of Metabolism
Syntex Research Centre
Research Park, Riccarton
Edinburgh, United Kingdom

Gerald Salen, M.D.
Professor
Department of Medicine
Veterans Administration Hospital
East Orange, New Jersey

Sarah Shefer, Ph.D.
Professor
Department of Medicine
University of Medicine and Dentistry of New Jersey
Newark, New Jersey

James Shepherd, M.B.Ch.B., Ph.D.
Reader in Pathological Biochemistry
Royal Infirmary
Glasgow, United Kingdom

M. T. R. Subbiah, Ph.D.
Professor of Experimental Medicine
Department of Internal Medicine
University of Cincinnati Medical Center
Cincinnati, Ohio

Z. Reno Vlahcevic, M.D.
Professor of Medicine
Chairman, Division of Gastroenterology
Department of Medicine
Medical College of Virginia
Richmond, Virginia

Kjell Wikvall, M.D., Ph.D.
Assistant Professor
Department of Pharmaceutical Biochemistry
University of Uppsala
Uppsala, Sweden

TABLE OF CONTENTS

General Features, Distribution, and Role

Chapter 1

CHOLESTEROL 7α-HYDROXYLASE: AN OVERVIEW

John R. Sabine

TABLE OF CONTENTS

I. FUNCTION

Any enzyme has two functions — the formation of a product and the removal of a substrate. The microsomal enzyme cholesterol 7α-hydroxylase (EC 1.14.13.17)* removes cholesterol and produces 7α-hydroxycholesterol (Figure 1).

II. IMPORTANCE

The importance of any enzyme, i.e., importance both to the cell and to the organism harboring that enzyme, likewise depends upon the relative significance of one or both of these primary functions. As far as is known at present neither cells nor organisms have any specific use for 7-hydroxycholesterol per se. This compound is, however, invaluable to all mammalian species in that it is the first step on the metabolic pathway from cholesterol to bile salts, which in turn are critical factors in mammalian digestion. Indeed, as the enzyme responsible for this step, cholesterol 7α-hydroxylase is now commonly conceded as being the rate-limiting enzyme for the entire pathway, and thus the enzyme critically determining the rate of formation of bile salts.[1] The relevance of the study of this enzyme to our understanding of such important disorders as gallstone formation and lipid malabsorption is thus obvious (compare Chapter 13).

On the other hand, while the formation of bile salts is undoubtedly important, it may well be that the more important role for cholesterol 7α-hydroxylase is the removal of cholesterol — from cells, from the blood stream, and ultimately from the body. Bile salts are by far the major excretory products of cholesterol catabolism, and losses by this route approximately equal direct cholesterol excretion, at least for the human.[2]

Currently it is probably impossible to determine in any specific situation which of these two metabolic roles, i.e., anabolic or catabolic, is the more important. The production of 7α-hydroxycholesterol, and thus ultimately of bile salts, is clearly important in the liver, from whence these salts can be passed by way of the bile into the intestine. But what role does the enzyme play in nonhepatic tissues, such as leukocytes,[3] or adipose tissue?[4] In these situations regulation of cholesterol levels may be important, but little work so far has appeared on the fate of the hydroxylated sterol product in these extrahepatic sites.

Any consideration of the importance of this enzyme must recognize the validity of yet another question. C7H can hydroxylate cholesterol, but does it — in vivo, *in situ*? Depending heavily upon isolation and assay conditions, the tissue preparation used to measure cholesterol 7α-hydroxylation in vitro will also hydroxylate a selection of other compounds, nonsterols as well as sterols. This question is covered in considerable depth later in this text and has significant implication in any discussion of control mechanisms (see below).

III. ISOLATION AND ASSAY

Another consideration that impinges heavily upon the measurement of the activity of this enzyme, and thus upon our understanding of its role and control, is the major variation possible in the actual numbers obtained because of differences in isolation and assay procedures.

Two major factors contribute to this variability. First, it appears that under normal conditions in vivo at least four proteins contribute to, and are even essential for, the enzyme's activity — the enzyme itself, cytochrome P-450,[5,6] NADPH - cytochrome P-450 reductase[5,6] and a cytosolic protein.[7,8] Different isolation procedures — from crude microsomes to purified preparations — will affect the proportions and the activities of these various components. Second, what is the size, location, and accessibility of the substrate pool? Is the entire

* Abbreviated for this paper as C7H.

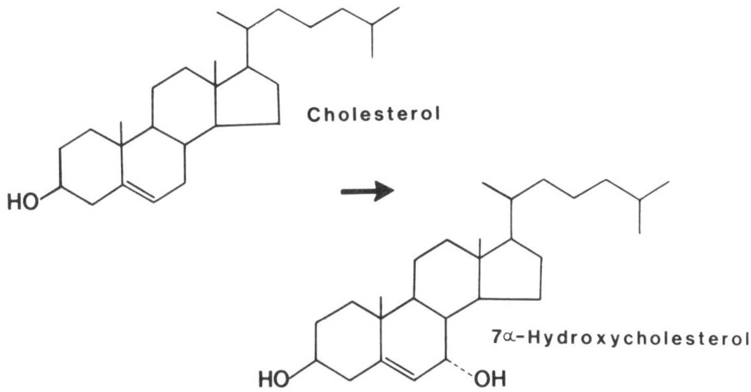

FIGURE 1. The enzymatic step catalyzed by cholesterol 7α-hydroxylase (EC 1.14.13.17).

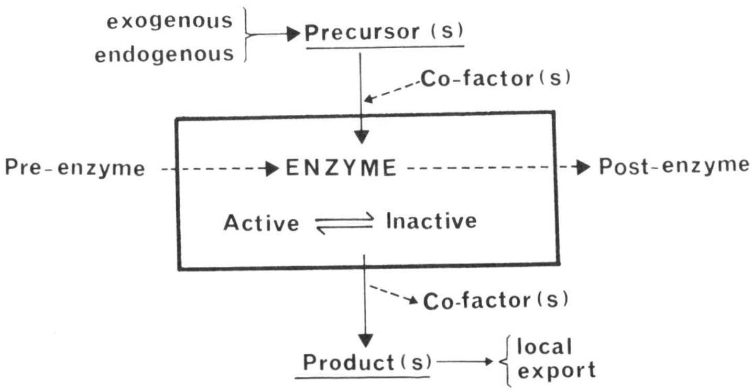

FIGURE 2. Schematic representation of possible sites for control of enzyme activity.

microsomal cholesterol content available as substrate, or only some of it?[1] Is the substrate pool derived only, predominantly, partially or not at all from that synthesized within the cell, or does cholesterol derived from outside the cell also play a role? And to what extent is the exogenous cholesterol that is added, in various ways, to an assay mixture miscible with the endogenous substrate pool already present? Obviously the answers to these questions will influence greatly not only the measured values we obtain in a specific assay system, but also our whole understanding of the physiological function of the enzyme and of the regulation of that function.

These and other questions relative to enzyme assay are covered more completely in Chapter 4.

IV. CONTROL

All the possibilities for mechanisms to control an enzyme, at the level of the enzyme itself, can be depicted schematically as in Figure 2.[9] Basically, there are five interactive control systems possible:

1. Rates of supply of substrate and of removal of product
2. The presence or absence of necessary cofactors and of any endogenous or exogenous activators or inhibitors

3. The relative rates of enzyme synthesis and degradation
4. Conversion between active and inactive forms of the enzyme protein
5. For membrane-bound enzymes, the fluid status of the membrane lipids

We must recognize that any attempt at control — either devised by nature or proposed by science — might operate through one or more of these mechanisms. Moreover, any apparent lack of control, as indicated by lack of change in measured enzyme activity, does not necessarily mean that these control features are not operating. They may cancel each other out.

As far as is currently known, there are two major controls regulating the activity of hepatic cholesterol 7α-hydroxylase in vivo. The first of these is the rate of return of bile salts to the liver.[1] In other words, there is a form of negative feedback in operation, such that bile salts absorbed from the intestine and transported back to the liver in portal blood inhibit the activity of the enzyme. The mechanism of this inhibition is not known. Likewise unknown is the mechanism of action of, and indeed the relative extent of, any feedforward stimulus to enzyme activity, i.e., the extent to which the supply of substrate, i.e., cholesterol, stimulates hydroxylase activity.

Superimposed on this continual feedback modulation by bile salts is a second control feature — a diurnal rhythm.[10-13] The activity of the enzyme is higher, under some conditions manyfold higher, at night than during the day. The precise mechanism by which this rhythm occurs is not known, nor is its physiological significance. Diurnal variation in the rate of synthesis and/or degradation of enzyme protein is an important, but not the only, factor.[10] Corticosteroid levels in the blood are highly correlated with the diurnal variation in enzyme activity,[10,12,13] but other regulatory stimuli cannot be excluded.

There is, perhaps, yet another way this whole question of physiological regulation of enzyme activity should be approached. Generally we tend to look for *change* in enzyme activity, and then equate that with some regulatory phenomena. It is possible, however, that constancy, i.e., no change at all, in enzyme activity might indeed represent a more tightly controlled situation. This potential dilemma highlights also another control question that is rarely asked, let alone answered, namely the question of priorities. With cholesterol 7α-hydroxylase, as with probably most enzymes, we do not know just what aspect(s) of the enzyme's function the cell and/or the body regards as critical. Is the rate of bile salt production critical? Is cholesterol removal critical? And how critical, compared with other possible priorities? The importance we assign to any particular "control" phenomena we observe will be greatly influenced by our answers, perceived or implied, to these questions.

A good example of this problem is shown by the relationship between cholesterol 7α-hydroxylase activity and vitamin C status. Given that considerable early work had suggested that the activity of the enzyme was seriously depressed by vitamin C deficiency, it is now clear[14] that enzyme active is low in *both* deficiency and excess. Too much vitamin C is just as inhibitory as too little!

V. CHOLESTEROL 7α-HYDROXYLASE AND PLASMA CHOLESTEROL

Given the possibility, as mentioned above, that a major role for cholesterol 7α-hydroxylase might be the removal of cholesterol, it is thus reasonable to ask what is the relationship between the activity of this enzyme and the size of the body's cholesterol pool. Indeed, one of the most frequent reasons why C7H activity is measured at all is in an attempt to explain why plasma cholesterol has gone up or down as a result of some experimental maneuver.

While the hypothesis for a significant removal role for the enzyme may hold in some circumstances, the cumulative evidence is far from convincing. An extensive, though not necessarily complete, survey of the literature since 1977 reveals that in a wide variety of experimental conditions — designed to examine the effects of various diets, drugs and diseases on animals of various species, strains, ages and sex — plasma cholesterol levels were negatively correlated with hepatic cholesterol 7α-hydroxylase activity (i.e., in support of the hypothesis) in 17 situations, but positively correlated in 24 and not correlated at all in 25!

The other major enzyme that has been studied in order to attempt to explain a change in the level of plasma cholesterol is 3-hydroxy-3-methylglutaryl coenzyme A reductase (HMGR), which catalyzes the primary rate-controlling step in the biosynthesis of cholesterol.[15] But in those same papers, when HMGR was also measured it was negatively correlated with C7H (and thus directly correlated with plasma cholesterol levels) in only 6 situations, but positively correlated in 29 and not correlated in 12.

As I see it the major stumbling block to our understanding of this complex issue is that we do not know, under any specific circumstances, what determines *in physiological terms* the level of cholesterol in the plasma. Obviously any change from one level to another must involve a temporary imbalance between input and output, but what determines the equilibrium level — and maintains it at equilibrium — is quite unknown. My guess[2] is that there is some form of "cholesterostat" mechanism, akin perhaps to the insulin/glucagon regulation of glucose levels, but this has not been demonstrated. Yet until we know at least some details of the homeostatic mechanism regulating plasma cholesterol level, we have no starting point for analyzing the effects of drugs, diets, or anything else, enzymes included.

This whole controversy is exemplified by studies of day/night variability. In every species studied there is a considerable diurnal variation in the measured activities of both cholesterol 7α-hydroxylase and 3-hydroxy-3-methylglutaryl coenzyme A reductase — they are both higher at night — yet there is no comparable variation in the level of plasma cholesterol. Moreover, the regular daily changes in the activities of these enzymes are not necessarily in phase[11] and the magnitude of the variations can be differentially affected by a wide range of factors, again all without necessarily influencing the level of plasma cholesterol.

Chapters 9 and 11 cover this area in considerably greater detail.

VI. THE APPROPRIATE ANIMAL MODEL

As I have already mentioned the study of cholesterol 7α-hydroxylase derives most of its impetus from two areas of biomedical concern — the synthesis of bile salts in normal and abnormal conditions, and the dynamics of cholesterol physiology. Thus a question of major importance is the appropriateness or otherwise of the various animal models that are used for these studies.

Obviously the only perfect model for human studies is the human, but both ethics and the law prevent the use of human subjects in most direct experimental procedures. Therefore, the question becomes one of how closely can an animal model approximate the human situation. The study of cholesterol gallstone disease clearly demands the use of animals that can suffer from the same disorder, but one might also argue that since humans do suffer from this obviously gross disturbance to their cholesterol metabolism, then the only appropriate comparisons that can be made in studies of normal cholesterol dynamics would be with these same animals. But do these animals also suffer from atherosclerosis? and cancer? and nervous disorders? i.e., other conditions in the human also characterized by gross malfunction of cellular and whole-body cholesterol metabolism.

Many recent papers have highlighted this dilemma. The mouse is different from the rat[13] which is in turn different from the monkey[12] and, moreover, not all strains of mice[16] nor rats[17] nor monkeys are the same.[12] Old animals differ from young ones,[18] males differ from females[11] and old and young, male and female, all differ when fed different basal diets,[19] or when previously challenged during neonatal development[20,21] or, obviously, when measured during the day rather than during the night.[10-13] The critical problem is not just that baseline values vary, but rather that responses to change — age, diet, disease, drugs, and so on — also vary. The problem will not be easily solved, but while it remains all experimental results on cholesterol 7α-hydroxylase must be interpreted with caution when applied to human metabolism.

REFERENCES

1. **Myant, N. B. and Mitropoulos, K. A.,** Cholesterol 7α-hydroxylase, *J. Lipid Res.,* 18, 135, 1977.
2. **Sabine, J. R.,** *Cholesterol,* Marcel Dekker, New York, 1977.
3. **Dodd, N. K., Sizer, C. E., and Dupont, J.,** Cholanic acids and cholesterol 7α hydroxylase activity in human leucocytes, *Biochem. Biophys. Res. Commun.,* 106, 385, 1982.
4. **Qureshi, A. A., Abuirmeileh, N., Fitch, M., Elson, C. E., Burger, W. C., and McGibbon, W. H.,** Regulation of lipid metabolism in restricted ovulator chicken *Gallus domesticus* by dietary supplementation with high protein barley, flour and culture filtrate, *Nutr. Rep. Int.,* 27, 87, 1983.
5. **Ahlsten, M., Hansson, R., Holmberg, I., Ljungdahl, I., and Wikvall, K.,** Reconstitution of cholesterol 7α-hydroxylase system from human liver microsomes, *Biochem. Med.,* 26, 307, 1981.
6. **Li, J. R. and Kim, D. N.,** Effects of superoxide dismutase on cholesterol 7α-hydroxylation in swine, *Steroids,* 35, 459, 1980.
7. **Spence, J. T. and Gaylor, J. L.,** Investigation of regulation of microsomal hydroxymethylglutaryl coenzyme A reductase and methyl sterol oxidase of cholesterol biosynthesis, *J. Biol. Chem.,* 252, 5852, 1977.
8. **Danielsson, H., Kalles, I., and Wikvall, K.,** Regulation of hydroxylations in biosynthesis of bile acids. Isolation of a protein from rat liver cytosol stimulating reconstituted cholesterol 7-α-hydroxylase activity, *J. Biol. Chem.,* 259, 4258, 1984.
9. **Sabine, J. R.,** Distribution and importance of HMGCoA reductase, in *3-Hydroxy-3-Methylglutaryl Co-Enzyme A Reductase,* Sabine, J. R., Ed., CRC Press, Boca Raton, 1983, chap. 1.
10. **Mitropoulos, K. A. and Balasabramaniam, S.,** The role of glucocorticoids in the regulation of the diurnal rhythm of hepatic β-hydroxy-β-methylglutaryl coenzyme A reductase and cholesterol 7-α-hydroxylase, *Biochem. J.,* 160, 49, 1976.
11. **Carlson, S. E., Mitchell, A. D., and Goldfarb, S.,** Sex-related differences in diurnal activities and development of hepatic microsomal 3-hydroxy-3-methylglutaryl coenzyme A reductase and cholesterol 7α-hydroxylase, *Biochim. Biophys. Acta,* 531, 115, 1978.
12. **Hulcher, F. H., Margolis, R. D., and Bowman, D. J.,** Circadian rhythm of cholesterol 7α-hydroxylase and cortisol in the African green monkey Cercopithecus aethiops, *Biochim. Biophys. Acta,* 529, 409, 1978.
13. **Van Cantford, J. and Gielen, J. E.,** Comparison of rat and mouse circadian rhythm of cholesterol 7α-hydroxylase activity, *J. Steroid Biochem.,* 10, 647, 1979.
14. **Holloway, D. E., Peterson, F. J., Prigge, W. F., and Gebhard, R. L.,** Influence of dietary ascorbic acid upon enzymes of sterol biosynthesis in the guinea pig, *Biochem. Biophys. Res. Commun.,* 102, 1283, 1981.
15. **Sabine, J. R.,** HMGCoA reductase and cholesterol, in *3-Hydroxy-3-Methylglutaryl Co-Enzyme A Reductase,* Sabine, J. R., Ed., CRC Press, Boca Raton, 1983, chap. 2.
16. **Rush, W. R., Fears, R., Green, J., and Parke, D. V.,** Liver microsomal mixed-function oxidase activity (MFO) and cholesterol metabolism in the obese, arthritic STR/1N strain of mouse, *Comp. Biochem. Physiol.,* 69B, 493, 1981.
17. **Story, J. A., Gomolinski, E., Czarnecki, S. K., Tepper, S. A., and Kritchevsky, D.,** Age strain interrelations in lipid metabolism of rats, *Lipids,* 16, 87, 1981.
18. **Story, J. A., Tepper, S. A., and Kritchevsky, D.,** Age related changes in the lipid metabolism of Fisher 344 rats, *Lipids,* 11, 623, 1976.
19. **O'Brien, B. C. and Reiser, R.,** Comparative effects of purified and human-type diets on cholesterol metabolism in the rat, *J. Nutr.,* 109, 98, 1979.
20. **Li, J. R., Bale, L. K., and Kottke, B. A.,** Effect of neo-natal modulation of cholesterol homeostasis on subsequent response to cholesterol challenge in adult guinea pigs, *J. Clin. Invest.,* 65, 1060, 1980.
21. **Naseem, S. M., Khan, M. A., Heald, F. P., and Nair, P. P.,** The influence of cholesterol and fat in maternal diet of rats on the development of hepatic cholesterol metabolism in the offspring, *Atherosclerosis,* 36, 1, 1980.

Chapter 2

BIOSYNTHESIS OF BILE ACIDS

Henry Danielsson and Kjell Wikvall

TABLE OF CONTENTS

I. INTRODUCTION

The biosynthesis of bile acids represents the major pathway for the metabolism and excretion of cholesterol in mammals. The principal functions of bile acids are to promote intestinal absorption of dietary fat, including fat-soluble vitamins and drugs, and to keep bile cholesterol in solution. The primary bile acids are those formed from cholesterol in the liver. They are secreted as conjugates with taurine or glycine by way of bile into the duodenum. The secondary bile acids are those formed from the primary bile acids in the intestine through the action of intestinal microorganisms. The predominant part of the bile acids is reabsorbed, mainly in the distal ileum, and transported back to the liver with portal blood. Prior to resecretion into bile, unconjugated bile acids are reconjugated, ketonic bile acids may be reduced, and bile acids may be hydroxylated. Only a small part of the bile acids circulating enterohepatically is lost daily with the feces.

The main primary bile acids in most mammalian species are cholic acid and chenodeoxycholic acid. The ratio between the two bile acids may vary with species. There are a few primary bile acids which can be regarded as species specific even if they may be present in trace amounts in bile of other species. For instance, rat and mouse synthesize in addition to cholic and chenodeoxycholic acids, α-muricholic and β-muricholic acids. In the pig, hyocholic acid (3α,6α,7α-trihydroxy-5β-cholanoic acid) is synthesized instead of cholic acid.[1,2]

Mammals synthesize small amounts of allo bile acids (5α-cholanoic acids). In primitive species, allo bile acids and/or allo bile alcohols may predominate. Pathways for synthesis of allo bile acids and bile alcohols will not be discussed in this chapter. The reader is referred to recent reviews of this topic.[2-4]

A common denominator for most primary bile acids is that they have a 7α-hydroxyl group. In fact, only very few primary bile acids not having a 7α-hydroxyl group have been discovered in nature. One example is β-muricholic acid carrying a 7β-hydroxyl group. However, this bile acid is formed from α-muricholic acid through an epimerization of the 7α-hydroxyl group.

The present chapter will summarize briefly pathways for the biosynthesis of the principal primary bile acids. A short discussion of the regulation of these processes is included.

II. SEQUENCE OF REACTIONS IN BIOSYNTHESIS OF BILE ACIDS

The major pathways for the biosynthesis of cholic acid and chenodeoxycholic acid in the rat were established by the late 1960s (Figure 1). Experimental work in man in vivo and with preparations from human liver has shown pathways for bile acid biosynthesis similar to those in the rat.[5]

The conversion of cholesterol to bile acids involves reactions catalyzed by microsomal, cytosolic, mitochondrial and peroxisomal enzymes. The 7α-hydroxylation of cholesterol is the first step. The reaction is catalyzed by the microsomal cholesterol 7α-hydroxylase system involving a cytochrome P-450 and NADPH-cytochrome P-450 reductase. The purification and properties of the enzyme system are discussed in Chapter 5. The 7α-hydroxylated 3β-hydroxy-Δ^5-steroid, i.e., 5-cholestene-3β,7α-diol, is converted into the corresponding 3-oxo-Δ^4-steroid, 7α-hydroxy-4-cholesten-3-one, by an NAD-dependent microsomal 3β-hydroxy-Δ^5-C_{27}-steroid oxidoreductase. The enzyme has been purified from rabbit liver microsomes and appears to be specific for the C_{27}-steroid.[6] Cytosolic oxidoreductases saturate the Δ^4-double bond and reduce the 3-oxo group into a 3α-hydroxyl group.[1] An NADPH-dependent Δ^4-3-ketosteroid 5β-reductase has been purified to homogeneity from rat liver cytosol.[7] The enzyme reduced the Δ^4-double bond of C_{19}-, C_{21}-, as well as C_{27}-steroids.

There may be two major branch points in the biosynthesis of cholic acid and chenodeox-

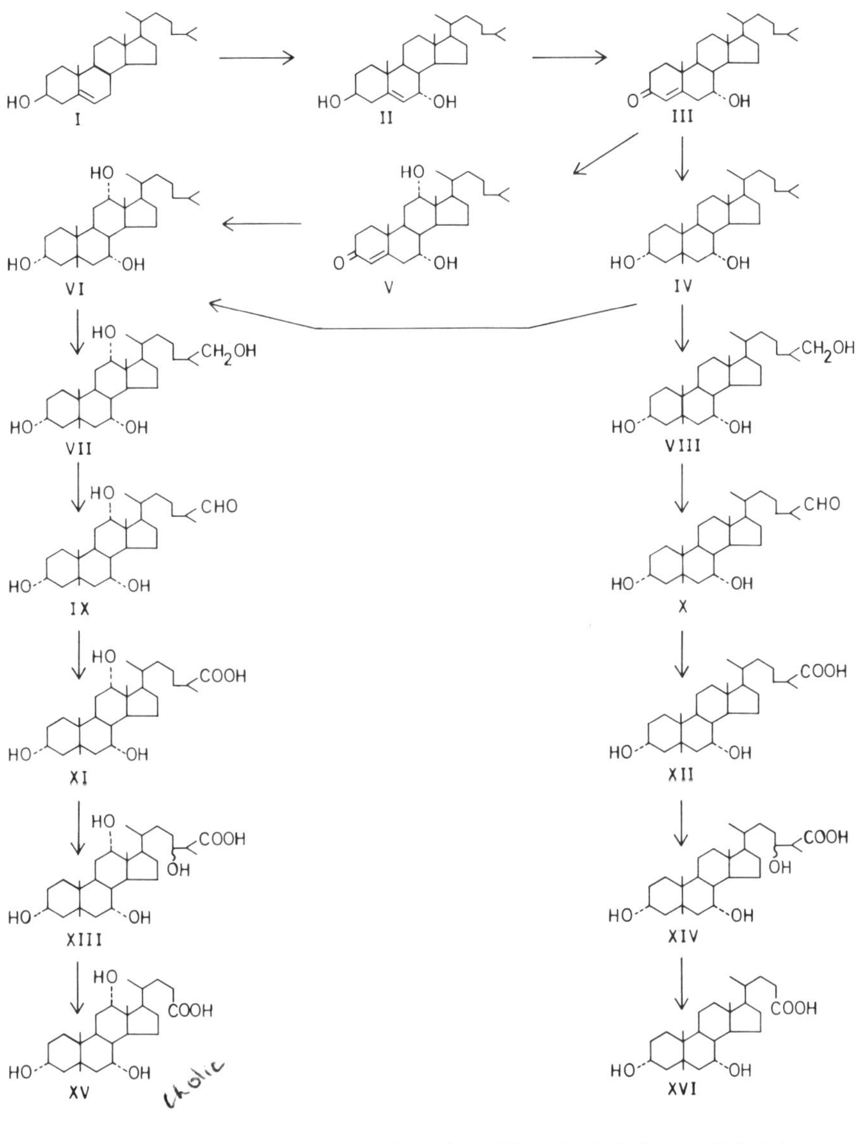

FIGURE 1. Major pathways for biosynthesis of primary bile acids. I, cholesterol; II, 5-cholestene-3β,7α-diol; III, 7α-hydroxy-4-cholesten-3-one; IV, 5β-cholestane-3α,7α-diol; V, 7α,12α-dihydroxy-4-cholesten-3-one; VI, 5β-cholestane-3α,7α,12α-triol; VII, 5β-cholestane-3α,7α,12α,26-tetrol; VIII, 5β-cholestane-3α,7α,26-triol; IX, 3α,7α,12α-trihydroxy-5β-cholestan-26-al; X, 3α,7α-dihydroxy-5β-cholestan-26-al; XI, 3α,7α,12α-trihydroxy-5β-cholestanoic acid; XII, 3α,7α-dihydroxy-5β-cholestanoic acid; XIII, 3α,7α,12α,24ξ-tetrahydroxy-5β-cholestanoic acid; XIV, 3α, 7α, 24ξ-trihydroxy-5β-cholestanoic acid; XV, cholic acid; XVI, chenodeoxycholic acid.

ycholic acid. Both 7α-hydroxy-4-cholesten-3-one and 5β-cholestane-3α,7α-diol serve as substrates for the microsomal, cytochrome P-450-dependent 12α-hydroxylase system.[8,9] Studies with whole microsomes have indicated that 7α-hydroxy-4-cholesten-3-one is the favored substrate in rabbit, rat, and man.[10-12] Studies with reconstituted systems have shown that the cytochrome P-450 catalyzing 12α-hydroxylation is different from that catalyzing 7α-hydroxylation of cholesterol and have provided strong evidence for the involvement of a specific species of cytochrome P-450 in the reaction (compare Chapter 5).[9,13] The 12α-hydroxylase has been ascribed a role in the regulation of the ratio of cholic acid to chenodeoxycholic acid.

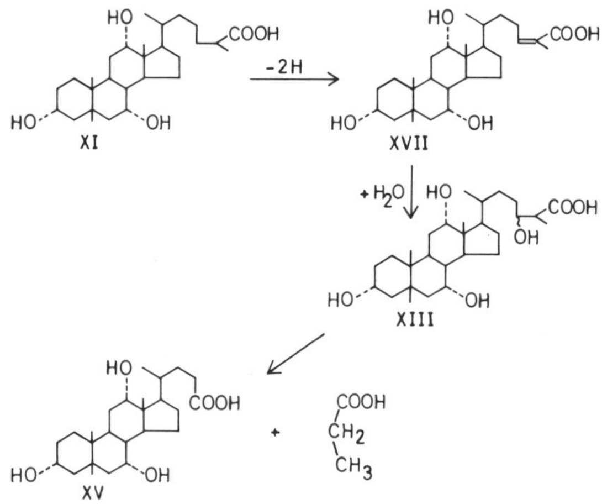

FIGURE 2. Mechanism for side chain oxidation. XI, 3α,7α,12α-trihydroxy-5β-cholestanoic acid; XIII, 3α,7α,12α, 24ξ-tetrahydroxy-5β-cholestanoic acid; XV, cholic acid; XVII, 3α,7α,12α-trihydroxy-5β-cholest-24-enoic acid.

It has been postulated that in the rat all of the changes of the steroid nucleus precede oxidation of the side chain in the biosynthesis of both chenodeoxycholic acid and cholic acid.[1] The main substrates for cleavage of the C_{27}-steroid side chain would then be 5β-cholestane-3α,7α-diol and 5β-cholestane-3α,7α,12α-triol. The major pathway for side chain degradation starts with an ω-hydroxylation, i.e., a 26-hydroxylation. In man, 7α-hydroxy-4-cholesten-3-one may be the main substrate for 26-hydroxylation in chenodeoxycholic acid formation.[12,14] Saturation of the Δ^4-double bond and reduction of the oxo group would then occur at a later, as yet not defined stage. In rat liver, two different 26-hydroxylating systems with different stereospecificity are present, one microsomal and one mitochondrial. Both systems are cytochrome P-450-dependent.[15,16] In rabbit and human liver, the mitochondrial 26-hydroxylase system is by far the most active.[12] The specific mitochondrial 26-hydroxylating cytochrome P-450 has been purified from rabbit liver.[17]

The 26-hydroxylated steroids are oxidized into the corresponding C_{27}-acids by liver alcohol (ethanol) and acetaldehyde dehydrogenase.[4,18] The role of liver alcohol dehydrogenase in bile acid biosynthesis has been recently questioned. Deermice with a genetic lack of liver alcohol dehydrogenase show a normal bile acid pattern and pool size.[19]

The final steps in side chain degradation involve a β-oxidation process starting with a 24-hydroxylation of the C_{27}-acids (Figure 2). The reaction, which involves desaturation and hydration, has been reported to be catalyzed by the mitochondrial, microsomal, as well as the peroxisomal fractions of rat liver homogenate.[20-22] Cleavage of the side chain results in the elimination of propionic acid and the formation of cholic acid and chenodeoxycholic acid. Neither the intermediates nor the enzymes involved in the β-oxidation process have yet been well characterized.

Alternative pathways for bile acid biosynthesis may exist. A mechanism for side chain degradation in cholic acid synthesis not involving mitochondrial 26-hydroxylation has been described (Figure 3). In this pathway, an initial, cytochrome P-450-dependent 25-hydroxylation in microsomes followed by a 24β-hydroxylation in the same subcellular fraction results in the elimination of acetone. This pathway was described some years ago by Shefer et al.[23] in connection with studies on the metabolic defect in the rare, inherited disease cerebrotendinous xanthomatosis. In cholic acid formation, the probable substrate for the 25-hydroxylase is 5β-cholestane-3α,7α,12α-triol and in chenodeoxycholic acid formation 5β-

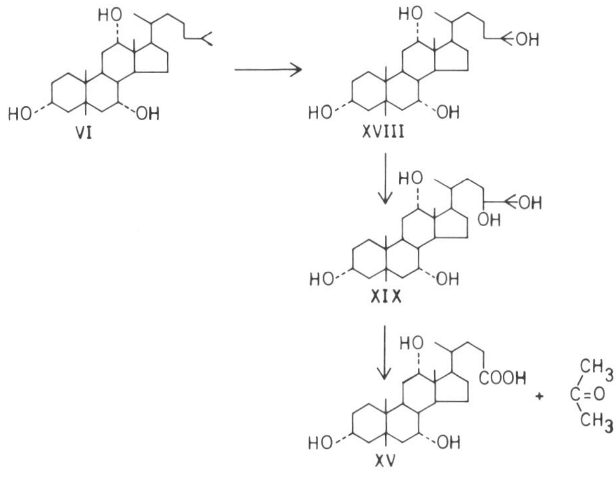

FIGURE 3. Alternative mechanism for side chain oxidation. VI, 5β-cholestane-3α,7α,12α-triol; XV, cholic acid; XVIII, 5β-cholestane-3α,7α,12α,25-tetrol; XIX, 5β-cholestane-3α,7α,12α,24β,25-pentol.

cholestane-3α,7α-diol. It is doubtful that the 25-hydroxylation pathway plays a role in chenodeoxycholic acid formation. Studies of 25-hydroxylation of various substrates by human liver microsomes show that 5β-cholestane-3α,7α-diol is a very poor substrate.[12,24] It should be mentioned that already in 1961 Whitehouse et al. reported the formation of labeled acetone from 26-^{14}C cholesterol after incubation with mitochondrial preparations from rat liver.[25] No studies of mechanisms of acetone formation were included in this report.

A mitochondrial pathway for the synthesis of chenodeoxycholic acid in the rat has been described (Figure 4A).[26] This pathway involves complete oxidation of the side chain before the changes in the steroid nucleus. Intermediates in this pathway are thus 26-hydroxycholesterol, 3β-hydroxy-5-cholenoic acid and lithocholic acid. Another pathway has also been described for chenodeoxycholic acid synthesis in the rat in which some early steps, including cholesterol 7α-hydroxylation, are catalyzed by liver mitochondria (Figure 4B).[27,28] In this connection it might be mentioned that a mitochondrial 12α-hydroxylase system in human fetal liver has been recently discovered.[29] In the rat, insertion of a 26-hydroxyl group apparently hinders 12α-hydroxylation. This is not the case in man. Thus, labeled 26-hydroxycholesterol, 7α,26-dihydroxy-4-cholesten-3-one and 5β-cholestane-3α,7α,26-triol have been shown to be precursors of not only chenodeoxycholic acid but also cholic acid.[5,30-32] Results suggesting a pathway in man from cholesterol to cholic acid not involving an initial 7α-hydroxylation of cholesterol have been reported.[32] No information is available concerning intermediates in such a pathway.

The quantitative importance of the different, alternative pathways has not been definitely established but they appear to play a minor role under normal conditions. However, they might be important during development and in certain pathological conditions (reviews, see References 4, 5, and 33).

III. REGULATION OF BILE ACID BIOSYNTHESIS

A. Regulation of Overall Bile Acid Biosynthesis

7α-Hydroxylation of cholesterol is the first and rate-limiting step in the main pathways for bile acid biosynthesis. In the rat and probably most other mammals, biosynthesis of bile acids appears to be regulated homeostatically by the amount of bile acids returning to the

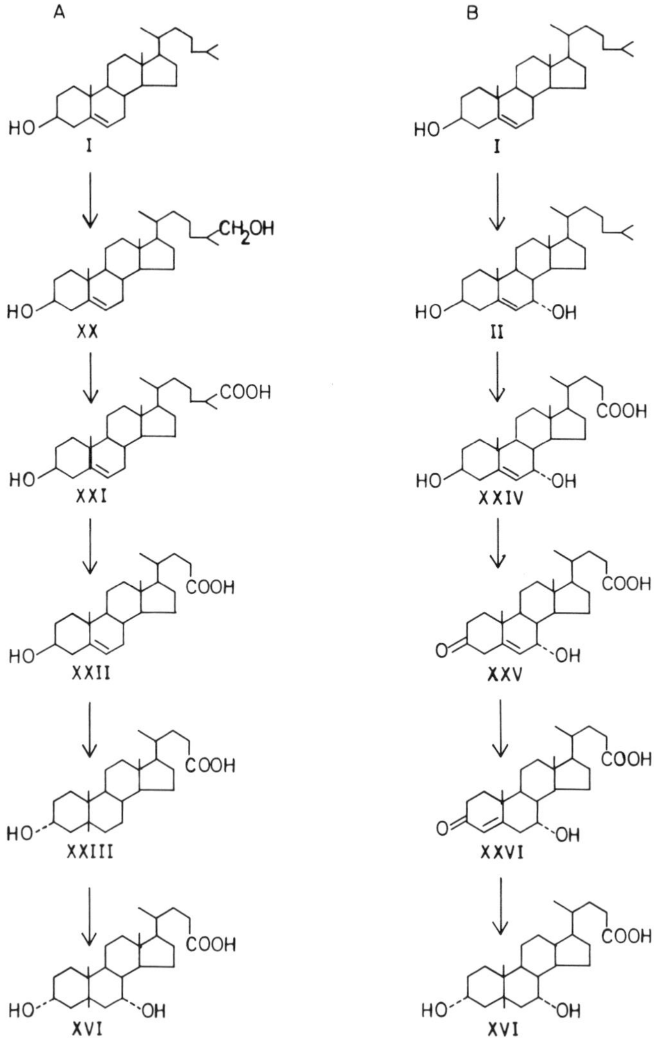

FIGURE 4. Alternative pathways for biosynthesis of chenodeoxycholic acid. A, mitochondrial pathway as suggested by Mitropoulos and Myant.[26] B, pathway as suggested by Yamasaki and collaborators.[27,28] I, cholesterol; II, 5-cholestene-3β,7α-diol; XVI, chenodeoxycholic acid; XX, 5-cholestene-3β,26-diol; XXI, 3β-hydroxy-5-cholestenoic acid; XXII, 3β-hydroxy-5-cholenoic acid; XXIII, lithocholic acid; XXIV, 3β,7α-dihydroxy-5-cholenoic acid; XXV, 3-oxo-7α-hydroxy-5-cholenoic acid; XXVI, 3-oxo-7α-hydroxy-4-cholenoic acid.

liver via the portal vein.[1] Interruption of the enterohepatic circulation by drainage of bile or by oral administration of bile acid-binding resins such as cholestyramine increase the rate of bile acid synthesis manyfold. Duodenal infusion or feeding of bile acids leads to a decrease in bile acid formation in both normal animals and those subjected to drainage of bile.[1,4] It was shown in the late 1960s that biliary drainage in the rat selectively increased microsomal cholesterol 7α-hydroxylase activity in parallel with the increased rate of bile acid synthesis.[34] Later, effects of cholestyramine feeding and bile acid feeding on the rate of bile acid biosynthesis were also shown to be correlated with changes in cholesterol 7α-hydroxylase activity. These findings are to be expected if 7α-hydroxylation of cholesterol is a rate-limiting reaction in the formation of bile acids. The increased rate of cholesterol 7α-hydroxylation in animals with biliary drainage can be prevented by administration of actinomycin D or puromycin, indicating that the increase in enzyme activity is due to increased synthesis of some protein component of the enzyme system rather than to activation of preexisting enzyme.[4,5,35] The rate of decrease of the 7α-hydroxylase activity after administration of actinomycin or puromycin to a bile fistula or control rat suggests that the half-life time of cholesterol 7α-hydroxylase is short, 2 to 3 hours.[36] Other enzyme activities in bile acid biosynthesis, e.g., the 12α-hydroxylase, have considerably longer half-life times. The short half-life time of cholesterol 7α-hydroxylase is consistent with a regulatory function for the enzyme system. It should be mentioned that there is a lag phase between start of biliary drainage and increased cholesterol 7α-hydroxylase activity which is considerably longer than the calculated turnover of the enzyme. Thus, in the rat cholesterol 7α-hydroxylase activity and bile acid biosynthesis start to increase about 24 hours after initiation of biliary drainage.[34] No explanation for this phenomenon has been presented.

The mechanism by which bile acids repress bile acid synthesis is not understood. The specificity of the effect of bile acids on bile acid formation has been studied to some extent. In the rat, feeding of taurocholic acid, taurochenodeoxycholic acid, and taurodeoxycholic acid inhibits cholesterol 7α-hydroxylase activity, whereas taurohyodeoxycholic acid has no effect. Feeding with taurocholate has been shown to inhibit the biosynthesis of both chenodeoxycholic acid and cholic acid; taurochenodeoxycholic acid inhibits biosynthesis of taurocholic acid. In man, chenodeoxycholic acid inhibits cholic acid synthesis and cholic acid inhibits chenodeoxycholic acid synthesis.[4,37]

The possibility of a direct allosteric inhibition of the enzyme system has been studied by addition of bile acids to microsomes.[38] An inhibition was observed. However, this inhibition could be due to nonspecific detergent effects of bile acids.

The effect of bile acids on bile acid biosynthesis has been recently studied in isolated hepatocytes and in cultured hepatocytes. No inhibitory effect on bile acid biosynthesis was observed when various bile acids were added to the hepatocytes.[39,40] It can not be excluded that the hepatocytes had lost an original ability to respond to bile acids.

The activity of cholesterol 7α-hydroxylase exhibits a diurnal rhythm in the rat with a maximum at midnight and a minimum during day time. The rise in enzyme activity can be prevented by treating the rat with actinomycin D or cycloheximide. The amplitude of the variations as well as the timing can be influenced by time for feeding and time of darkness and light. The diurnal cycle of the cholesterol 7α-hydroxylase activity is probably influenced by variations in secretion of glucocorticoids. Adrenalectomy depresses the cholesterol 7α-hydroxylase activity and causes an almost total loss of the diurnal variations. Other hormones, such as thyroid hormones, pituitary hormones, and insulin, have been reported to influence cholesterol 7α-hydroxylase activity.[4,35]

Generally, changes in the rate of bile acid biosynthesis are paralleled by changes in the rate of cholesterol biosynthesis in the liver. The biosynthetic rates are reflected in variations in the activities of the rate-limiting enzymes, cholesterol 7α-hydroxylase and HMG-CoA reductase, respectively (Table 1). Parallel changes have been observed in biliary drainage, in the administration of cholestyramine, in bile acid feeding, upon portacaval anastomosis, and in lymphatic drainage. The two enzymes have the same diurnal rhythm and are influenced in the same way by glucocorticoids, adrenalectomy, thyroid hormones, thyroidectomy, starvation, and by feeding glucose after starvation. In certain cases, there is a dissociation between effects on cholesterol 7α-hydroxylase and HMG-CoA reductase. Feeding tomatine, which interferes with the absorption of cholesterol in the intestine, increases the activity of HMG-CoA reductase, whereas the cholesterol 7α-hydroxylase activity remains unchanged or is only slightly increased. Feeding rats a diet rich in cholesterol decreases the rate of HMG-CoA reduction whereas cholesterol 7α-hydroxylation is increased or unchanged. A

Table 1

Treatment	Cholesterol 7α-hydroxylase activity	HMG-CoA reductase activity
Biliary fistula or feeding of cholestyramine	↑	↑
Feeding of bile acids	↓	↓
Diurnal rhythm	Yes	Yes
Glucocorticoids	↑	↑
Adrenalectomy	↓	↓
Thyroid hormone	↑	↑
Thyroidectomy	↓	↓
Feeding of glucose after a fast	↑	↑
Feeding of tomatine	→	↑
Feeding of cholesterol	↑→	↓
Vitamin C deficiency	↓	↑→

↑ = increase; ↓ = decrease; → = no change.

diminished cholesterol 7α-hydroxylase activity has been observed in guinea pigs deficient in vitamin C. HMG-CoA reductase activity is unaffected. Administration of vitamin C restores the cholesterol 7α-hydroxylase activity.[4,35] With respect to the interrelationship between the two enzyme activities it is not known whether both activities are influenced simultaneously or one is influenced before the other.

There is some information concerning the mechanisms by which certain factors influence HMG-CoA reductase activity. It has been recently demonstrated that the stimulation of the enzyme activity in response to treatment with cholestyramine is associated with an increase of the specific mRNA for HMG-CoA reductase and increased enzyme synthesis.[41,42] The inhibition of HMG-CoA reductase activity in response to cholesterol feeding has been shown to be associated with decreased enzyme synthesis.[41] The variation in diurnal activity has also been correlated with variations in mRNA and HMG-CoA reductase concentrations.[43] Thus, long-term regulation of cholesterol biosynthesis, mediated by changes in HMG-CoA reduction, is primarily due to changes in enzyme quantity. It is now well established that the activity of HMG-CoA reductase can be modulated by means of a dephosphorylation-phosphorylation mechanism. This mechanism is apparently involved in short-term regulation of HMG-CoA reductase.[44]

The mechanisms by which different factors influence the cholesterol 7α-hydroxylase activity have not been established. Long-term regulation of bile acid synthesis might also be mediated by changes in enzyme quantity. A role of a dephosphorylation-phosphorylation mechanism in short-term regulation of cholesterol 7α-hydroxylase activity has been proposed (compare Chapter 7).[45,46]

B. Regulation of the Ratio between Biosynthesis of Cholic Acid and Chenodeoxycholic Acid

The molecular mechanisms for the regulation of the proportion between cholic acid and chenodeoxycholic acid formed in the liver are not fully understood. Since microsomal 12α-hydroxylation is a step unique for cholic acid biosynthesis it is possible that the 12α-hydroxylase system plays a role in this regulation. In fact, there are positive correlations between the species-specific ratios of the two primary bile acids formed and 12α-hydroxylase activity in liver microsomes of certain animals. For instance, the low ratio of cholic acid to chenodeoxycholic acid formed in the guinea pig is correlated with a low 12α-hydroxylase activity. The rabbit has a high 12α-hydroxylase activity and produces cholic acid almost exclusively as a primary bile acid. Recently, a positive correlation has been found in the hamster between the 12α-hydroxylase activity and the sex-linked differences in the ratio of cholic acid to chenodeoxycholic acid.[47] It has been suggested that the reduced cholic acid biosynthesis in patients with alcoholic cirrhosis is due to defective 12α-hydroxylation.[5] It appears that microsomal 12α-hydroxylase activity is subject to a feedback control by bile acids. At least in the rat and hamster 12α-hydroxylase activity is inhibited by feeding bile acids.[37,48]

Since introduction of a 26-hydroxyl group seems to prevent 12α-hydroxylation in the rat, it is possible that the mitochondrial or microsomal 26-hydroxylase system could have a regulatory role in combination with the 12α-hydroxylase system. It has been reported that rats with alloxan-induced diabetes show an increased cholic acid biosynthesis but a normal chenodeoxycholic acid biosynthesis.[49] Unpublished experiments have shown an increased microsomal 12α-hydroxylase activity and a decreased microsomal 26-hydroxylase activity in alloxan-treated rats.[50]

Administration of thyroid hormone to rats results in a decreased ratio between cholic acid and chenodeoxycholic acid. The change in ratio is reflected in a decreased 12α-hydroxylation and an increased microsomal 26-hydroxylation. Conversely, thyroidectomy is associated with an increase of 12α-hydroxylase activity and a decrease of microsomal 26-hydroxylase activity. Thus, thyroid hormone status in the rat affects the activities of the microsomal 12α-hydroxylase and the 26-hydroxylase systems in the same direction as it affects the ratio between cholic acid and chenodeoxycholic acid. It should be mentioned that the mitochondrial 26-hydroxylase activity is not affected by thyroid hormone status.[4,51]

The situation in man may be different. Introduction of a 26-hydroxyl group does not prevent a subsequent 12α-hydroxylation. Microsomal 26-hydroxylase activity is very low in human liver and thyroid hormone has no effect on the ratio between cholic acid to chenodeoxycholic acid.[4] In a recent in vivo study, Björkhem et al. did not find a correlation between the ratio of cholic acid to chenodeoxycholic acid and the apparent 12α-hydroxylase activity.[52] These workers studied the 12α-hydroxylation of 7α-hydroxy-4-cholesten-3-one which is believed to be the major physiological substrate for the 12α-hydroxylase system. It cannot be excluded, however, that 12α-hydroxylation of 5β-cholestane-3α,7α-diol plays a role in cholic acid biosynthesis and that this 12α-hydroxylation might be affected. In this connection, it could be mentioned that Okuda et al. have suggested that 12α-hydroxylation of 7α-hydroxy-4-cholesten-3-one and 12α-hydroxylation of 5β-cholestane-3α,7α-diol are catalyzed by two separate cytochrome P-450 isozymes.[53]

REFERENCES

1. **Danielsson, H.,** Mechanisms of bile acid biosynthesis, in *The Bile Acids,* Vol. 2, Nair, P. P. and Kritchevsky, D., Eds., Plenum Press, New York, 1973, 1, 305.
2. **Haslewood, G. A. D.,** *The Biological Importance of Bile Salts,* Elsevier-North-Holland, Amsterdam, 1978.
3. **Danielsson, H. and Sjövall, J.,** Bile acid metabolism, *Annu. Rev. Biochem.,* 44, 233, 1975.
4. **Björkhem, I.,** Mechanism of bile acid biosynthesis in mammalian liver, in *New Comprehensive Biochemistry,* Danielsson, H. and Sjövall, J., Eds., Elsevier, Amsterdam, 1985, in press.
5. **Danielsson, H. and Gustafsson, J.,** Biochemistry of bile acids in health and disease, in *Pathobiology Annual,* Vol. 11, Ioachim, H. L., Ed., Raven Press, New York, 1981, 259.
6. **Wikvall, K.,** Purification and properties of a 3β-hydroxy-Δ^5-C_{27}-steroid oxidoreductase from rabbit liver microsomes, *J. Biol. Chem.,* 256, 3376, 1981.
7. **Okuda, A. and Okuda, K.,** Purification and characterization of Δ^4-3-ketosteroid 5β-reductase, *J. Biol. Chem.,* 259, 7519, 1984.

8. **Bernhardsson, C., Björkhem, I., Danielsson, H., and Wikvall, K.,** 12α-Hydroxylation of 7α-hydroxy-4-cholesten-3-one by a reconstituted system from rat liver microsomes, *Biochem. Biophys. Res. Commun.*, 54, 1030, 1973.
9. **Hansson, R. and Wikvall, K.,** Hydroxylations in biosynthesis of bile acids: cytochrome P-450 LM_4 and 12α-hydroxylation of 5β-cholestane-3α,7α-diol, *Eur. J. Biochem.*, 125, 423, 1982.
10. **Nicolau, G., Cohen, B. I., Salen, G., and Mosbach, E.,** Studies on the 12α and 26-hydroxylation of bile alcohols by rabbit liver microsomes, *Lipids*, 11, 148, 1976.
11. **Einarsson, K.,** On the properties of the 12α-hydroxylase in cholic acid biosynthesis. Bile acids and steroids 198, *Eur. J. Biochem.*, 5, 101, 1968.
12. **Björkhem, I., Gustafsson, J., Johansson, G., and Persson, B.,** Biosynthesis of bile acids in man: hydroxylation of the C_{27}-steroid side chain, *J. Clin. Invest.*, 55, 478, 1975.
13. **Boström, H. and Wikvall, K.,** Hydroxylations in biosynthesis of bile acids: isolation of subfractions with different substrate specificity from cytochrome P-450 LM_4, *J. Biol. Chem.*, 257, 11755, 1982.
14. **Björkhem, I., Oftebro, H., Skrede, S., and Pedersen, J.,** Assay of intermediates in bile acid biosynthesis using isotope dilution-mass spectrometry: hepatic levels in the normal state and in cerebrotendinous xanthomatosis, *J. Lipid Res.*, 22, 191, 1981.
15. **Björkhem, I., Danielsson, H., and Wikvall, K.,** Side chain hydroxylations in biosynthesis of cholic acid: 25- and 26-hydroxylation of 5β-cholestane-3α,7α,12α-triol by reconstituted systems from rat liver microsomes, *J. Biol. Chem.*, 251, 3495, 1976.
17. **Wikvall, K.,** Hydroxylations in biosynthesis of bile acids: isolation of a cytochrome P-450 from rabbit liver mitochondria catalyzing 26-hydroxylation of C_{27}-steroids, *J. Biol. Chem.*, 259, 3800, 1984.
18. **Okuda, A. and Okuda, K.,** Physiological function and kinetic mechanism of human liver alcohol dehydrogenase as 5β-cholestane-3α,7α,12α,26-tetrol dehydrogenase, *J. Biol. Chem.*, 258, 2899, 1983.
19. **Sjovall, J., Andersson, S. H. G., and Lieber, C. S.,** Bile acids in deermice lacking liver alcohol dehydrogenase, *Biochim. Biophys. Acta*, 836, 8, 1985.
20. **Masui, T. and Staple, E.,** The formation of bile acids from cholesterol. The conversion of 5β-cholestane-3α,7α,12α-triol-26-oic acid to cholic acid via 5β-cholestane-3α,7α,12α,24-tetraol-26-oic acid by rat liver, *J. Biol. Chem.*, 241, 3889, 1966.
21. **Gustafsson, J.,** Biosynthesis of cholic acid in rat liver: 24-hydroxylation of 3α,7α,12α-trihydroxy-5β-cholestanoic acid, *J. Biol. Chem.*, 250, 8243, 1975.
22. **Kase, F., Björkhem, I., and Pedersen, J. I.,** Formation of cholic acid from 3α,7α,12α-trihydroxy-5β-cholestanoic acid by rat liver peroxisomes, *J. Lipid Res.*, 24, 1560, 1983.
23. **Shefer, S., Cheng, F. W., Dayal, B., Hauser, S., Trut, G. S., Salen, G., and Mosbach, E. H.,** A 25-hydroxylation pathway of cholic acid biosynthesis in man and rat, *J. Clin. Invest.*, 57, 897, 1976.
24. **Ahlsten, M., Hansson, R., Holmberg, I., Ljungdahl, I., and Wikvall, K.,** Reconstitution of cholesterol 7α-hydroxylase system from human liver microsomes, *Biochem. Med.*, 26, 307, 1981.
25. **Whitehouse, M. W., Staple, E., and Gurin, S.,** Catabolism *in vitro* of cholesterol. II. Further studies on the oxidation of cholesterol by rat liver mitochondria, *J. Biol. Chem.*, 236, 68, 1961.
26. **Mitropoulos, K. A. and Myant, N. B.,** The formation of lithocholic acid, chenodeoxycholic acid and α- and β-muricholic acids from cholesterol incubated with rat-liver mitochondria, *Biochem. J.*, 103, 472, 1967.
27. **Ayaki, Y. and Yamasaki, K.,** *In vitro* conversion of 7α-hydroxycholesterol to some natural C_{24}-bile acids with special reference to chenodeoxycholic acid biogensis, *J. Biochem.*, 68, 341, 1970.
28. **Ikawa, S., Ayaki, Y., Ogura, M., and Yamasaki, K.,** The metabolism *in vivo* and *in vitro* of 3-oxo-7α-hydroxychol-4-enoic acid-$24^{14}C$ as an intermediate of chenodeoxycholic acid biogenesis, *J. Biochem.*, 71, 579, 1972.
29. **Gustafsson, J.,** Bile acid synthesis during development. Mitochondrial 12α-hydroxylation in human fetal liver, *J. Clin. Invest.*, 75, 604, 1985.
30. **Javitt, N. B. and Emerman, S.,** Metabolic pathways of bile acid formation in the rat, *Mt. Sinai J. Med.*, 37, 477, 1970.
31. **Andersson, K. E., Kok, E., and Javitt, N. B.,** Bile acid synthesis in man: metabolism of 7α-hydroxycholesterol-^{14}C and 26-hydroxycholesterol-^{3}H, *J. Clin. Invest.*, 51, 112, 1972.
32. **Vlahcevic, Z. R., Schwartz, C. C., Gustafsson, J., Halloran, L. G., Danielsson, H., and Swell, L.,** Biosynthesis of bile acids in man: multiple pathways to cholic and chenodeoxycholic acid, *J. Biol. Chem.*, 255, 2925, 1980.
33. **Salen, G. and Mosbach, E. H.,** The metabolism of sterols and bile acids in cerebrotendinous xanthomatomatosis, in *The Bile Acids*, Vol. 3, Nair, P. P. and Kritchevsky, D., Eds., Plenum Press, New York, 1976, 115.
34. **Danielsson, H., Einarsson, K., and Johansson, G.,** Effect of biliary drainage on individual reactions in the conversion of cholesterol to taurocholic acid, *Eur. J. Biochem.*, 2, 44, 1967.
35. **Myant, N. B. and Mitropoulos, K. A.,** Cholesterol 7α-hydroxylase, *J. Lipid Res.*, 18, 135, 1977.

36. **Einarsson, K. and Johansson, G.,** Bile acids and steroids 206. Effect of actinomycin D and puromycin on the conversion of cholesterol into bile acids in bile-fistula rats, *FEBS Lett.*, 1, 219, 1968.
37. **Danielsson, H.,** Influence of dietary bile acids on formation of bile acids in rat, *Steroids,* 22, 667, 1973.
38. **Shefer, S., Hauser, S., and Mosbach, E. H.,** 7α-Hydroxylation of cholestanol by rat liver microsomes, *J. Lipid Res.*, 9, 328, 1968.
39. **Botham, K. M., Lawson, M. E., Beckett, G. J., Percy-Robb, I. W., and Boyd, G. S.,** The effect of portal blood bile salt concentrations on bile salt synthesis in rat liver: studies with isolated hepatocytes, *Biochim. Biophys. Acta,* 666, 238, 1981.
40. **Davis, R. A., Highsmith, W. E., McNeal, M. M., Schexnayder, J. A., and Kuan, C. W.,** Bile acid synthesis by cultured hepatocytes: inhibition by mevinolin but not by bile acids, *J. Biol. Chem.*, 258, 4079, 1983.
41. **Liscum, L., Luskey, K. L., Chin, D. J., Ho, Y. K., Goldstein, J. L., and Brown, M. S.,** Regulation of 3-hydroxy-3-methylglutaryl Coenzyme A reductase and its mRNA in rat liver as studied with a monoclonal antibody and a cDNA probe, *J. Biol. Chem.*, 258, 8450, 1983.
42. **Clarke, C. F., Edwards, P. A., Lan, S.-F., Tanaka, R. D., and Fogelman, A. M.,** Regulation of 3-hydroxy-3-methylglutaryl-coenzyme A reductase mRNA levels in rat liver, *Proc. Natl. Acad. Sci. U.S.A.*, 80, 3305, 1983.
43. **Clarke, C. F., Fogelman, A. M., and Edwards, P. A.,** Diurnal rhythm of rat liver mRNAs encoding 3-hydroxy-3-methylglutaryl Coenzyme A reductase: correlation of functional and total mRNA levels with enzyme activity and protein, *J. Biol. Chem.*, 259, 10439, 1984.
44. **Ingebritsen, T. S. and Gibson, D. M.,** Reversible phosphorylation of hydroxymethylglutaryl-CoA reductase, *Mol. Asp. Cell Reg.*, 1, 63, 1980.
45. **Sanghvi, A., Grassi, E., Warty, V., Diven, W., Wight, C., and Lester, R.,** Reversible activation-inactivation of cholesterol 7α-hydroxylase possibly due to phosphorylation-dephosphorylation, *Biochem. Biophys. Res. Commun.*, 103, 886, 1981.
46. **Goodwin, C. D., Cooper, B. W., and Margolis, S.,** Rat liver cholesterol 7α-hydroxylase: modulation of enzyme activity by changes in phosphorylation state, *J. Biol. Chem.*, 257, 4469, 1982.
47. **Kuroki, S., Muramoto, S., Kuramoto, T., and Hoshita, T.,** Sex differences in gallbladder bile acid composition and hepatic steroid 12α-hydroxylase activity in hamsters, *J. Lipid Res.*, 24, 1543, 1983.
48. **Kuroki, S. and Hoshita, T.,** Effect of bile acid feeding on hepatic steroid 12α-hydroxylase activity in hamsters, *Lipids,* 18, 789, 1983.
49. **Uchida, K., Takase, H., Kadowaki, M., Nomura, Y., Matsubara, T., and Takeuchi, N.,** Altered bile acid metabolism in alloxan diabetic rats, *Jpn. J. Pharmacol.*, 29, 553, 1979.
50. **Wikvall, K.,** unpublished data.
51. **Björkhem, I., Danielsson, H., and Gustafsson, J.,** On the effect of thyroid hormone on 26-hydroxylation of C_{27}-steroids in rat liver, *FEBS Lett.*, 31, 20, 1973.
52. **Björkhem, I., Eriksson, M., and Einarsson, K.,** Evidence for a lack of regulatory importance of the 12α-hydroxylase in formation of bile acids in man: an in vivo study, *J. Lipid Res.*, 24, 1451, 1983.
53. **Murakami, K., Okada, Y., and Okuda, K.,** Purification and characterization of 7α-hydroxy-4-cholesten-3-one 12α-monooxygenase, *J. Biol. Chem.*, 257, 8030, 1982.

Chapter 3

INTRODUCTION TO CHOLESTEROL 7α-HYDROXYLASE: ROLE AS RATE-LIMITING ENZYME IN BILE ACID SYNTHESIS

Kathleen M. Botham

TABLE OF CONTENTS

I. HISTORICAL INTRODUCTION

A scheme showing 7α-hydroxycholesterol as the first intermediate in the biosynthetic pathway from cholesterol to bile acids was proposed by Danielsson in 1963.[1] This proposal was based on earlier in vivo and in vitro work on the metabolism of cholesterol and other hypothetical intermediates in the pathway.[2-4] Experimental proof showing that the 7α-hydroxylation of cholesterol is the first step in bile acid synthesis soon followed from studies using liver homogenates.[5-7] The currently accepted pathway for bile acid formation is shown in Figure 1. A rate-limiting role for cholesterol 7α-hydroxylase in the pathway was postulated as early as 1958,[8] but the first experimental evidence for this was not forthcoming until the late 1960s when several laboratories showed that the activity of the enzyme was increased markedly by biliary drainage and by feeding the bile acid sequestrant, cholestyramine[9-12] (both procedures known to cause large increases in the amount of bile acid produced by the liver), while the activities of other enzymes involved in the synthesis of bile acids were increased only slightly or not at all.[13-14] Recognition of the rate-limiting nature of cholesterol 7α-hydroxylase coupled with the knowledge gained from earlier work that excretion of bile acids in the feces is a major route for the elimination of cholesterol from the body[15,16] established the importance of the enzyme and extensive studies have been carried out in order to gain information about its nature and regulation. A number of different experimental systems have been used including the whole animal, isolated hepatocytes, isolated perfused liver, liver microsomal preparations, and partially purified enzyme preparations, but despite the intensive efforts of scientists over two decades the membrane-associated character and essentially insoluble substrate of cholesterol 7α-hydroxylase have made the understanding of the enzyme a slow and difficult process, and many questions remain unanswered. Nevertheless, considerable progress has been made and in the remainder of this chapter I shall give a broad outline of the current knowledge of cholesterol 7α-hydroxylase, with particular reference to its rate-limiting role, and show how each of the experimental systems mentioned above has contributed to different aspects of the overall picture.

II. SYSTEMS USED FOR THE STUDY OF CHOLESTEROL 7α-HYDROXYLASE

A. Whole Animal

Ideally the whole animal is the best system for the study of cholesterol 7α-hydroxylase as all the factors outside the liver that may influence the system such as the enterohepatic circulation of bile acids and cholesterol, plasma lipoproteins, circulating hormones, etc., are present. It is clearly difficult, however, to study one reaction in a biosynthetic pathway directly using the whole animal. Another disadvantage lies in the very complexity of the system, which often leads to problems in interpretation of the results obtained. The overall process of bile acid synthesis can be investigated in vivo, however, and much information about the regulation of the pathway, and therefore indirectly the regulation of cholesterol 7α-hydroxylase, has been gained in this way. Isotopic equilibrium methods have been used by Chevallier[17] and others[18-20] to determine bile acid turnover under various conditions. Radioactive labeling techniques have been employed to investigate the pool size, origin, and compartmentation of the substrate for cholesterol 7α-hydroxylase in both bile fistula and intact rats,[21-23] while Björkhem and Lewenhaupt have used the nonradioactive isotope ^{18}O for study of the origin of the cholesterol used in bile acid synthesis.[24] In vivo experiments, particularly with bile fistula rats, have also contributed information about the regulation of cholesterol 7α-hydroxylase (see Section V).

B. Isolated Hepatocytes

A method for the preparation of isolated rat liver parenchymal cells in high yield using continuous recirculating perfusion of rat liver with collagenase was described by Berry and

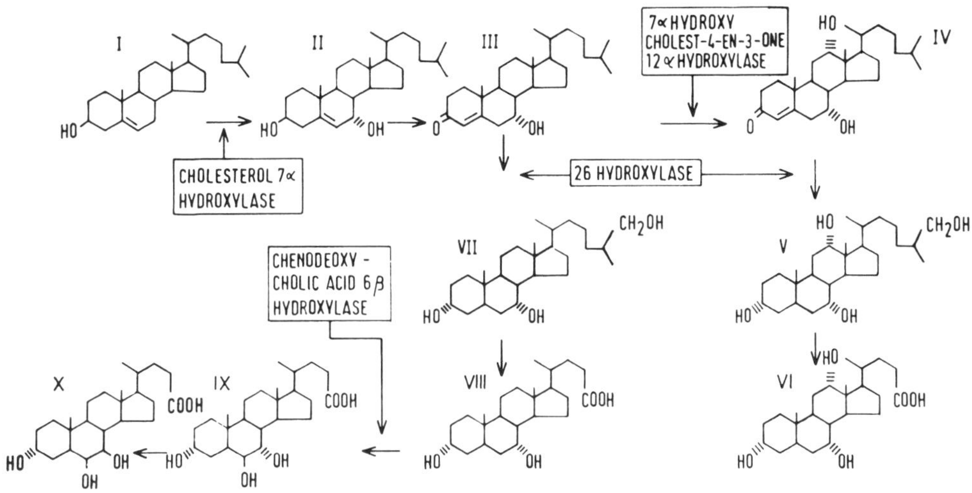

FIGURE 1. Pathways for bile acid synthesis in the rat. I, cholesterol: II, 7α-hydroxycholesterol: III, 7α-hydroxycholest-4-en-3-one: IV, 7α, 12α-dihydroxycholest-4-en-3-one: V, 5β-cholestane-3α, 7α, 12α, 26 tetrol: VI, cholic acid: VII, 5β-cholestane-3α, 7α, 26 triol: VIII, chenodeoxycholic acid: IX, 3α, 6β, 7α-trihydroxycholanic acid (α-muricholic acid): X, 3α,6β,7β-trihydroxycholanic acid (β-muricholic acid).

Friend in 1969,[25] and, with minor modifications, this technique has been used extensively for the investigation of all aspects of liver function since then. Isolated intact cells have the advantage of an easily manipulated simple in vitro system, while retaining the intracellular organization that is essential for the regulation of enzyme activity. First attempts to study cholesterol 7α-hydroxylase directly by measuring its activity in isolated hepatocytes were not encouraging as this parameter decreased to about 50% of its original value after 2 hr of incubation of the cells no matter what conditions were used.[26] Experiments in which the mass of bile acids produced by the cells was determined by gas-liquid chromatography,[27,28] an enzymatic luminometric method,[29] or radioimmunoassay,[30] however, have proved more successful. As cholesterol 7α-hydroxylase is the rate-limiting enzyme for bile acid synthesis, the amount of bile acid produced by isolated hepatocytes is directly related to its activity. Botham et al. showed that the synthesis of bile acids during short term (4 hr) incubation of isolated hepatocytes in suspension reflected that found in vivo in being higher in cells from rats fed cholestyramine and when the hepatocytes were prepared during the dark phase of the rats' diurnal cycle[30] (Figure 2). These findings were confirmed by Kempen et al.[29]

A disadvantage of the isolated hepatocyte system is that cells suspension are not polarized as they are in vivo so that bile acids are taken up on one side and secreted on the other. One consequence of this is that a relatively large amount of β-muricholic acid, a metabolite of chenodeoxycholic acid in rat liver[31] (see Figure 1), is synthesized by isolated hepatocytes.[29,32] It is necessary, therefore, to take the synthesis of β-muricholic acid into account when studying bile acid synthesis in isolated rat hepatocytes. Production of this bile acid has been shown to be increased in cells from cholestyramine-fed rats.[29,32]

Much information about the regulation of cholesterol 7α-hydroxylase has been obtained with the use of isolated hepatocytes in suspension and these studies will be discussed in Section V. Experimental systems using liver cells in suspension, however, have the disadvantage of being short-lived so that only short-term regulation can be studied. In addition it is possible that some receptors on the surface of cells may be temporarily damaged during isolation with collagenase. For some studies, therefore, longer term maintenance of hepatocytes than can be obtained with cell suspensions is necessary. Isolated hepatocytes in monolayer culture have been shown to produce bile acids for periods up to 48 hr.[33,34]

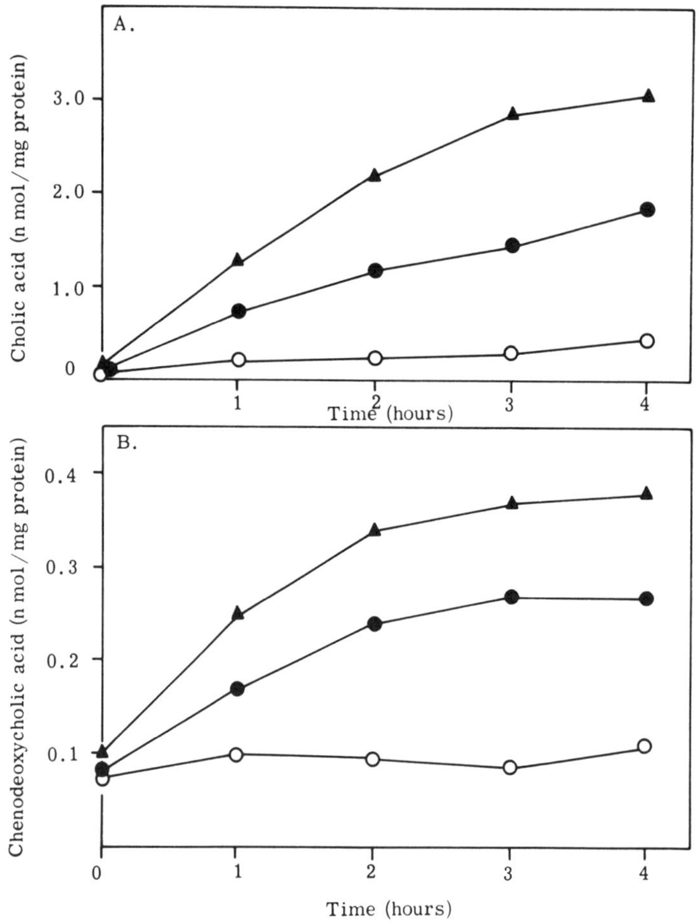

FIGURE 2. Synthesis of bile acids by isolated hepatocytes in suspension. Synthesis of (A) conjugated cholic acid and (B) conjugated chenodeoxycholic acid by hepatocytes isolated from rats fed a control diet (○—○) or a diet containing 4% cholestyramine (●—●) or rats fed a diet containing 4% cholestyramine and killed in the mid-dark period of the diurnal cycle (▲—▲) (From Botham, K. M., Beckett, G. J., Percy-Robb, I. W., and Boyd, G. S., *Eur. J. Biochem.*, 103, 299, 1980. With permission.)

Furthermore, bile acid production by the cells has been found to follow a similar pattern to that found in vivo in being increased in hepatocytes prepared from rats given cholestyramine in the diet compared to those fed a low fat soft diet,[26] while cells from rats given a commercial pellet diet show an intermediate level of synthesis[106] (Figure 3). Isolated hepatocytes maintained in monolayer culture have been used successfully in the investigation of the longer-term regulation of cholesterol 7α-hydroxylase and the importance of lipoprotein cholesterol as a substrate for the enzyme (see Section V).

C. Isolated Perfused Liver

One way of avoiding the problems of the complexity of whole animal experiments while maintaining the integrity of the liver is to use the isolated perfused organ. This system has the advantage over isolated hepatocytes of retaining the polarity of the liver cells with respect to the circulation of bile acids (see Section II.B) and avoids the difficulties of possible damage to receptors. Only short-term experiments can be carried out, however, and the

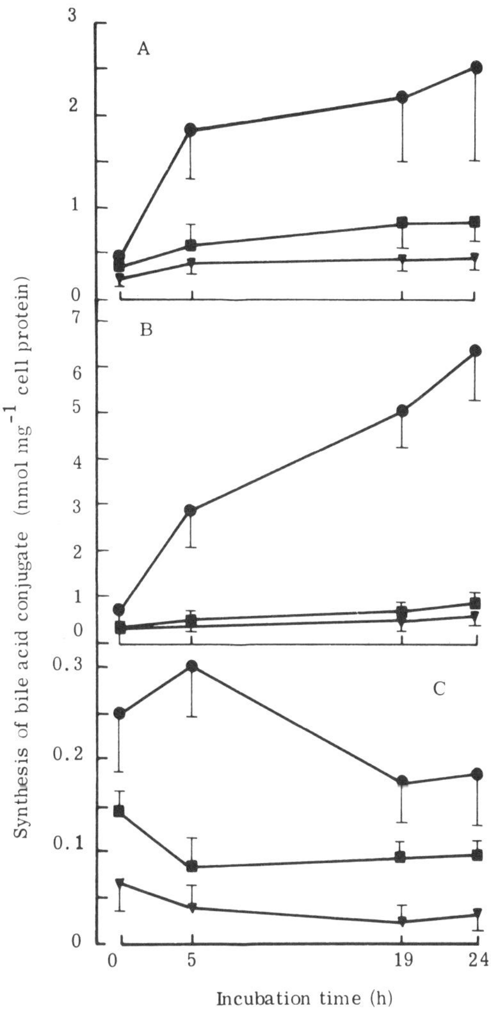

FIGURE 3. Synthesis of bile acids in isolated hepatocytes in monolayer culture. Synthesis of (A) conjugated cholic acid, (B) conjugated β-muricholic acid and (C) conjugated chenodeoxycholic acid by hepatocytes isolated from rats fed a low fat soft diet (▼ - ▼) a commercial pellet diet (■ - ■) or a diet containing 4% cholestyramine (● - ●) during 24 hr in monolayer culture. (From Ford, R. P., Ph.D. thesis, University of Edinburgh, 1984. With permission.)

system lacks the ease of manipulation characterizing experiments with isolated hepatocytes. Isolated perfused livers have been used by Ayaki et al.[22] to investigate the contribution of lipoproteins to precursor pools for cholesterol 7α-hydroxylase and have also contributed evidence for the rate-limiting role of the enzyme (see Section IV).

D. Microsomal Preparations

Isolated liver microsomal preparations provide a simple easily manipulable in vitro system for the study of cholesterol 7α-hydroxylase under conditions in which further metabolism of 7α-hydroxycholesterol can be eliminated.[35] In this system the rate of conversion of [4-^{14}C]cholesterol to 7α-hydroxy[^{14}C]cholesterol is related to the activity of cholesterol 7α-hydroxylase. The formation of 7α-hydroxycholesterol by nonenzymatic autoxidation was a problem in early studies[36] but this was soon found to be minimized by homogenizing the liver in the presence of EDTA[9] or by incubating the microsomes with β-mercaptoethylamine.[10] Although widely used in the study of cholesterol 7α-hydroxylase the [4-^{14}C] method has some drawbacks. These have been reviewed in detail by Myant and Mitropoulos,[35] and include; uncertainty as to whether the exogenously added [4-^{14}C]cholesterol is fully equilibrated with the enzyme substrate pool and how this is dependent on the method of addition; uncertainty as to the size of the substrate pool as changes in this would alter the measured activity of the enzyme; and the impossibility of attaining concentrations of cholesterol sufficient to saturate the enzyme. It is clear that all these problems except the last would be overcome if the mass of 7α-hydroxycholesterol formed during the incubation of microsomes could be determined. In order to achieve this Mitropoulos and Balasubramaniam[37] devised a method based on ^{3}H-acetylation of the 7α-hydroxy[^{14}C]cholesterol isolated from microsomal incubations. In this way the mass of 7α-hydroxycholesterol could be estimated from the specific activity of the ^{3}H acetyl group and losses during extraction and isolation calculated from the ^{3}H:^{14}C ratio. Other workers have used gas chromatography in conjunction with mass spectrometry to measure the mass of 7α-hydroxycholesterol.[38-40] This technique has the advantage of being less time-consuming than the ^{3}H-acetylation method, but the disadvantage of requiring large, expensive equipment.

Isolated microsomal preparations have been used to obtain information about the nature of cholesterol 7α-hydroxylase,[41-43] its substrate specificity[9,44,45] and substrate pool size and origin.[39,46] Many studies on the rate-limiting role and regulation of the enzyme have also made use of this system either alone or after conditions in vivo have been changed (see Sections IV and V).

E. Partially Purified Preparations

In order to carry out detailed investigations of the nature, kinetics, and mechanism of action of cholesterol 7α-hydroxylase, isolation and purification of the enzyme protein is required, but because of the close association of cholesterol 7α-hydroxylase with microsomal membranes and the labile nature of the cytochrome P-450 component of the enzyme system this has proved extremely difficult to achieve.

Partially purified preparations of varying efficiency in the 7α-hydroxylation of cholesterol have been obtained however, by persistent studies notably in the laboratories of Danielsson, Wikvall and co-workers, and Boyd and co-workers.[47-50]

Cholesterol 7α-hydroxylase is known to consist of at least two parts, NADPH cytochrome P-450 reductase and cytochrome P-450, and the stimulatory effect of cholestyramine has been found to be associated with the cytochrome P-450 fraction.[47] The NADPH cytochrome P-450 reductase component has proved relatively easy to purify and can be used together with purified cytochrome P-450 to reconstitute the cholesterol 7α-hydroxylase system. The 7α-hydroxylating activity of the cytochrome P-450 fractions obtained in different laboratory in this type of reconstituted system varies depending on the method used for the solubilization and separation of the hemoproteins. A combination of a low concentration of cholate with nonionic detergents such as Renex® 690 or Emulgen 913 has been used by some workers,[47,48] but as cholesterol 7α-hydroxylase is highly sensitive to bile acids and detergents, the 7α-hydroxylating activity of cytochrome P-450 fractions obtained by this method is relatively low. Ozasa and Boyd, using the nonionic detergent Nonidet, have isolated a cytochrome P-450 fraction which is able to 7α-hydroxylate cholesterol at a rate three-to fourfold higher than that found in cytochrome P-450 fractions solubilized and separated by any other method.[50] In view of this fact it is perhaps surprising that the techniques described by Ozasa and Boyd have not become more widely used.

Studies using purified cytochrome P-450 fractions in the reconstituted cholesterol 7α-hydroxylase system have assisted in the investigation of the kinetics,[48,50] specificity,[50-53] and system components[48,54] of the enzyme.

III. THE NATURE OF CHOLESTEROL 7α-HYDROXYLASE

Studies with liver microsomal preparations have shown that cholesterol 7α-hydroxylase requires NADPH and molecular oxygen for full activity,[37,42,43] and is inhibited by carbon monoxide and antibody to NADPH cytochrome P-450 reductase.[10,11,37] Furthermore, the inhibition by carbon monoxide is reversed by monochromatic light at 450 nm.[43] These findings indicate that cholesterol 7α-hydroxylase belongs to a group of enzymes known as mixed function oxidases,[55] which involve NADPH cytochrome P-450 reductase as a flavoprotein and cytochrome P-450 as a terminal oxidase. Cholesterol 7α-hydroxylase has been shown to consist of at least these two components.[12,43,47]

The mixed function oxidase system of rat liver microsomes contains a number of forms of cytochrome P-450 inducible by drugs such as phenobarbitone and 3-methylcholanthrene and differing in substrate specificity, subunit molecular weight, and electrophoretic behavior.[56] The activity of cholesterol 7α-hydroxylase in rat liver microsomes, however, has usually been found to be unaffected by phenobarbitone treatment,[10,57,58] and conversely, increasing the activity of the enzyme by interrupting the enterohepatic circulation does not result in any detectable change in total hepatic cytochrome P-450 or NADPH cytochrome P-450 reductase content.[58,69] Cholesterol 7α-hydroxylase activity also seems to be more sensitive to detergents and salts than other cytochrome P-450-dependent reactions.[60] More recently the work of Böstrom and Wikvall[52] and Ozasa and Boyd[50] with purified cytochrome P-450 fractions from rat and rabbit liver microsomes in the reconstituted enzyme system has shown that cholesterol 7α-hydroxylating activity can be separated from other cytochrome P-450-dependent hydroxylating activities. Danielsson and Wikvall have also found that the cytochrome P-450 catalyzing the 7α-hydroxylation of cholesterol has a shorter half-life than those catalyzing other hydroxylations.[61] These findings indicate that the cytochrome P-450 involved in the 7α-hydroxylation of cholesterol is distinct not only from those concerned with drug metabolism, but also from those catalyzing other hydroxylations in bile acid synthesis, and probably comprises only a small part of the total liver cytochrome P-450 content.

Current knowledge about the specificity of cholesterol 7α-hydroxylase for its substrate, cholesterol, is limited mainly because of the difficulties involved in the interpretation of experiments with cholesterol analogues using isolated microsomal preparations.[35] Boyd et al. have shown shown that the length of the side chain of the sterol is important for efficient 7α-hydroxylation in liver microsomes[59] and even more important in the reconstituted enzyme system.[51] Recent advances in the isolation of specific cytochrome P-450 fractions for cholesterol 7α-hydroxylation[48-53] should facilitate experiments on the ability of the enzyme to catalyze the 7α-hydroxylation of various analogues of cholesterol and lead to progress in this area in the near future.

A role for a cytosolic protein in the activation of cholesterol 7α-hydroxylase was suggested by early work showing that 20,000 g supernatant prepared from rat liver 7α-hydroxylated cholesterol more efficiently than microsomes and that this increase in efficiency could be reproduced by adding 100,000 g supernatant to the microsomal preparations.[5,41] More recently, several groups have reported the isolation of partially purified proteins from rat liver 100,000 g supernatant which activate cholesterol 7α-hydroxylase either in isolated microsomes[62-65] or in the reconstituted enzyme complex.[48,54] The physiological function of such proteins, however, remains unclear. Some work with microsomal preparations has suggested that the cytosolic protein is involved in a dephosphorylation-phosphorylation mechanism for the regulation of cholesterol 7α-hydroxylase activity,[64,65] but this has not been found to be the case in experiments with the partially purified enzyme.[48,54] Modulation of cholesterol 7α-hydroxylase activity by sulfydryl reagents, however, has been observed in both experimental systems[40,41,66] and Danielsson et al. have proposed that the most likely function of the protein is to facilitate the reduction of protein disulfide bonds in the cytochrome P-450 molecule by glutathione and/or thioredoxin, both biologically important sulfydryl reagents.[54]

IV. THE RATE-LIMITING ROLE OF CHOLESTEROL 7α-HYDROXYLASE

It was recognized at an early stage that the position of cholesterol 7α-hydroxylase in catalyzing the first step in the metabolism of cholesterol to bile acids and thus determining the ultimate fate of the cholesterol molecule involved in the reaction makes it an obvious candidate for the rate-limiting enzyme for the pathway.[8] If cholesterol 7α-hydroxylase is rate-limiting for the formation of bile acids, then two requirements that can be demonstrated experimentally should be fulfilled. First, the activity of the enzyme should be increased or decreased in situations where bile acid synthesis is known to be raised or lowered while the activities of other enzymes in the pathway remain unaffected, and second, the formation of bile acids from the product of the 7α-hydroxylation reaction, 7α-hydroxycholesterol, should be unregulated.

Interruption of the enterohepatic circulation of bile acids by biliary drainage[67] or by feeding cholestyramine[14] is known to cause a large increase in bile acid synthesis and experiments with microsomal preparations have shown that cholesterol 7α-hydroxylase activity increases concurrently with bile acid production in bile fistula and cholestyramine-fed rats, but the activities of some other enzymes involved in the pathway are not changed.[13,14] The 12α-hydroxylation of 7α-hydroxy-4-cholesten-3-one was also found to be stimulated but to a lesser extent than the 7α-hydroxylation of cholesterol in biliary drained rats.[13,14] The significance of this will be discussed in Section VI.B. Mosbach and co-workers, in studies in vivo with bile fistula rats, have found that although duodenal infusion of taurocholate inhibits the incorporation of radioactive acetate, mevalonate, and cholesterol into bile acids, incorporation of 7α-hydroxy[^{14}C]cholesterol is not affected.[68] Conversely, cholestyramine feeding was found to lead to stimulation of the incorporation of radioactive acetate, mevalonate, and cholesterol, but not 7α-hydroxy[^{14}C]cholesterol into bile acids in perfused rabbit liver.[69] More direct evidence has been obtained from experiments with isolated hepatocytes which have shown that addition of 7α-hydroxycholesterol to the incubation medium increases the amount of bile acid produced by the cells[30] (Figure 4). Addition of other oxygenated derivatives of cholesterol to the hepatocyte incubations either had no effect or inhibited bile acid synthesis, showing that the stimulation is specific for 7α-hydroxycholesterol.

Enzymes that have a rate-limiting function are often rapidly turned over with half-lives of the order of a few hours. Work with isolated microsomes prepared from the livers of rats given the protein synthesis inhibitors cycloheximide, actinomycin D, or puromycin has indicated that cholesterol 7α-hydroxylase has a half-life of 2 to 3 hr.[45,70]

Taken together, these studies from experimental systems using the whole animal, perfused liver, isolated hepatocytes, and microsomal preparations, provide a substantial body of evidence for the rate-limiting nature of cholesterol 7α-hydroxylase under most physiological conditions. We can state with some confidence, therefore, that the body normally regulates the amount of bile acid synthesized in the liver by regulating the activity of cholesterol 7α-hydroxylase. Our understanding of the mechanisms by which this is brought about, however, is far from complete. The extensive studies that have been carried out in this area indicate that several different processes may be involved in the regulation of cholesterol 7α-hydroxylase activity.

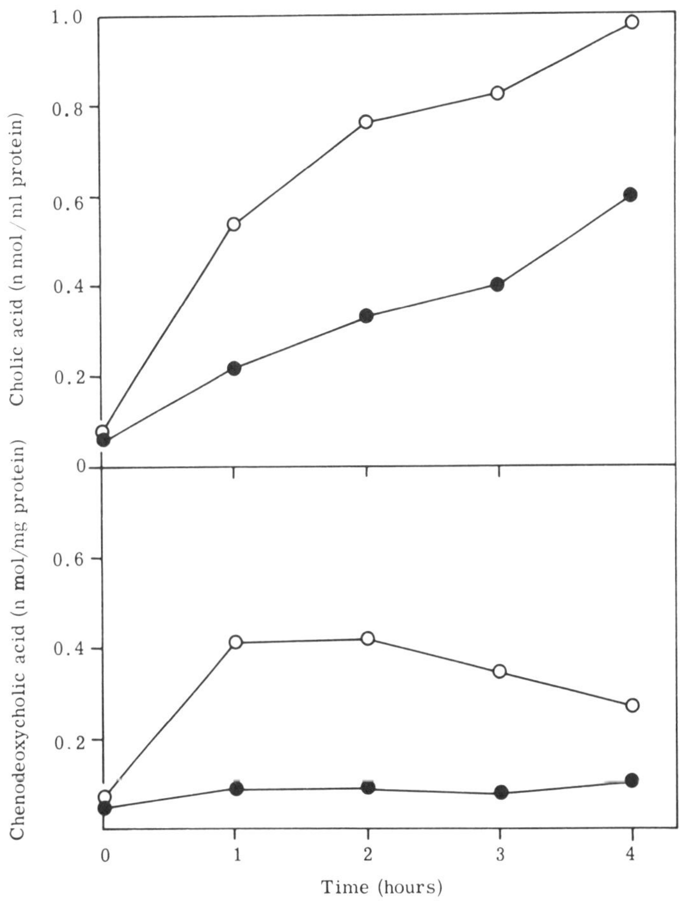

FIGURE 4. The effect of 7α-hydroxycholesterol on bile acid synthesis in isolated hepatocytes. Synthesis of bile acids by isolated hepatocytes in suspension in the presence (○—○) and absence (●—●) of 7α-hydroxycholesterol. (From Botham, K. M., Beckett, G. J., Percy-Robb, I. W., and Boyd, G. S., *Eur. J. Biochem.*, 103, 299, 1980. With permission.)

V. THE REGULATION OF CHOLESTEROL 7α-HYDROXYLASE

The major regulatory mechanisms which are thought to act on cholesterol 7α-hydroxylase include feedback inhibition by bile acids returning to the liver in the enterohepatic circulation, supply of the substrate, cholesterol, and phosphorylation-dephosphorylation of the enzyme. Before discussing each of these in detail it is important to point out that the large (five to tenfold) increase in cholesterol 7α-hydroxylase activity that occurs when animals are fed cholestyramine or biliary drained is not observed until after a relatively long lag period of approximately 24 hr.[9,13] In contrast to this, small changes in the activity of the enzyme take place during the rats normal 24 hr diurnal cycle with a minimum occurring during the day and a maximum (approximately twofold higher) during the night under normal light conditions.[71] Mitropoulos et al. have shown that this diurnal variation persists during biliary drainage and when rats are fed cholestyramine and have suggested that the diurnal variation in cholesterol 7α-hydroxylase activity and the longer term stimulatory effect caused by interrupting the enterohepatic circulation are regulated by two independent processes.[72]

A. Feedback Inhibition

In the normal animal bile acids are synthesized at a relatively low rate in the liver, secreted into bile, and then pass into the intestine where 95% is reabsorbed and returns to the liver in the portal blood. When bile acids are diverted from this enterohepatic circulation by biliary drainage, bile acid synthesis increases by about five to tenfold,[67] while raising the concentration of bile acids in the portal blood by oral administration of cholate leads to a decrease in their synthesis.[73,74] These observations suggest that bile acids returning to the liver in the portal blood may exert a feedback inhibition on their own synthesis. In support of this, the activity of cholesterol 7α-hydroxylase in microsomal preparations has been shown to be increased after feeding cholestyramine and after portacaval anastomosis in the rat.[9,75] In addition, Shefer et al. have reported that duodenal infusion of taurocholate at a rate greater than 10 mg/100 g/hr into bile fistula rats restores bile acid synthesis to its normal low level.[74] Another study, however, did not detect any repression of bile acid synthesis after infusion of similar concentrations of taurocholate into similarly treated rats.[76] Cronholm and Sjövall could not detect any effect of cholestyramine feeding (1%) on portal blood bile acid concentrations,[77] but more recently, Botham et al. have found that feeding a higher level of cholestyramine (4%) to rats caused a small, but significant decrease in bile acid levels in portal blood.[78] In spite of the inconsistencies in the evidence, it has been accepted for many years that feedback inhibition by bile acids plays a major role in the regulation of cholesterol 7α-hydroxylase.[35] The results of some recent studies, particularly those with isolated hepatocytes, however, have begun to cast doubt on this idea.

That the diurnal variation in cholesterol 7α-hydroxylase activity is not under feedback control has been demonstrated by Botham et al. who have found; first, that bile acid production in hepatocytes isolated from cholestyramine-fed rats is not inhibited by concentrations of taurocholate similar to those found in the portal blood of rats with a low rate of bile acid synthesis during a 4 hr incubation period;[79] second, that the portal blood concentrations of bile acids in bile fistula rats reach their lowest point about 6 to 8 hr before hepatocytes isolated from such rats begin to show increased bile acid production;[78] and third, that the maximum level of bile acids in the portal blood during the diurnal variation coincides with or follows the peak in activity of cholesterol 7α-hydroxylase, rather than precedes it.[78] These results indicate that the response of bile acid synthesis to changes in portal blood bile acid concentrations are too slow to explain the diurnal variation in cholesterol 7α-hydroxylase activity, but do not rule out the possibility that the longer term regulatory mechanism may involve feedback inhibition. Davis et al., however, in experiments with isolated hepatocytes in monolayer culture, have shown that physiological concentrations of exogenous taurocholate do not inhibit the incorporation of [^{14}C] mevalonic acid into bile acids over 48 hr of incubation (Figure 5).[80] Furthermore, these workers have been unable to demonstrate any repression of bile acid synthesis by intrajugular infusion of taurocholate into bile fistula rats.[81] The results from the studies with isolated hepatocytes both in suspension and in monolayer culture[79,80] suggest, then, that bile acids do not interact directly with liver cells to inhibit their own synthesis. It is possible, however, that bile acid concentrations in the intestine play some part in the initiation of feedback inhibition of bile acid synthesis so that lack of some intestinal factor might explain the results obtained with isolated hepatocytes.[80] If this were the case, then the discrepancy between the results of Shefer et al.[74] and Davis[81] would be explained as infusion of bile acid into the duodenum, but not into the jugular vein, would be expected to inhibit bile acid production in bile fistula rats. Pries et al.,[82] however, did find inhibition of bile acid synthesis with intrajugular infusions of cholate in biliary drained rats.

The accepted view of bile acids returning to the liver in the portal blood regulating cholesterol 7α-hydroxylase activity has been questioned by these studies and clearly more work is needed to establish whether intestinal factors and/or bile acids are involved in a

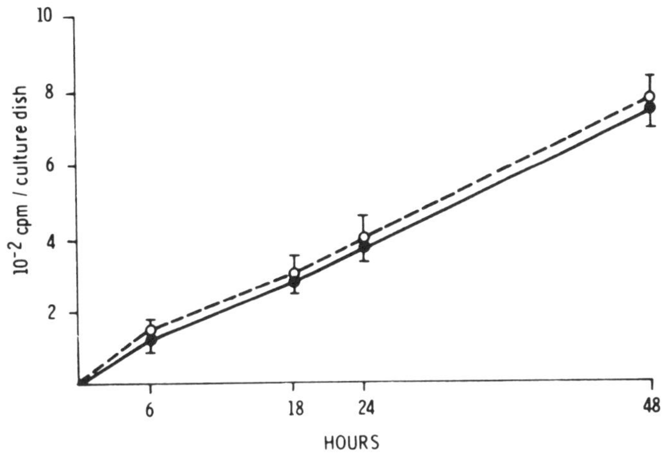

FIGURE 5. The effect of taurocholate on bile acid synthesis in hepatocytes in monolayer culture. Incorporation of [2-^{14}C] mevalonic acid into bile acids by monolayer cultures of hepatocytes incubated in the presence (○ - ○) and absence (●—●) of 100 μM taurocholate for 48 hr. (From Davis, R. A., Highsmith, W. E., McNeal, M. M., Schexnayder, J. A., and Kuan, J. W., *J. Biol. Chem.*, 258, 4079, 1983. With permission.)

negative feedback process and the importance of this type of mechanism in the overall control of cholesterol 7α-hydroxylase activity.

B. Substrate Supply

The possible sources for the substrate for cholesterol 7α-hydroxylase include newly synthesized cholesterol, preformed cholesterol existing in various functional pools within the cell and extracellular cholesterol in the form of plasma lipoproteins containing cholesterol arising from the diet (chylomicron remnants) or from other tissues in the body (LDL and HDL). These distinctions, of course, are somewhat artificial as all the sources of cholesterol are interchangeable with each other and all the exchangeable cholesterol in the body is potentially accessible to the cholesterol 7α-hydroxylase enzyme.[35] They do provide a useful device, however, for the discussion of how the supply of substrate may be involved in the regulation of the enzyme.

Myant and Mitropoulos,[35] based on studies on the relative specific activities of bile acids and cholesterol after incorporation of radioactive precursors in vitro in isolated microsomes and in vivo in bile fistula rats, have suggested that the preferred substrate for cholesterol 7α-hydroxylase is newly synthesized cholesterol. Similar experiments by Ogura et al. using [^{14}C]mevalonic acid and [^{3}H]-cholesterol[21] and Björkhem and Lewenhaupt using the inhalation of ^{18}O,[24] and the finding that HMGCoA reductase, the major rate-limiting enzyme in cholesterol biosynthesis, and cholesterol 7α-hydroxylase are located in the same submicrosomal fraction,[83] support this view. It is important to point out, however, that these experiments do not show that newly synthesized cholesterol is the sole substrate for cholesterol 7α-hydroxylase, but only that it is preferred. Nevertheless, it is possible that in certain physiological conditions the rate of bile acid synthesis might be determined by the rate of cholesterol biosynthesis rather than directly by the activity of cholesterol 7α-hydroxylase. The activities of HMG CoA reductase and cholesterol 7α-hydroxylase are often, but not always, changed in parallel and it has been suggested that there is an obligatory link between the regulation of the two enzymes.[35] This has implications in the overall control of cholesterol metabolism and will be discussed in Section VI.A.

Preformed intracellular unesterified cholesterol certainly can act as a substrate for cholesterol 7α-hydroxylase. Long et al. in experiments with bile fistula rats, have calculated that animals with a relatively intact enterohepatic circulation (i.e., 2 hr after making the bile fistula) synthesize 75% of bile acids from preformed cholesterol.[84] Similar results have been obtained by Stange et al.[23] using infusion of ^3H_2O into biliary drained rats. These workers have found that the proportion of bile acid derived from the different precursor pools depends on the individual bile acid studied, but after 12 hr biliary drainage at least 76% originates from preformed cholesterol. Both groups have reported that increasing the time of biliary drainage leads to an increased proportion of bile acid arising from newly synthesized cholesterol.[23,84] Kempen et al.,[85] in experiments on the incorporation of ^3H_2O into cholesterol and bile acids by isolated hepatocytes from cholestyramine-fed rats, have calculated that during the first hour of incubation all the bile acid is formed from preexisting cholesterol. After this time, however, the rate of bile acid synthesis dropped and newly synthesized cholesterol became more important as a substrate, presumably because the sources of preformed cholesterol available to the enzyme were then exhausted and the rate of bile acid production became dependent on the rate of cholesterol synthesis.

Davis et al.[81] have found a significant correlation between the cholesterol concentration of microsomal membranes from control, mid-dark cycle, and bile fistula rats and cholesterol 7α-hydroxylase activity and have proposed that microsomal cholesterol is primarily responsible for the regulation of bile acid synthesis. Unesterified cholesterol is known to modulate the fluidity of cell membranes in a way that can affect their permeability and the properties of membrane-bound enzymes and transport systems.[86] It is possible, therefore, that microsomal membrane cholesterol concentrations may regulate cholesterol 7α-hydroxylase activity by causing physical changes in the membrane as well as by acting as a supply of substrate.

The hypothesis that this mechanism plays a major role in the control of cholesterol 7α-hydroxylase activity, however, requires more evidence before it can be substantiated.

The effect of exogenous cholesterol from the diet on cholesterol 7α-hydroxylase activity is unresolved. Some workers have found increased activity of the enzyme on feeding cholesterol to rats,[10,20,72] while others, including those studying the mass of 7α-hydroxycholesterol formed in microsomal preparations, have found no significant change.[38,87] In contrast to this, bile acid synthesis as measured by fecal excretion of bile acids in isotope equilibrium studies has shown a consistent two to threefold increase on supplementation of the diet with cholesterol.[17-20] To add to the confusion, Botham and Boyd reported no significant difference in bile acid production between cholesterol-fed and control rats either in vitro with isolated hepatocytes in suspension or in vivo during biliary drainage,[88] whereas Davis et al. found that cholesterol-feeding increased the amount of bile acid formed by hepatocytes in monolayer culture.[34]

One approach that may help to resolve the problem of these conflicting results is to use isolated hepatocytes to investigate the effect of exogenous cholesterol in the form of lipoproteins on bile acid synthesis. A subfraction of HDL from rat plasma (HDL_2) has been shown to increase bile acid production in monolayer cultures of hepatocytes from cholestyramine-fed, but not control diet, rats.[89] These results suggest that the cells from rats with a basal rate of bile acid synthesis have sufficient cholesterol from intracellular sources to meet the demand of cholesterol 7α-hydroxylase, while those from rats with a stimulated rate of synthesis do not and are able to make use of cholesterol originating from an extracellular source. The work of Stange et al.[23] and Long et al.[84] showing that prolonged biliary drainage leads to an increase in the proportion of bile acid derived from newly synthesized cholesterol and that of Kempen et al.,[85] showing that the rate of synthesis of bile acids in hepatocytes from cholestyramine-fed rats falls after the first hour of incubation when it becomes more dependent on newly synthesized cholesterol, supports this idea. Davis et al., using monolayers of hepatocytes from rats with a basal rate of bile acid synthesis, have also

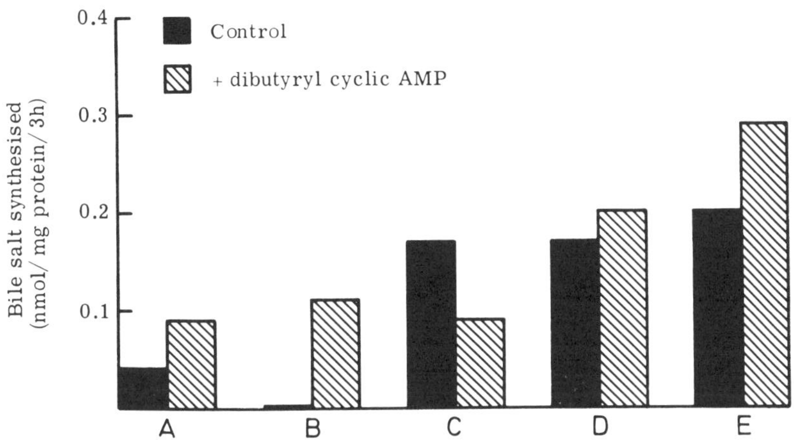

FIGURE 6. The effect of dibutyryl cyclic AMP on bile acid synthesis in isolated hepatocytes. Synthesis of (A) conjugated cholic acid, (B) conjugated chenodeoxycholic acid, (C) conjugated β-muricholic acid, (D) conjugated chenodeoxycholic acid + conjugated β-muricholic acid and (E) conjugated cholic acid + conjugated chenodeoxycholic acid + conjugated β-muricholic acid (total bile acids) in isolated hepatocytes incubated in suspension in the presence ▧ or absence ■ of 1m*M* dibutyryl cyclic AMP for 3h at 37°C. (From Botham, K. M. and Boyd, G. S., *Eur. J. Biochem.*, 136, 313, 1983. With permission.)

reported that LDL and HDL from cholesterol-fed rats did not affect the amount of bile acid formed. In this study, however, a lipoprotein fraction of d < 1.02 g thought to contain chylomicron remnants among other low density lipoprotein fractions, did increase bile acid synthesis in the cells.[34] This is the only information available as yet on the effect of chylomicron remnants, the form in which dietary cholesterol reaches the liver, on bile acid synthesis in hepatocyte monolayer cultures. The evidence obtained so far, then, suggests that exogenous extracellular cholesterol does not affect bile acid synthesis and, therefore, cholesterol 7α-hydroxylase activity, except in certain physiological conditions when a shortage of substrate may occur. Further studies, however, particularly on the effects of chylomicron remnants, should lead to rapid progress in this area.

C. Phosphorylation-Dephosphorylation

In recent years, evidence from studies with microsomal preparations and isolated hepatocytes has begun to accumulate suggesting that cholesterol 7α-hydroxylase activity can be changed by alterations in its phosphorylation state. Goodwin et al. found that the activity of the enzyme in microsomal preparations was increased in the presence of sodium fluoride, an inhibitor of phosphoprotein phosphatases, and decreased in the presence of bacterial alkaline phosphatase. The loss of activity in the latter case, however, was partially restored by ATP and magnesium.[90] Other laboratories have reported similar results.[65,91] These data suggest that cholesterol 7α-hydroxylase is activated by phosphorylation and inhibited by dephosphorylation. If this were the case, then cyclic AMP, which is known to promote the phosphorylation of enzymes, might be expected to have a role in the regulation of cholesterol 7α-hydroxylase. Investigation of this possibility has been facilitated by the use of isolated hepatocytes. Botham and Boyd have shown that dibutyryl cyclic AMP increases the synthesis of cholic and chenodeoxycholic acid, but decreases the production of β-muricholic acid (Figure 6A-C) in isolated hepatocytes in suspension. Total bile acid synthesis, however, was increased by 30 to 50%[92] (Figure 6E). (The differences in the synthesis of individual bile acids (Figure 6A-D) may be caused by effects of cyclic AMP on other enzymes in the bile acid pathway and will be discussed in Section VI.B.) Sundaram et al.[93] have reported similar results for cholic and chenodeoxycholic acid. A correlation between the intracellular cyclic AMP content of hepatocytes with and without induction by glucagon and bile acid synthesis has also been demonstrated.[94] These studies show that cyclic AMP is able to modulate the activity of cholesterol 7α-hydroxylase, presumably by a phosphorylation mechanism in isolated hepatocytes in vitro. The effects on bile acid synthesis, however, are observable after 1 hr of incubation of the cells[93,94] and are of the order of 30 to 50% (Figure 6).[92,93] These findings, therefore, cannot account for the stimulation of cholesterol 7α-hydroxylase activity that occurs when the enterohepatic circulation is interrupted by feeding cholestyramine or by biliary drainage as this is much larger (five to tenfold) and requires a time period of approximately 16 to 20 hr before it begins to be seen.[9,13] The diurnal variation in cholesterol 7α-hydroxylase activity has been shown to be regulated independently of the longer term effect,[94,78,79] and it is possible that cyclic AMP may increase the activity of the enzyme in the regulation of the relatively small, short-term changes that occur during the diurnal cycle. This is supported by the experiments of Mitropoulos and Balasubramaniam showing that the diurnal variation is regulated, at least partially, by glucocorticoid hormones.[95] There is little evidence available on the involvement of the cylic nucleotide in the longer term stimulation of cholesterol 7α-hydroxylase, but recent experiments in our laboratory with bile fistula rats suggest that the increase in bile acid synthesis, which normally occurs when rats are biliary drained, is inhibited by dibutyryl cyclic AMP.[107]

The regulation of cholesterol 7α-hydroxylase activity by a phosphorylation mechanism is an attractive idea from the point of view of the overall regulation of cholesterol metabolism as the activities of two other key enzymes in this process in the liver, HMG CoA reductase and acyl coenzyme A: cholesterol acyl transferase (ACAT) are known to be controlled by similar mechanisms (see Section VI.A). All the evidence so far, however, for such a process operating on cholesterol 7α-hydroxylase has come from in vitro experiments, and further work is needed to establish whether phosphorylation-dephosphorylation is important in the regulation of the enzyme in vivo.

VI. THE RELATIONSHIP OF CHOLESTEROL 7α-HYDROXYLASE TO OTHER ENZYMES OF CHOLESTEROL METABOLISM

The relationship of cholesterol 7α-hydroxylase to other enzymes in cholesterol metabolism can be considered in two different ways. First, from the point of view of the control of cholesterol homeostasis in the liver and second, by looking at its interaction with other enzymes involved in bile acid synthesis.

A. HMG CoA Reductase and ACAT

There is a great deal of evidence indicating that the concentration of unesterified cholesterol within the cell is tightly regulated within a narrow range.[96] Figure 7 shows the enzymes that are involved in this process in the liver.

That a close relationship exists between the major regulatory enzyme in cholesterol synthesis, HMG CoA reductase, and cholesterol 7α-hydroxylase has been known for some time.[35] More recently, the importance of ACAT, the enzyme which controls the esterification of cholesterol, in the regulation of cholesterol homeostasis in the liver has become apparent.[96] Experiments with microsomal preparations have shown that the activity of HMG CoA reductase is inhibited by phosphorylation and by increased cholesterol concentrations,[35,97] while hepatic ACAT is activated under both conditions.[98] These findings, together with the evidence that cholesterol 7α-hydroxylase activity is increased by phosphorylation (Section V.C), have led some workers to propose that the activities of the three enzymes are coordinately regulated in the liver,[86,98] so that when excess cholesterol enters the cell, cholesterol synthesis, controlled by HMG CoA reductase, would be reduced while cholesterol degra-

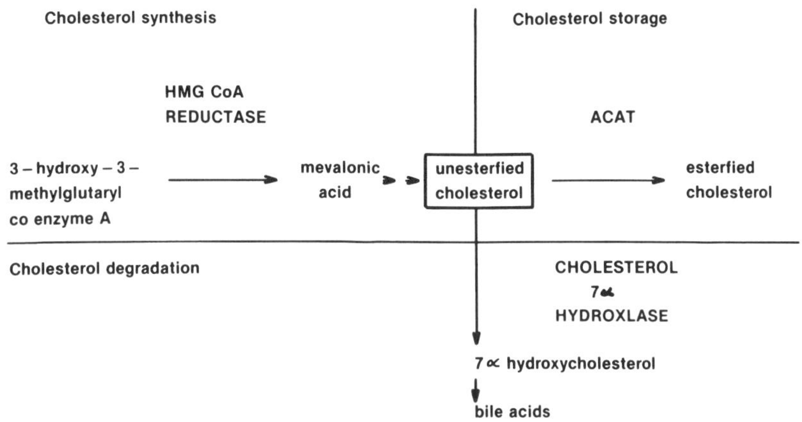

FIGURE 7. Enzymes regulating cholesterol homeostasis in the liver.

dation and storage, regulated by cholesterol 7α-hydroxylase and ACAT, respectively, would be enhanced. Conversely, a shortage of cholesterol within the cell would lead to activation of HMG CoA reductase and inhibition of cholesterol 7α-hydroxylase and ACAT. The regulation of these three enzymes by phosphorylation-dephosphorylation would provide an elegant short-term mechanism for the control of cholesterol homeostasis in the liver. There are difficulties, however, with this model. In most situations where cholesterol 7α-hydroxylase activity is increased, HMG CoA reductase activity is also activated rather than inhibited.[35] Even the short-term changes that take place during the diurnal variation in the activity of the two enzymes occur in parallel so that both are at a maximum in the mid-dark and at a minimum in the mid-light period of the cycle.[95] In addition, the amplitude of the diurnal rhythm of both enzymes is decreased by adrenalectomy and partially restored by glucocorticoids.[95] The relationship between HMG CoA reductase and cholesterol 7α-hydroxylase, therefore, cannot be fully accounted for in terms of a phosphorylation-dephosphorylation mechanism. The evidence that the preferred substrate for cholesterol 7α-hydroxylase is newly synthesized cholesterol,[21,24,35,83] however, suggests another link between the two enzyme systems. In conditions where cholesterol 7α-hydroxylase was not saturated with substrate, an increase in HMG CoA reductase activity might be expected to lead to increased 7α-hydroxylation of cholesterol, whereas in other conditions a rise in the activity of cholesterol 7α-hydroxylase may cause induction of HMG CoA reductase as a result of cholesterol removal.[35]

It has been established, then that a close relationship exists between the three key enzymes controlling cholesterol homeostasis in the liver. The exact nature of the interactions involved, however, are not yet completely understood.

B. Other Enzymes of Bile Acid Synthesis

Cholesterol 7α-hydroxylase is generally accepted to be the rate-limiting enzyme in the overall synthesis of bile acids. There is some evidence, however, that 7α-hydroxy-cholest-4-en-3-one-12α-hydroxylase (12α-hydroxylase) may play a secondary role in the regulation of the pathway. This enzyme catalyzes the conversion of 7α-hydroxy-cholest-4-en-3-one to 7α,12α-dihydroxy-cholest-4-en-3-one (see Figure 1), and once this reaction has occurred only cholic acid can be formed.[99] Experiments with microsomal preparations have shown that the activity of the 12α-hydroxylase enzyme is depressed by oral administration of bile acids[100] and increased after about 24 hr biliary drainage.[13] In isolated hepatocytes, 12α-hydroxylation has been reported to be increased by dibutyryl cyclic AMP.[92] Under the conditions used in these studies, then, the activity of 12α-hydroxylase varies in the same way as that of cholesterol 7α-hydroxylase, but to a lesser extent. On the other hand, 12α-hydroxylation in microsomes from cholesterol-fed and phenobarbitone-treated rats has been shown to be decreased while cholesterol 7α-hydroxylase activity is enhanced or unchanged,[101,102] and starvation increases the activity of 12α-hydroxylase, but inhibits cholesterol 7α-hydroxylase.[101] In some physiological circumstances, therefore, 12α-hydroxylase appears to be regulated independently from cholesterol 7α-hydroxylase.

The position of 12α-hydroxylase in the bile acid synthetic pathway suggests that it may be involved in the control of the relative amounts of cholic and chenodeoxycholic acids formed in the liver. Treatment of rats with thyroid hormones is known to lead to a reversal of the normal ratio between the two bile acids[99] and 12α-hydroxylase activity has been shown to be reduced in microsomes isolated from the livers of hyperthyroid rats and raised in those from hypothyroid animals.[103] The first step in the pathway leading specifically to chenodeoxycholic acid is catalyzed by a microsomal 26 hydroxylase (see Figure 1) and the activity of this enzyme has been found to be increased in hyperthyroid and decreased in hypothyroid rats.[104] Although there is some evidence from in vitro experiments to suggest that chenodeoxycholic acid can be synthesized in liver mitochondria in a pathway involving 26 hydroxycholesterol and lithocholic acid but not 7α-hydroxycholesterol,[105] it seems likely that this route is not quantitatively important in vivo.[101]

In the rat, chenodeoxycholic acid is further metabolized by the action of chenodeoxycholic acid 6β-hydroxylase to α-muricholic acid, which is then isomerized to β-muricholic acid[31] (see Figure 1). The ratio of the amount of chenodeoxycholic to β-muricholic acid formed in isolated hepatocytes in suspension has been shown to be increased markedly in the presence of dibutyryl cyclic AMP (see Figure 6),[92] suggesting that the activity of chenodeoxycholic acid 6β-hydroxylase may also be influenced by phosphorylation.

12α-Hydroxylase and microsomal 26 hydroxylase can be envisaged, then, as acting in concert to regulate the ratio of cholic to chenodeoxycholic acid synthesized in the liver, with chenodeoxycholic acid 6β-hydroxylase controlling the further metabolism of chenodeoxycholic acid and cholesterol 7α-hydroxylase determining the total amount of bile acid produced.

VII. CONCLUDING REMARKS

In this chapter I have attempted to give a broad introduction to current knowledge about cholesterol 7α-hydroxylase. It will be clear to the reader that for many aspects of the enzyme this knowledge is somewhat limited. There is good reason, however, to believe that the next few years will see much more rapid progress than has been the case in the past. The availability of new techniques for the study of the enzyme, particularly the isolation of highly purified preparations of the cytochrome P-450 component, which should help to elucidate the enzymology of cholesterol 7α-hydroxylase, and the use of isolated hepatocytes, the enormous potential of which has only just begun to be tapped, coupled with the emergence of new concepts concerning cholesterol fluxes within cells and coordinate regulation of the enzymes controlling cholesterol metabolism, should not only lead to a fuller understanding of cholesterol 7α-hydroxylase itself, but also allow us to place the enzyme in the context of the overall management of cholesterol homeostasis by liver cells.

REFERENCES

1. **Danielsson, H.,** Present status of research on catabolism and excretion of cholesterol, *Adv. Lipid Res.*, 1, 335, 1963.
2. **Bergström, S. and Lindstedt, S.,** The formation of cholic acid from 3α 7α dihydroxycoprostane in the rat, *Biochim. Biophys. Acta*, 19, 556, 1956.
3. **Lindstedt, S.,** The formation of bile acids from 7α hydroxycholesterol in the rat, *Acta Chem. Scand.*, 11, 417, 1957.
4. **Danielsson, H.,** Conversion of Δ^5-cholestene-3α-12α-diol to cholic acid in the rabbit, *Acta Chem. Scand.*, 16, 1534, 1962.
5. **Danielsson, H. and Einarsson, K.,** The enzymatic formation of 7α-hydroxycholesterol from cholesterol in rat liver homogenates, *Acta Chem. Scand.*, 18, 831, 1964.
6. **Mendelsohn, D., Mendelsohn, L., and Staple, E.,** The catabolism *in vitro* of cholesterol: formation of the 7 epimeric hydroxycholesterols from cholesterol in rat liver, *Biochim. Biophys. Acta*, 97, 379, 1965.
7. **Danielsson, H. and Einarsson, K.,** On the conversion of cholesterol to 7α, 12α dihydroxycholest-4-en-3-one, *J. Biol. Chem.*, 241, 1449, 1966.
8. **Bergström, S.,** Bile acids: formation and metabolism, in *Ciba Foundation Symposium: Biosynthesis of Terpenes and Sterols*, Wolstenholme, G. E. W. and O'Connor, R. M., Eds., Little, Brown, Boston, 1959, 185.
9. **Shefer, S., Hauser, S., and Mosbach, E. H.,** 7α Hydroxylation of cholestanol by rat liver microsomes, *J. Lipid Res.*, 9, 328, 1968.
10. **Boyd, G. S., Scholan, N. A., and Mitton, J. R.,** Factors influencing cholesterol 7α hydroxylase activity in the rat liver, *Adv. Exp. Biol. Med.*, 4, 443, 1969.
11. **Wada, F., Hirata, K., Nakao, K., and Sakamoto, Y.,** Participation of the microsomal electron transport system involving cytochrome P450 in 7α hydroxylation of cholesterol, *J. Biochem. (Tokyo)*, 66, 699, 1969.
12. **Johansson, G.,** Oxidation of cholesterol 3β hydroxy-5-pregnen-20-one and 3β hydroxy-5-androsten-17-one by rat liver microsomes, *Eur. J. Biochem.*, 21, 68, 1971.
13. **Danielsson, H., Einarsson, K., and Johansson, G.,** Effect of biliary drainage on individual reactions in the conversion of cholesterol to taurocholic acid, *Eur. J. Biochem.*, 2, 44, 1967.
14. **Johansson, G.,** Effect of cholestyramine and diet on hydroxylations in the biosynthesis and metabolism of bile acids, *Eur. J. Biochem.*, 17, 292, 1970.
15. **Siperstein, M. D., Jayko, M. E., Chaikoff, I. L., and Dauben, W. G.,** Nature of the metabolic products of [^{14}C] cholesterol excreted in bile and feces, *Proc. Soc. Exp. Biol. Med.*, 83, 71, 1952.
16. **Bergström, S. and Norman, A.,** Metabolic products of cholesterol in bile and feces of rat. Steroids and bile acids, *Proc. Soc. Exp. Biol. Med.*, 83, 71, 1953.
17. **Chevallier, F.,** Dynamics of cholesterol in rats, studied by the isotopic equilibrium method, *Adv. Lipid Res.*, 5, 209, 1967.
18. **Wilson, J. D.,** The quantification of cholesterol excretion and degradation in the isotopic steady state in the rat: the influence of dietary cholesterol, *J. Lipid Res.*, 5, 409, 1964.
19. **Beher, W. T., Casazza, K. K., Beher, M. E., Filus, A. M., and Bertasius, J.,** Effects of cholesterol on bile acid metabolism in the rat, *Proc. Soc. Exp. Biol. Med.*, 134, 595, 1970.
20. **Raicht, R. F., Cohen, B. I., Shefer, S., and Mosbach, E. H.,** Sterol balance studies in the rat. Effects of dietary cholesterol and β-sitosterol on sterol balance and rate-limiting enzymes of sterol metabolism, *Biochim. Biophys. Acta*, 388, 374, 1975.
21. **Ogura, M., Ayaki, Y., and Goto, M.,** Further studies on the possible compartmentation of the precursor pool of cholesterol for the biosynthesis of cholic acid in the rat, *J. Biochem. (Tokyo)*, 80, 537, 1976.
22. **Ayaki, Y., Tsumadate, T., Endo, S., and Ogura, M.,** Role of endogenous and exogenous cholesterol as the precursor for bile acids in the rat, *Steroids*, 38, 495, 1981.
23. **Stange, E. F., Spady, D. K., and Dietschy, J. M.,** Characterisation of the cholesterol precursor pools used for bile acid synthesis *in vivo*, in *Enterohepatic Circulation of Bile Acids and Sterol Metabolism*, Paumgartner, G., Stiehl, A., and Gerok, W., Eds., MTP Press, Lancaster, 1985, in press.
24. **Björkhem, J. and Lewenhaupt, A.,** Preferential utilisation of newly synthesised cholesterol as a substrate for bile acid synthesis, *J. Biol. Chem.*, 254, 5252, 1979.
25. **Berry, M. N. and Friend, D. S.,** High yield preparation of isolated rat liver parenchymal cells, *J. Cell. Biol.*, 43, 506, 1969.
26. **Botham, K. M. and Boyd, G. S.,** Cholesterol 7α hydroxylase in isolated rat liver cells, *Eur. J. Biochem.*, 95, 533, 1979.
27. **Yousef, I. M., Ho, J., and Jeejeebhoy, K. N.,** Bile acid synthesis in isolated rat hepatocytes, *Can. J. Biochem.*, 56, 780, 1978.
28. **Whiting, M. J. and Edwards, A. M.,** Measurement of cholic acid synthesis and secretion by isolated rat hepatocytes, *J. Lipid Res.*, 20, 914, 1979.

29. **Kempen, H. J. M., Vos-van Holstein, M. P. M., and de Lange, J.,** Bile acids and lipids in isolated rat hepatocytes: content, synthesis and release, as affected by cholestyramine treatment of donor rats, *J. Lipid Res.*, 23, 823, 1982.
30. **Botham, K. M., Beckett, G. J., Percy-Robb, I. W., and Boyd, G. S.,** Bile acid synthesis in isolated rat liver cells. The effect of 7α hydroxycholesterol, *Eur. J. Biochem.*, 103, 299, 1980.
31. **Mahowald, T. A., Matschiner, J. T., Hsia, S. L., Richter, R., Doisy, E. A., Elliot, W. H., and Doisy, E. A.,** Metabolism of deoxycholic acid-24-^{14}C and chenodeoxycholic acid-24-^{14}C in the rat, *J. Biol. Chem.*, 225, 781, 1957.
32. **Botham, K. M. and Boyd, G. S.,** The metabolism of chenodeoxycholic acid to β-muricholic acid in rat liver, *Eur. J. Biochem.*, 134, 191, 1983.
33. **Botham, K. M., Craig, I. F., and Boyd, G. S.,** Bile acid synthesis in isolated hepatocytes in primary culture, *Biochem. Soc. Trans.*, 9, 212P, 1981.
34. **Davis, R. A., Highsmith, W. E., Malone-McNeal, M., Archambault-Schexnayder, J., and Kuan, J. C. W.,** Bile acid secretion by cultured hepatocytes: regulation by cholesterol availability, *J. Biol. Chem.*, 258, 3661, 1983.
35. **Myant, N. B. and Mitropoulos, K. A.,** Cholesterol 7α hydroxylase, *J. Lipid Res.*, 18, 135, 1977.
36. **Danielsson, H.,** On the oxidation of cholesterol in liver mitochondrial preparations. Bile acids and sterols 99, *Acta Chem. Scand.*, 14, 846, 1960.
37. **Mitropoulos, K. A. and Balasubramaniam, S.,** Cholesterol 7α hydroxylase in rat liver microsomal preparations, *Biochem. J.*, 128, 1, 1972.
38. **Björkhem, I. and Danielsson, H.,** 7α Hydroxylation of exogenous and endogenous cholesterol in rat liver microsomes, *Eur. J. Biochem.*, 53, 63, 1975.
39. **Björkhem, I. and Danielsson, H.,** Assay of liver microsomal cholesterol 7α hydroxylase using deuterated carrier and gas chromatography mass spectrometry, *Anal. Biochem.*, 59, 508, 1974.
40. **Galli, G., Sanghvi, A., Kienle, M. G., and Galli, G.,** A rapid, sensitive method for simultaneous measurements of tissue levels of oxygenated derivatives of cholesterol, *Anal. Biochem.*, 85, 430, 1978.
41. **Mitton, J. R., Scholan, N. A., and Boyd, G. S.,** The oxidation of cholesterol in rat liver sub-cellular particles, *Eur. J. Biochem.*, 20, 569, 1971.
42. **Scholan, N. A. and Boyd, G. S.,** The cholesterol 7α hydroxylase enzyme system, *Hoppe-Seyler's Z. Physiol. Chem.*, 349, 1628, 1968.
43. **Boyd, G. S., Grimwade, A. M., and Lawson, M. E.,** Studies on rat liver microsomal cholesterol 7α hydroxylase, *Eur. J. Biochem.*, 37, 334, 1973.
44. **Boyd, G. S., Brown, M. J. G., Hattersley, N. G., and Suckling, K. E.,** Studies on the specificity of rat liver microsomal cholesterol 7α hydroxylase, *Biochim. Biophys. Acta*, 337, 132, 1974.
45. **Brown, M. J. G. and Boyd, G. S.,** The specificity of rat liver cholesterol 7α hydroxylase, *Eur. J. Biochem.*, 44, 37, 1974.
46. **Balasubramaniam, S., Mitropoulos, K. A., and Myant, N. B.,** Evidence for the compartmentation of cholesterol in rat liver microsomes, *Eur. J. Biochem.*, 34, 77, 1973.
47. **Björkhem, I., Danielsson, H., and Wikvall, K.,** 7α Hydroxylation of cholesterol by reconstituted systems from rat liver microsomes, *Biochem. Biophys. Res. Commun.*, 61, 934, 1974.
48. **Wikvall, K., Böstrom, H., Danielsson, H., Hansson, R., Kalles, I., and Lidström, B.,** Studies on properties of purified cholesterol 7α hydroxylase from rabbit and rat liver microsomes, in *Bile Acids and Lipids,* Paumgartner, G., Stiehl, A., and Gerok, W., Eds., MTP Press, Lancaster, 1981, 31.
49. **Hattersley, N. G. and Boyd, G. S.,** Solubilisation of rat liver microsomal cholesterol 7α hydroxylase, *FEBS Lett.*, 53, 80, 1975.
50. **Ozasa, S. and Boyd, G. S.,** Cholesterol 7α hydroxylase of rat liver. Studies on the solubilisation, resolution and reconstitution of the enzyme complex, *Eur. J. Biochem.*, 119, 263, 1981.
51. **Danielsson, H.,** Cytochrome P450 and biosynthesis of bile acids, in *Bile Acids and Cholesterol in Health and Disease,* Paumgartner, G., Stiehl, A., and Gerok, W., Eds., MTP Press, Lancaster, 1983, 3.
52. **Böstrom, H. and Wikvall, K.,** Hydroxylations in the biosynthesis of bile acids. Isolation of subfractions with different substrate specificity from cytochrome P450 LM_4, *J. Biol. Chem.*, 257, 11755, 1982.
53. **Chiang, J. Y. L., Malmer, M., and Hutterer, F.,** A form of rabbit liver cytochrome P450 that catalyses 7α hydroxylation of cholesterol, *Biochim. Biophys. Acta,* 750, 291, 1983.
54. **Danielsson, H., Kalles, I., and Wikvall, K.,** Regulation of hydroxylations in biosynthesis of bile acids. Isolation of a protein from rat liver cytosol stimulating reconstituted cholesterol 7α hydroxylase activity, *J. Biol. Chem.*, 259, 4258, 1984.
55. **Mason, H. S.,** Oxidases, *Annu. Rev. Biochem.*, 34, 595, 1965.
56. **Coon, M. J., Vermilion, J. L., Vatsis, K. P., French, J. S., Dean, W. L., and Haugen, D. A.,** Biochemical studies on drug metabolism: isolation of multiple forms of liver microsomal cytochrome P450, in *Drug Metabolism Concepts,* ACS Symp. Ser. No. 44, Jerina, D. M., Ed., American Chemical Society, Washington, D.C., 1977, 46.

57. **Einarsson, K. and Johansson, G.,** Effect of phenobarbitol on the conversion of cholesterol to taurocholic acid. Bile acids and steroids 204, *Eur. J. Biochem.*, 6, 293, 1968.
58. **Balasubramaniam, S. and Mitropoulos, K. A.,** The role of cytochrome P450 in the 7α hydroxylation of cholesterol, *Biochem. Soc. Trans.*, 3, 964, 1975.
59. **Boyd, G. S., Brown, M. J. G., Hattersley, N. G., Lawson, M. E., and Suckling, K. E.,** Rat liver cholesterol 7α hydroxylase, in *Advances in Bile Acid Research,* Matern, S., Hackenschmidt, J., Back, P., and Gerok, W., Eds., F. K. Schattauer Verlag, New York, 1975, 45.
60. **Björkhem, I. and Danielsson, H.,** Heterogeneity of hepatic mixed function oxidases, *Biochem. Biophys. Res. Commun.*, 51, 766, 1973.
61. **Danielsson, H. and Wikvall, K.,** Evidence for a specific cytochrome P450 with a short half life catalysing 7α hydroxylation of cholesterol, *Biochem. Biophys. Res. Commun.*, 103, 46, 1981.
62. **Craig, I. F., Boyd, G. S., McLeod, R. J., and Suckling, K. E.,** Partial characterization of a protein from rat liver 100,000g supernatant which stimulates cholesterol 7α hydroxylase, *Biochem. Soc. Trans.*, 7, 967, 1979.
63. **Spence, J. T. and Gaylor, J. L.,** Investigation of the regulation of microsomal 3-hydroxy-3-methylglutaryl co-enzyme A reductase and methyl sterol oxidase of cholesterol biosynthesis, *J. Biol. Chem.*, 252, 5852, 1977.
64. **Kwok, C. T., Burnett, W., and Hardie, I. R.,** Regulation of rat liver microsomal cholesterol 7α hydroxylase: presence of a cytosolic activator, *J. Lipid Res.*, 22, 570, 1981.
65. **Sanghvi, A., Grassi, E., Warty, V., Diven, W., Wight, C., and Lester, R.,** Reversible activation-inactivation of cholesterol 7α hydroxylase possibly due to phosphorylation-dephosphorylation, *Biochem. Biophys. Res. Commun.*, 103, 886, 1981.
66. **Kalles, I. and Wikvall, K.,** Role of sulphydryl groups in the catalytic activity of the purified cholesterol-7α-hydroxylase system, *Biochem. Biophys. Res. Commun.*, 100, 1361, 1981.
67. **Ericksson, S.,** Biliary excretion of bile acids and cholesterol in bile fistula rats. Bile acids and steroids, *Proc. Soc. Exp. Biol. Med.*, 94, 578, 1957.
68. **Shefer, S., Hauser, S., Bekersky, I., and Mosbach, E. H.,** Biochemical site of regulation of bile acid synthesis in the rat, *J. Lipid Res.*, 11, 404, 1970.
69. **Mosbach, E. H., Rothschild, M. A., Bekersky, I., Oratz, M., and Mongelli, J.,** Bile acid synthesis in the isolated, perfused rabbit liver, *J. Clin. Invest.*, 50, 1720, 1971.
70. **Einarsson, K. and Johansson, G.,** Bile acids and steroids 206. Effect of actinomycin D and puromycin on the conversion of cholesterol into bile acids in bile-fistula rats, *FEBS Lett.*, 1, 219, 1968.
71. **Gielen, J., van Cantfort, J., Robaye, B., and Renson, J.,** Rythme circadien de la cholesterol-7α-hydroxylase chez la rat, *C.R. Acad. Sci. Ser. D.*, 269, 731, 1969.
72. **Mitropoulos, K. A., Balasubramaniam, S., and Myant, N. B.,** The effect of interruption of the enter-hepatic circulation of bile acids and of cholesterol feeding on cholesterol 7α hydroxylase in relation to the diurnal rhythm in its activity, *Biochim. Biophys. Acta,* 326, 428, 1973.
73. **Beher, W. T., Baker, G. T., and Anthony, W. L.,** Effects of bile acids on fecal excretion of end products of cholesterol metabolism, *Am. J. Physiol.*, 199, 736, 1960.
74. **Shefer, S., Hauser, S., Bekersky, I., and Mosbach, E. H.,** Feedback regulation of bile acid biosynthesis in the rat, *J. Lipid Res.*, 10, 646, 1969.
75. **Boyd, G. S. and Lawson, M. E.,** The effect of portacaval transposition on hepatic cholesterol 7α hydroxylase activity in the rat, *FEBS Lett.*, 64, 435, 1976.
76. **Wilson, J. D., Bentley, W. H., and Crowley, G. T.,** Regulation of bile acid formation in intact animals, in *Bile Salt Metabolism,* Schiff, L., Carey, J. B., and Dietschy, J. M., Eds., Charles C Thomas, Springfield, Ill., 1969, 140.
77. **Cronholm, T. and Sjövall, J.,** Bile acids in portal blood of rats fed different diets and cholestyramine. Bile acids and steroids 189, *Eur. J. Biochem.*, 2, 375, 1967.
78. **Botham, K. M., Lawson, M. E., Beckett, G. J., Percy-Robb, I. W., and Boyd, G. S.,** Portal blood concentration of conjugated cholic and chenodeoxycholic acids; relationship to bile salt synthesis in liver cells, *Biochim. Biophys. Acta,* 665, 81, 1981.
79. **Botham, K. M., Lawson, M. E., Beckett, G. J., Percy-Robb, I. W., and Boyd, G. S.,** The effect of portal blood bile salt concentrations on bile salt synthesis in rat liver; studies with isolated hepatocytes, *Biochim. Biophys. Acta,* 666, 238, 1981.
80. **Davis, R. A., Highsmith, W. E., Malone-McNeal, M., Archambault-Schexnayder, J., and Kuan, J. C. W.,** Bile acid synthesis by cultured hepatocytes, *J. Biol. Chem.*, 258, 4079, 1983.
81. **Davis, R. A.,** Regulation of bile acid synthesis by microsomal membrane cholesterol, in *Enterohepatic Circulation of Bile Acids and Sterol Metabolism,* Paumgartner, G., Stiehl, A., and Gerok, W., Eds., MTP Press, Lancaster, 1985, in press.
82. **Pries, J. M., Gustafson, A., Wiegand, D., and Duane, W. C.,** Taurocholate is more potent than cholate in suppression of bile salt synthesis in rat, *J. Lipid Res.*, 24, 141, 1983.

83. **Mitropoulos, K. A., Venkatesan, S., Balasubramaniam, S., and Peters, S. T. J.,** The sub-microsomal localisation of 3-hydroxy-3-methylglutaryl co-enzyme A reductase, cholesterol 7α hydroxylase and cholesterol in rat liver, *Eur. J. Biochem.*, 82, 419, 1978.
84. **Long, T. T., Jakoi, L., Stevens, R., and Quarfordt, S.,** The sources of rat biliary cholesterol and bile acid, *J. Lipid Res.*, 19, 872, 1978.
85. **Kempen, H. J., Vos-van Holstein, M., and de Lange, J.,** Bile acids and lipids in isolated rat hepatocytes: sources of cholesterol for bile acid formation, estimated by incorporation of tritium from ^3H_2O and by the effect of ML 236B, *J. Lipid Res.*, 24, 316, 1983.
86. **Scallen, T. J. and Sanghvi, A.,** Regulation of three key enzymes in cholesterol metabolism by phosphorylation/dephosphorylation, *Proc. Natl. Acad. Sci. U.S.A.*, 80, 2477, 1983.
87. **Shefer, S., Hauser, S., Lapar, V., and Mosbach, E. H.,** Regulatory effects of sterols and bile acids on hepatic 3-hydroxy-3-methylglutaryl co-enzyme A reductase and cholesterol 7α hydroxylase in the rat, *J. Lipid Res.*, 14, 573, 1973.
88. **Botham, K. M. and Boyd, G. S.,** The effect of dietary cholesterol on bile acid synthesis in rat liver, *Biochem. Soc. Trans.*, 13, 873, 1985.
89. **Ford, R. P., Botham, K. M., Suckling, K. E., and Boyd, G. S.,** The effect of a rat plasma high density lipoprotein subfraction [HDL_2] on the synthesis of bile salts by rat hepatocyte monolayers, *FEBS Lett.*, 179, 177, 1985.
90. **Goodwin, C. D., Cooper, B. W., and Margolis, S.,** Rat liver cholesterol 7α hydroxylase. Modulation of enzyme activity by changes in phosphorylation state, *J. Biol. Chem.*, 257, 4469, 1982.
91. **Kwok, C. T., Pillay, S. P., and Hardie, I. R.,** Regulation of rat liver microsomal cholesterol 7α hydroxylase by magnesium and ATP, effect of pH, *Biochem. Biophys. Res. Commun.*, 116, 966, 1983.
92. **Botham, K. M. and Boyd, G. S.,** The effect of dibutyryl adenosine 3′5′-monophosphate on the synthesis of bile acids in isolated hepatocytes from rat, *Eur. J. Biochem.*, 136, 313, 1983.
93. **Sundaram, G. S., Rothman, V., and Margolis, S.,** Stimulation of bile acid synthesis by dibutyryl cyclic AMP in isolated rat hepatocytes, *Lipids*, 18, 443, 1983.
94. **Botham, K. M., Suckling, K. E., and Boyd, G. S.,** The effect of glucagon-induced adenosine 3′5′-monophosphate concentrations on bile acid synthesis in isolated rat liver cells, *FEBS Lett.*, 168, 317, 1984.
95. **Mitropoulos, K. A. and Balasubramaniam, S.,** The role of glucocorticoids in the regulation of the diurnal rhythm of 3-hydroxy-3-methylglutaryl coenzyme A reductase and cholesterol 7α hydroxylase, *Biochem. J.*, 160, 49, 1976.
96. **Spector, A. A., Mathur, S. N., and Kaduce, T. L.,** Role of acylcoenzyme A: cholesterol acyl transferase in cholesterol metabolism, *Prog. Lipid Res.*, 18, 31, 1979.
97. **Ingebritsen, T. S., Lea, H. S., Parker, R. A., and Gibson, D. M.,** Reversible modulation of the activities of both liver microsomal hydroxymethyl-glutaryl co-enzyme A reductase and its inactivated enzyme. Evidence for regulation by phosphorylation-dephosphorylation, *Biochem. Biophys. Res. Commun.*, 81, 1268, 1978.
98. **Suckling, K. E., Stange, E. F., and Dietschy, J. M.,** Dual modulation of hepatic and intestinal acyl-coA:cholesterol acyltransferase activity by (de-phosphorylation and substrate supply *in vivo, FEBS Lett.*, 151, 111, 1983.
99. **Danielsson, H.,** Mechanisms of bile acid synthesis, in *The Bile Acids*, Vol. 2, Nair, P. P. and Kritchevsky, D., Eds., Plenum Press, New York, 1971.
100. **Danielsson, H.,** Effect of biliary obstruction on formation and metabolism of bile acids in rat, *Steroids*, 22, 667, 1973.
101. **Danielsson, H. and Sjövall, J.,** Bile acid metabolism, *Annu. Rev. Biochem.*, 44, 233, 1975.
102. **Gustafsson, B. E., Einarsson, K., and Gustafsson, J. A.,** Influence of cholesterol feeding on liver microsomal metabolism of steroids and bile acids in conventional and germ-free rats, *J. Biol. Chem.*, 250, 8496, 1975.
103. **Mitropoulos, K. A., Suzuki, M., Myant, N. B., and Danielsson, H.,** Effect of thyroidectomy and thyroxine treatment on the activity of 12α hydroxylase and of some components of microsomal electron transport chains in rat liver. *FEBS Lett.*, 1, 13, 1968.
104. **Björkhem, I., Danielsson, H., and Gustafsson, J.,** On the effect of thyroid hormone on 26-hydroxylation of C_{27} steroids in rat liver, *FEBS Lett.*, 31, 20, 1973.
105. **Mitropoulos, K. A., Avery, M. D., Myant, N. B., and Gibbons, E. F.,** The formation of cholest-5-ene-3β, 26 diol as an intermediate in the conversion of cholesterol into bile acids by liver mitochondria, *Biochem. J.*, 130, 363, 1972.
106. **Ford, R. P.,** unpublished results.
107. **Botham, K. M.,** unpublished results.

Assay and Biochemical Characterization

Chapter 4

CHOLESTEROL 7α-HYDROXYLASE (7α-MONOOXYGENASE EC 1.14.13.17): METHODS OF ASSAY

Sarah Shefer, Gerald Salen, and Ashok K. Batta

TABLE OF CONTENTS

I. INTRODUCTION

It is well established that 7α-hydroxylation of cholesterol is the first and most important reaction in the bile acid synthetic pathway.[1,2] This reaction provides 7α-hydroxycholesterol, which is totally committed to bile acid synthesis, and hence the enzyme that catalyzes this reaction, cholesterol 7α-hydroxylase (EC 1.14.13.17), determines the rate of formation of bile acids.[3,4] It has been suggested that cholesterol 7α-hydroxylase exists in phosphorylated (active) and dephosphorylated (inactive) forms and that an activation/inactivation mechanism might exist in the short-term regulation of this enzyme,[5-7] but recent studies have failed to support this hypothesis.[8] This enzyme is an NADPH-dependent mixed-function oxidase consisting of a specific species of cytochrome P-450 and NADPH cytochrome P-450 reductase.[9,10]

Although purification and measurement of enzyme protein have not been fully accomplished, considerable information about the properties and regulation of cholesterol 7α-hydroxylase has been obtained through the measurement of in vivo enzyme activity. In the following section, the commonly used assays for the measurement of cholesterol 7α-hydroxylase activity in man and other animals will be discussed and the advantages and disadvantages of the individual procedures will be emphasized.

II. ISOTOPE INCORPORATION PROCEDURE

The isotope incorporation procedure for measuring cholesterol 7α-hydroxylase activity utilizes radiolabeled cholesterol as the substrate and is based on the following assumptions:

1. Enzyme activity correlates with the amount of enzyme protein.
2. The radioactive substrate equilibrates with the endogenous substrate pool and is 7α-hydroxylated in a similar manner. It is obvious that in order to introduce the substrate cholesterol into the aqueous incubation media, detergents, liposomes, or low-density lipoproteins are necessary to solubilize the cholesterol.[11-14]
3. It is important to emphasize that the in vitro assay should be carried out under conditions where the substrate is not rate-limiting. In other words, an abundance of substrate must be present to saturate the available enzyme.

The most commonly used method to measure cholesterol 7α-hydroxylase activity is the isotope incorporation method.[11,15,16] In this method, [4-^{14}C]cholesterol is incubated with hepatic microsomes and cofactors and the activity of cholesterol 7α-hydroxylase is expressed as the amount of 7α-hydroxycholesterol formed per milligram of microsomal protein per unit time.

The procedure for measuring cholesterol 7α-hydroxylase activity is briefly described below: liver tissue (0.05 to 4.0 g) is chilled on ice after removal and homogenized in a Potter-Elvehjem homogenizer with 5 volumes of a solution containing 0.05 *M* Tris buffer (pH 7.4) and 1.15% KCl. The whole homogenate is centrifuged at 10,000 × g for 20 min to sediment nuclei, cell debris, and the mitochondrial fraction. The microsomes are isolated by centrifugation at 100,000 × g for 90 min and the pellet is washed by resuspending in the same volume of a solution containing 1.15% KCl and 10 m*M* EDTA (pH 7.0), and centrifuged for an additional 60 min to remove any ribosomes. The final pellet is resuspended in 0.25 *M* sucrose solution so that the protein concentration ranges from 25 to 30 mg protein per milliliter. When stored at −62°C, this microsomal suspension is stable for a period of one month without appreciable loss of cholesterol 7α-hydroxylase activity. All subsequent operations are carried out at 0 to 5°C. The complete assay system contains in a volume of 0.5 mℓ: 75 m*M* potassium phosphate buffer (pH 7.4), 80 m*M* EDTA, 2.4 m*M* $MgCl_2$, 1.0

Table 1
EFFECT OF DRUGS ON CHOLESTEROL 7α-HYDROXYLASE ACTIVITY AS DETERMINED BY THE ISOTOPE INCORPORATION METHOD

Treatment	7α-Hydroxycholesterol formed ($pmol \cdot mg^{-1}$ $protein \cdot min^{-1}$)
None	9.11 ± 0.33[20]
Sodium taurocholate (1.0% of diet)	3.08 ± 0.18[20]
Cholestyramine (2.5% of diet)	32.5 ± 1.9[21]

m*M* 1,4-dithiotreitol, 100 nmol [4-^{14}C]cholesterol solubilized with a detergent (Cutscum or Tween-80) or liposomes and is preincubated for 2 min with the washed microsomes containing 0.5 to 1.0 mg protein. The reaction is initiated by the addition of NADPH (0.6 μmol). The incubation is carried out with air as the gas phase at 37°C for 20 min, care being taken to exclude light.[17] Zero-time and boiled enzyme controls are run with each experiment. The reaction is terminated by the addition of 15 to 20 volumes of methylene chloride: ethanol (5:1), and the sterols are extracted after the addition of 3.0 mℓ water. The organic solvent is evaporated at 60°C under a current of nitrogen. Under these conditions 85 to 90% of the steroids are recovered from the incubation media. The dried steroid fraction is dissolved in 40 μℓ of benzene and applied as a spot to a 250μ-thick silica-gel G plate. Unlabeled 7-keto-cholesterol and 7α- and 7β-hydroxycholesterol (30 μg each) are applied at the same spot. The plate is developed with diethyl ether and the radioactivity of the pertinent spots is determined in a liquid-scintillation counter. The R_f values of the sterols are: cholesterol, 0.88; 7α-hydroxycholesterol, 0.40, 7β-hydroxycholesterol 0.54; and 7-ketocholesterol, 0.63.[11]

The data can be expressed as the percentage of radioactive substrate transformed to 7α-hydroxycholesterol. In general, the amount of exogenous cholesterol greatly exceeds the endogenous cholesterol in the microsomes, but to date there is no conclusive evidence that the concentration of the exogenous cholesterol is sufficient to saturate the enzyme.[12,18] The use of various solubilizers has been shown to differently affect the accessibility of the exogenous cholesterol to the enzyme. Therefore, variations in the measured activity of cholesterol 7α-hydroxylase by the isotope incorporation procedure might be brought about not only by changes in the amount of enzyme protein but also by usage of different vehicles for the cholesterol substrate. For example, exogenous cholesterol is more accessible to the enzyme system if added in Tween-80 than in acetone.[12] Despite these drawbacks, extensive experience by many investigators suggests that this method is valid for comparative purposes, even though it fails to give an absolute measurement of enzyme activity.[19] It should be emphasized that the substrate used, [4-^{14}C]cholesterol, must be free from contamination of 7-oxygenated cholesterol derivatives.

Typical data that illustrate the application of the isotope incorporation method demonstrating the effect of drugs on cholesterol 7α-hydroxylase activity in the rat are shown in Table 1.[20,21]

Another method for measuring cholesterol 7α-hydroxylase activity in which [7α-^{3}H]cholesterol is used as the substrate, has been developed by Van Cantfort et al.,[22] and Johnson et al.[23] This method utilizes the stereospecificity of the enzyme to hydroxylate cholesterol only at the 7α-position. Tritiated water released into the incubation medium is collected by distillation under reduced pressure and its radioactivity reflects accurately the rate of cholesterol 7α-hydroxylase activity. This method gives results comparable to those obtained with [4-^{14}C]cholesterol as substrate, and it is advantageous over the isotope incorporation method described above, in its complete independence from further metabolism of the 7α-hydroxycholesterol formed. However, this method has not yet been widely adopted, partly because it is difficult to synthesize [7α-^{3}H]cholesterol in which the tritium is present exclusively in the 7α-position.

III. MASS MEASUREMENT OF THE BIOSYNTHETIC 7α-HYDROXYCHOLESTEROL

Because of inaccuracies relating to the measurement of cholesterol 7α-hydroxylase activity by the isotope incorporation methods described above, specifically, due to the effect of the endogenous cholesterol pool, other methods have been developed that measure the mass of the biosynthetic 7α-hydroxycholesterol formed. Mitropoulos and Balasubramaniam have developed a double isotope derivative dilution procedure in which the radioactive 7α-hydroxycholesterol formed during the incubation of [4-^{14}C]cholesterol is separated by thin-layer chromatography, the radioactive band is located by radioautography and acetylated with tritiated acetic anhydride.[24] The [^{3}H,^{14}C]7α-hydroxycholesterol diacetate formed is then separated from the monoacetylated and unacetylated 7α-hydroxycholesterol and from other radioactive contaminants by thin-layer chromatography and its amount calculated from the known specific radioactivity of the tritiated acetic anhydride. A zero time control is performed in order to take into account the endogenous 7α-hydroxycholesterol present. Although this assay is not influenced by the substrate pool and measures the actual mass of 7α-hydroxycholesterol formed, this procedure is very cumbersome and time consuming and does not overcome the problem of enzyme saturation.

A modification of this method has been developed by Shefer et al.[21] The method is based on the quantitative acetylation of the whole incubation mixture with tritiated acetic anhydride/pyridine at 100°C for 6 hr. The 7α-hydroxycholesterol diacetate is then separated from all other incubation products by a one step thin-layer chromatography on alumina plate using a solvent system of diethyl ether: hexane,1:1 (v/v); R_f: 7α-hydroxycholesterol diacetate, 0.61; 7β-hydroxycholesterol diacetate, 0.51; 7-ketocholesterol acetate, 0.37; and cholesterol acetate, 0.88. In this assay, there is no need to isolate the 7α-hydroxycholesterol by thin-layer chromatography prior to acetylation, the overall recovery is higher and amounts as low as 0.1 nmol can be measured.

IV. SELECTED-ION MONITORING-MASS SPECTROMETRY

Björkhem and Danielsson have developed a method to measure the mass of 7α-hydroxycholesterol by an isotope dilution technique using deuterated carrier and gas chromatography-mass spectrometry.[12,25] Standard incubation conditions are employed using endogenous cholesterol as substrate. At the end of the incubation period, a known amount of [24,25,7β-$^{2}H_3$]7α-hydroxycholesterol is added, and after extraction and thin-layer chromatography, the 7α-hydroxycholesterol fraction is subjected to combined gas chromatography-mass spectrometry as the trimethylsilyl ether derivative. The amount of the nondeuterated 7α-hydroxycholesterol present in the extraction products is calculated from the ratio between the relative intensities of the peaks due to the nondeuterated molecules (m/e 456) and the trideuterated molecules (m/e 459), no correction for incomplete recovery being required. In a typical assay, the degree of stimulation of cholesterol 7α-hydroxylase activity by biliary drainage in the rat as measured by this method was shown to be 320% (control, 41.0 vs. biliary fistula, 139.0 pmol/mg protein per minute).[12] This technique is sensitive and less time consuming than that of Mitropoulos et al.,[24] and Shefer et al.[21] However, endogenous

Table 2
EFFECT OF DIET ON RAT HEPATIC CHOLESTEROL 7α-HYDROXYLASE ACTIVITY AS MEASURED BY THE "ACETONE POWDER" ASSAY

Treatment	7α-Hydroxycholesterol formed (pmol·mg^{-1} protein·min^{-1})	Total cholesterol in acetone powder (nmol·mg^{-1} protein)
None	11.5 ± 0.95	0.42 ± 0.12
Cholesterol (2% of diet)	11.7 ± 0.90	0.45 ± 0.15
Cholic acid (1% of diet)	5.5 ± 0.48	0.47 ± 0.13
Cholestyramine (5% of diet)	19.9 ± 1.7	0.38 ± 0.08

microsomal cholesterol is used as the only substrate and therefore the incubation is probably performed under conditions where the enzyme is not saturated with the substrate.[12,18] Furthermore, [24,25,7β-$^2H_3$]7α-hydroxycholesterol has to be synthesized and the method requires a specially adapted mass-spectrometer that is not often available.

In a modification of this selected-ion monitoring method, Kienle et al. have used 5α-cholestane instead of the trideuterated 7α-hydroxycholesterol as the internal standard and injected the total lipid extract, after trimethylsilylation, onto a gas chromatograph-mass spectrometer.[26] The enzyme activity was calculated from the ratio between the intensities of the ions at m/e 456, due to the 7α-hydroxycholesterol molecules and at m/e 372, due to 5α-cholestane.

V. MEASUREMENT OF CHOLESTEROL 7α-HYDROXYLASE ACTIVITY IN THE ABSENCE OF ENDOGENOUS CHOLESTEROL

In the above-described methods the activity of cholesterol 7α-hydroxylase is measured in the presence of endogenous cholesterol where one has to assume equilibration with exogenous substrate. Furthermore, measurements are carried out under conditions where saturation of the enzyme might be incomplete. In order to overcome these difficulties a method for measuring cholesterol 7α-hydroxylase activity has been developed by Shefer et al.[27] In this method the microsomal suspension is precipitated into 40 volumes of acetone at -10°C.[28,29] The precipitate thus formed ("acetone powder") is virtually free from endogenous cholesterol and retains the cholesterol 7α-hydroxylase activity.[27] A partial removal of endogenous cholesterol from lyophilized hepatic microsomes by extraction with *n*-butanol at −30°C has also been reported.[30] Using the "acetone powder" of the microsomal preparation as the enzyme source and [4-^{14}C]cholesterol as the substrate one can employ the simple isotope incorporation method. Since all of the substrate is provided exogenously, problems relating to equilibration of endogenous with exogenous substrate are eliminated. Incubations are carried out under optimal conditions and enzyme activity is expressed in terms of mass of 7α-hydroxycholesterol formed per milligram protein per unit time with great reproducibility and accuracy. Shown in Table 2 is the effect of different dietary conditions on rat hepatic cholesterol 7α-hydroxylase activity as measured by the "acetone powder" assay.[27]

The method can be rapidly performed, is not affected by the preexisting microsomal cholesterol pool and can measure cholesterol 7α-hydroxylase activity under conditions where hepatic cholesterol content is varied. Furthermore, the affinity of different sterols to the 7α-hydroxylating enzyme can now be determined in the absence of endogenous cholesterol, whereas previous measurements were carried out in the presence of microsomal cholesterol.[31]

REFERENCES

1. **Danielsson, H. and Einarsson, K.,** Formation and metabolism of bile acids, in *The Biological Basis of Medicine,* Vol. 5, Bittar, E. E. and Bittar, N., Eds., Academic Press, London, 1969, 279.
2. **Danielsson, H.,** Mechanism of bile acid biosynthesis, in *The Bile Acids, Chemistry, Physiology and Metabolism,* Vol. 2, Nair, P. P. and Kritchevsky, D., Eds., Plenum Press, New York, 1973, 1.
3. **Shefer, S., Hauser, S., Bekersky, I., and Mosbach, E. H.,** Biochemical site of regulation of bile acid biosynthesis in the rat, *J. Lipid Res.,* 11, 404, 1970.
4. **Einarsson, K., Hellström, K., and Kalner, M.,** Feedback regulation of bile acid formation in man, *Metabolism,* 22, 1477, 1973.
5. **Sanghvi, A., Grassi, E., Warty, V., Diven, W., Wright, C., and Lester, R.,** Reversible activation-inactivation of cholesterol 7α-hydroxylase possibly due to phosphorylation-dephosphorylation, *Biochem. Biophys. Res. Commun.,* 103, 886, 1981.
6. **Kwok, C. T., Burnell, W., and Hardie, I. R.,** Regulation of rat liver microsomal cholesterol 7α-hydroxylase: presence of a cytosolic activator, *J. Lipid Res.,* 22, 570, 1981.
7. **Goodwin, C. D., Cooper, B. W., and Margolis, S.,** Rat liver cholesterol 7α-hydroxylase. Modulation of enzyme activity by changes in phosphorylation state. *J. Biol. Chem.,* 257, 4469, 1982.
8. **Wikvall, K., Boström, H., Danielsson, H., Hansson, R., Kalles, I., and Lidström, B.,** Studies on properties of purified cholesterol 7α-hydroxylase from rabbit and rat liver microsomes, in *Bile Acids and Lipids,* Paumgartner, G., Stiehl, A., and Gerok, W., Eds., MTP Press, Lancaster, 1981, 31.
9. **Salen, G. and Shefer, S.,** Bile acid synthesis, *Annu. Rev. Physiol.,* 45, 679, 1983.
10. **Danielsson, H. and Wikvall, K.,** Evidence for a specific cytochrome P-450 with short half-life catalyzing 7α-hydroxylation of cholesterol, *Biochem. Biophys. Res. Commun.,* 103, 46, 1981.
11. **Shefer, S., Hauser, S., and Mosbach, E. H.,** 7α-Hydroxylation of cholestanol by rat liver microsomes, *J. Lipid Res.,* 9, 328, 1968.
12. **Björkhem, I. and Danielsson, H.,** 7α-Hydroxylation of exogenous and endogenous cholesterol in rat liver microsomes, *Eur. J. Biochem.,* 53, 63, 1975.
13. **Sanghvi, A., Grassi, E., Bartman, C., and Diven, W. F.,** Effects of detergents on in vitro 7α-hydroxycholesterol formation by rat liver microsomes, *Lipids,* 17, 644, 1981.
14. **Sanghvi, A., Diven, W. F., Seltman, H. J., Paul, R., and Rizk, M.,** Properties of cholesterol 7α-hydroxylase in rat liver microsomal acetone powder. *Metabolism,* 33, 443, 1984.
15. **Danielsson, H., Einarsson, K., and Johansson, G.,** Effect of biliary drainage on individual reactions in the conversion of cholesterol to taurocholic acid. Bile acids and steroids. 180, *Eur. J. Biochem.,* 2, 44, 1967.
16. **Nicolau, G., Shefer, S., Salen, G., and Mosbach, E. H.,** Determination of hepatic cholesterol 7α-hydroxylase activity in man, *J. Lipid Res.,* 15, 146, 1974.
17. **Hais, I. M. and Myant, N. B.,** Photolysis of cholesterol during biological experiments, *Biochem. J.,* 94, 85, 1965.
18. **Balasubramaniam, S., Mitropoulos, K. A., and Myant, N. B.,** Evidence for the compartmentation of cholesterol in rat-liver microsomes, *Eur. J. Biochem.,* 34, 77, 1973.
19. **Myant, N. B. and Mitropoulos, K. A.,** Cholesterol 7α-hydroxylase, *J. Lipid Res.,* 18, 135, 1977.
20. **Shefer, S., Hauser, S., Lapar, V., and Mosbach, E. H.,** Regulatory effects of sterols and bile acids on hepatic 3-hydroxy-3-methylglutaryl CoA reductase and cholesterol 7α-hydroxylase in the rat, *J. Lipid Res.,* 14, 573, 1973.
21. **Shefer, S., Nicolau, G., and Mosbach, E. H.,** Isotope derivative assay of microsomal cholesterol 7α-hydroxylase, *J. Lipid Res.,* 16, 92, 1975.
22. **Van Cantfort, J., Renson, J., and Gielen, J.,** Rat liver cholesterol 7α-hydroxylase. 1. Development of a new assay based on the enzymatic exchange of tritium located on the 7α-position of the substrate, *Eur. J. Biochem.,* 55, 23, 1975.
23. **Johnson, D. B., Tyor, M. P., and Lack, L.,** The use of [7α-^{3}H]-and[7α, 7β-^{3}H] cholesterol in the enzymatic assay of cholesterol 7α-hydroxylase, *J. Lipid Res.,* 17, 353, 1976.
24. **Mitropoulos, K. A. and Balasubramaniam, S.,** Cholesterol 7α-hydroxylase in rat liver microsomal preparations, *Biochem. J.,* 128, 1, 1972.
25. **Björkhem, I. and Danielsson, H.,** Assay of liver microsomal cholesterol 7α-hydroxylase using deuterated carrier and gas chromatography-mass spectrometry, *Anal. Biochem.,* 59, 508, 1974.
26. **Kienle, M. G., Galli, G., Bosisio, E., Cighetti, G., and Paoletti, R.,** Evaluation of enzyme activities by gas chromatography-mass spectrometry: HMG CoA reductase and cholesterol 7α-hydroxylase, *J. Chromatogr.,* 289, 267, 1984.
27. **Shefer, S., Cheng, F. W., Hauser, S., Batta, A. K., and Salen, G.,** Regulation of bile acid synthesis. Measurement of cholesterol 7α-hydroxylase activity in rat liver microsomal preparations in the absence of endogenous cholesterol, *J. Lipid Res.,* 22, 532, 1981.

28. **Scallen, T. J., Dean, W. J., and Schuster, M. W.,** Enzymatic conversion of squalene to cholesterol by an acetone powder of rat liver microsomes, *J. Biol. Chem.*, 243, 5202, 1968.
29. **Boyd, G. S., Grimwalde, A. M., and Lawson, M. E.,** Studies on rat liver microsomal cholesterol 7α-hydroxylase, *Eur. J. Biochem.*, 37, 334, 1973.
30. **Boyd, G. S., Hattersley, N. G., Ozasa, S., and Suckling, K. E.,** Cholesterol 7α-hydroxylase in rat liver microsomes-characteristics, solubilization, resolution and substrate specificity, in *Microsomes, Drug Oxidations and Chemical Carcinogenesis,* Vol. II, Coon, M. J., Conney, A. H., Estabrook, R. W., Gelboin, H. V., Gillette, J. R., and O'Brien, P. J., Eds., Academic Press, New York, 1980, 699.
31. **Aringer, L.,** Conversion of 7α-hydroxycholesterol and 7α-hydroxy-β-sitosterol to 3α,7α-dihydroxy-and 3α,7α,12α-trihydroxy-5β-steroids in vitro, *J. Lipid Res.*, 16, 426, 1975.

Chapter 5

PURIFICATION AND PROPERTIES OF CHOLESTEROL 7α-HYDROXYLASE

Ronnie Hansson and Kjell Wikvall

TABLE OF CONTENTS

I. INTRODUCTION

The 7α-hydroxylation of cholesterol is the first and major rate-limiting step in the biosynthesis of bile acids. The reaction is catalyzed by a microsomal monooxygenase system involving a cytochrome P-450 and NADPH-cytochrome P-450 reductase.[1-4] The cholesterol 7α-hydroxylase system has been extensively studied over the years by several laboratories.[3,5-9] The activity is usually assayed by measuring the extent of conversion of added labeled cholesterol into 7α-hydroxycholesterol. One of the problems in the assay of cholesterol 7α-hydroxylase activity using whole microsomes is the presence of endogenous cholesterol, which together with the exogenous cholesterol, can serve as substrate for the enzyme system (compare Chapter 4). It is also difficult to investigate properties of cholesterol 7α-hydroxylase such as substrate specificity and mode of regulation with crude microsomal fraction as source of enzyme.

The aim of the present chapter is to summarize the present status of knowledge concerning purification and properties of cholesterol 7α-hydroxylase.

II. CYTOCHROME P-450

A. Purification of Multiple Forms of Cytochrome P-450

Cytochrome P-450 is the collective name of a group of membrane-bound heme proteins, found not only in liver microsomes and mitochondria, but also in the microsomal and mitochondrial fractions of various other tissues.[10-13] Cytochrome P-450 in liver microsomes has a broad substrate specificity, catalyzing monooxygenation of various exogenous substrates as well as of endogenous substrates such as fatty acids, prostaglandins, and steroids.[13-15] This broad substrate specificity of cytochrome P-450 is partly explained by the presence of multiple species of cytochrome P-450 in the microsomal fraction. Certain compounds such as phenobarbital and polycyclic hydrocarbons are able to increase the amount of cytochrome P-450 and NADPH-cytochrome P-450 reductase in liver microsomes.[14,16,17]

The first report on successful solubilization and fractionation of the microsomal monooxygenase system was published by Lu and Coon in 1968.[18] They achieved a partial separation of three components: cytochrome P-450, NADPH-cytochrome P-450 reductase, and phospholipid. Using these components they were able to reconstitute ω-hydroxylation of lauric acid. During the last ten years several groups have made significant advances in the purification of cytochrome P-450. Electrophoretically homogeneous preparations of cytochrome P-450 have been obtained from liver of for example phenobarbital- and 3-methylcholanthrene-treated rats and rabbits and β-naphthoflavone-treated rabbits (for recent reviews, see References 15, 17, and 19).

The cytochrome P-450 system is solubilized from liver microsomes by the use of a detergent. Sodium cholate is often used in this procedure, always in the presence of the cytochrome P-450-protecting agent glycerol. The solubilized microsomes are subjected to different chromatographic procedures such as ion-exchange chromatography on DEAE-cellulose, unspecific adsorption chromatography on hydroxylapatite, and hydrophobic interaction chromatography on octylamine-Sepharose. The chromatographic procedures are often performed in the presence of a nonionic detergent. Fractionation of the solubilized microsomes with polyethylene glycol and/or ammonium sulfate may be done prior to or in between various chromatographic procedures.

Two criteria of purity of a cytochrome P-450 preparation were used in earlier work. For each cytochrome P-450 species there will be a calculated specific content of cytochrome P-450, based on minimum molecular weight and assuming one heme per molecule of protein. The specific content is expressed as nanomoles of cytochrome P-450 per milligram of protein. Generally, the maximal specific content of cytochrome P-450 will be between 18 and 21

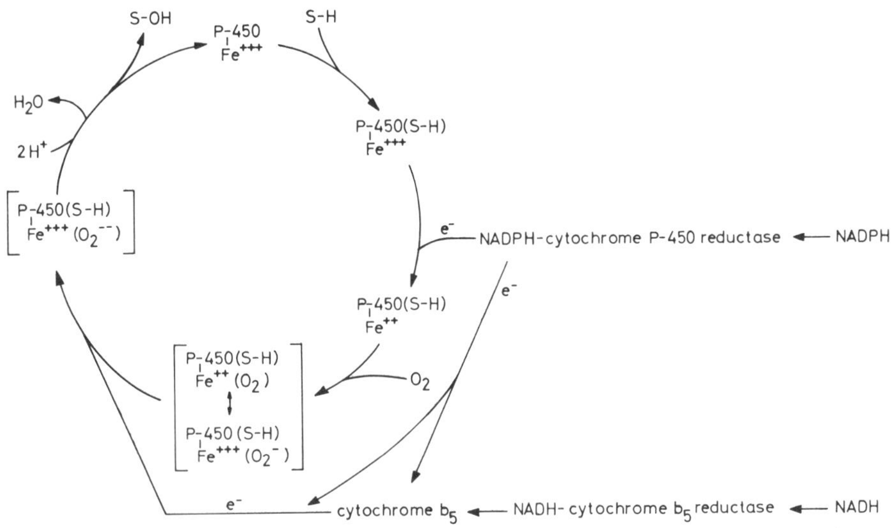

FIGURE 1. Mechanism of cytochrome P-450-catalyzed reactions. S-H represents substrate.

nmol/mg of protein.[20-23] The second criteria of purity is the presence of only one protein band upon sodium dodecyl sulfate (SDS)-polyacrylamide gel electrophoresis. Recently, several additional criteria of purity have been used such as one spot on two-dimensional gel electrophoresis,[24] one N-terminal amino acid sequence[25] and immunological homogeneity.[26] As research in the cytochrome P-450 field progresses, an additional criterion of homogeneity may be the specific catalytic properties with regard to endogenous substrates.

B. Mechanism of Cytochrome P-450-Catalyzed Reactions

The mechanism of cytochrome P-450-catalyzed reactions has been studied mainly with drugs as substrates. Figure 1 shows the generally accepted sequence of events.[27,28] The first step is the binding of the substrate to the oxidized form of cytochrome P-450 (Fe^{3+}). This enzyme-substrate complex is reduced by an electron transferred from NADPH by NADPH-cytochrome P-450 reductase. Molecular oxygen now binds to the reduced enzyme-substrate complex and this complex is reduced by a second electron. This electron may be transferred from NADPH in the same way as the first electron or by a mechanism involving participation of another microsomal hemoprotein, cytochrome b_5.[29-31] The oxygenated complex is rearranged, decomposed, and the hydroxylated substrate is released together with a water molecule and cytochrome P-450 in the oxidized form (see also Chapter 6).

As mentioned above, early studies of the cytochrome P-450 systems showed a requirement for phosphatidylcholine. The lipid may be stabilizing the enzyme complex and/or may enhance the rate of electron transfer from NADPH to cytochrome P-450.[13] Some reactions catalyzed by purified cytochromes P-450 show low degree of phospholipid dependency.[13]

III. EARLY STUDIES ON PURIFICATION OF CYTOCHROME P-450 ACTIVE IN 7α-HYDROXYLATION OF CHOLESTEROL

The first report showing 7α-hydroxylation of cholesterol in a reconstituted system consisting of partially purified cytochrome P-450, NADPH-cytochrome P-450 reductase, NADPH, and phosphatidylcholine was published in 1974 by Björkhem et al.[32] The cholesterol 7α-hydroxylase system was solubilized from rat liver microsomes using sodium cholate as detergent and cytochrome P-450 and NADPH-cytochrome P-450 reductase were partially purified as described by Lu et al.[33] for drug hydroxylase systems. The stimulatory effect of cholestyramine feeding on microsomal cholesterol 7α-hydroxylase activity was shown to be associated with the cytochrome P-450 fraction.[32] Several laboratories have subsequently reported the solubilization and reconstitution of cholesterol 7α-hydroxylase with protein components of varying degree of purity.[34-38] From studies with homogeneous NADPH-cytochrome P-450 reductase and highly purified cytochrome P-450 fractions, clearcut results have been obtained showing an absolute requirement for both protein components. There was, however, no marked effect upon omission of the phospholipid.[39] In 1979 to 1981, at least partial separations of cytochrome P-450 fractions with cholesterol 7α-hydroxylase activity from fractions with other hydroxylase activities in bile acid biosynthesis were reported[35,39-41] (see also Chapter 6).

A cytochrome P-450 fraction with a specific content of 14.5 nmol of cytochrome P-450 per milligram of protein and with the ability to catalyze an efficient 7α-hydroxylation of cholesterol was described in 1980 by Hansson and Wikvall.[40] This cytochrome P-450 fraction was obtained from liver microsomes of cholestyramine-treated rats. The purified fraction revealed one major protein band upon SDS-gel electrophoresis, and catalyzed in addition to 7α-hydroxylation of cholesterol, also 12α-hydroxylation of 5β-cholestane-3α,7α-diol, 25-hydroxylation of 5β-cholestane-3α,7α,12α-triol, and 7α-hydroxylation of taurodeoxycholic acid. However, all these hydroxylase activities were lower with the purified cytochrome P-450 fraction than with the microsomal fraction.

Ozasa and Boyd[35] reported in 1981 the properties of a cytochrome P-450 fraction prepared from liver microsomes of cholestyramine-treated rats, which catalyzed 7α-hydroxylation of cholesterol more efficiently than the microsomal fraction. This fraction also catalyzed effient 12α- and 25-hydroxylations of C_{27}-steroids as well as 7α-hydroxylation of taurodeoxycholic acid. The preparation showed several protein bands upon gel electrophoresis and the specific content of cytochrome P-450 was 1 to 2 nmol/mg protein.

Work with whole microsomes as well as with partially purified cytochrome P-450 preparations indicates that there is a specific species of cytochrome P-450 in liver microsomes catalyzing 7α-hydroxylation of cholesterol. This contention is further strengthened by a report showing that a cytochrome P-450 fraction from rat liver microsomes catalyzing 7α-hydroxylation of cholesterol had a shorter half-life time than the bulk of cytochrome P-450.[42]

IV. CATALYTIC PROPERTIES OF HIGHLY PURIFIED SPECIES OF CYTOCHROME P-450

A. Rabbit Liver Microsomal Cytochrome P-450

Seven different isozymes of cytochrome P-450 have been isolated from rabbit liver microsomes by Coon and associates and Johnson, Muller-Eberhard, and associates.[23,25,43-48] The various isozymes have been designated according to their relative mobilities upon SDS-polyacrylamide gel electrophoresis. The protein with the lowest molecular weight and the greatest electrophoretic mobility is called LM_1 (LM = liver microsomal) and the other isozymes are numbered according to their decreasing mobility and increasing molecular weight. Cytochromes P-450 LM_1 and LM_{3c} appear to be constitutive forms, whereas cytochromes P-450 LM_2, LM_{3a}, LM_4, and LM_6 are inducible. Cytochrome P-450 LM_4 is the major species in microsomes from untreated rabbits.[15,43,49]

Cytochrome P-450 LM_2 is induced by phenobarbital treatment and the concentration of this isozyme in untreated animals is very low.[43,49] Cytochrome P-450 LM_{3b} has been suggested to be a constitutive form, as it was not possible to find any agent that could induce this form. However, it has been recently reported that cytochrome P-450 LM_{3b} is induced by triacetyloleandomycin, an antibiotic.[50] Cytochrome P-450 LM_4 can be induced by various compounds such as β-naphthoflavone, 3-methylcholanthrene, and isosafrole.[15,43,51]

Preparation of NADPH-cytochrome P-450 reductase to homogeneity has been described by several laboratories.[52-54] The purified reductase molecule contains one mole of FMN and one mole of FAD.

The purified cytochrome P-450 isozymes have different but overlapping substrate specificity. Cytochrome P-450 LM_2 seems to be primarily involved in oxidation of exogenous substrates and cytochrome P-450 LM_4 in oxidation of endogenous substrates.

The involvement of cytochromes P-450 LM_2, LM_{3b}, and LM_4 in bile acid biosynthesis has been studied by Hansson, Wikvall, and collaborators[40,55-57] and by Chiang et al.[38] Table 1 summarizes the physical properties of cytochromes P-450 LM_2, LM_{3b}, and LM_4 as well as the catalytic properties of these three forms towards cholesterol and intermediates in bile acid biosynthesis. All preparations from rabbit liver microsomes but one had a specific content of cytochrome P-450 of more than 10 nmol/mg protein and they were apparently homogeneous on SDS-polyacrylamide gel electrophoresis. Cytochrome P-450 LM_4 prepared by Chiang et al.[38] showed a higher molecular weight but a lower specific content of cytochrome P-450 than the cytochrome P-450 LM_4 prepared by Hansson, Wikvall et al.[40,55-57]

Cytochrome P-450 LM_2 did not catalyze hydroxylation of C_{27}-steroids. Cytochrome P-450 LM_{3b} prepared from untreated or cholestyramine-treated rabbits did not catalyze 7α-hydroxylation of cholesterol and showed only low activity towards the other C_{27}-steroids. Cytochrome P-450 LM_4 from untreated and differently treated rabbits catalyzed 12α- and 25-hydroxylation of C_{27}-steroids more efficiently than the corresponding microsomes. Cytochrome P-450 LM_4 isolated from cholestyramine-treated rabbits catalyzed in addition 7α-hydroxylation of cholesterol. This hydroxylation was more efficient with the purified enzyme fraction than with microsomes.

These findings indicate that the stimulatory effect of cholestyramine treatment on 7α-hydroxylation of cholesterol found with whole microsomes is associated with the cytochrome P-450 LM_4 fraction. It should be pointed out that cholestyramine treatment does not increase, if anything decreases, the total amount of cytochrome P-450.[32,39] It is likely that 7α-hydroxylase activity in cytochrome P-450 LM_4 fractions prepared from rabbits treated otherwise is not zero but below the limit of detection.[55] The results in Table 1 show that cytochrome P-450 LM_4 plays a major role in the biosynthesis of bile acids, whereas cytochrome P-450 LM_{3b} plays at the most a minor role and cytochrome P-450 LM_2 is not involved in this metabolic pathway.

Cytochrome P-450 LM_4 has been considered to be homogeneous according to a number of criteria although some irregularities have been found.[26,43,58,59] Studies on the properties of the hydroxylations in bile acid biosynthesis indicate that cytochrome P-450 LM_4 contains more than one species of cytochrome P-450. It has been found that agents that induce cytochrome P-450 LM_4 do not induce the hydroxylase activities in bile acid biosynthesis.[56] On the other hand, starvation and cholestyramine treatment induce 12α- and 7α-hydroxylase activity, respectively.[55,56] These findings indicate the presence of at least two species of cytochrome P-450 in cytochrome P-450 LM_4 catalyzing 12α- and 7α-hydroxylation, respectively.

Studies by Chiang et al.[38] showed that antibodies raised against cytochrome P-450 LM_4 prepared from phenobarbital-treated rabbits were able to inhibit to some extent 7α-hydroxylation of cholesterol catalyzed by cytochrome P-450 LM_4 prepared from cholestyramine-treated rabbits. This finding and the results of two-dimensional peptide fingerprints of the two cytochrome P-450 LM_4 fractions indicate that these two fractions are not identical but very similar.

Efforts have been made to separate 7α-, 12α-, and 25-hydroxylase activities in cytochrome P-450 LM_4 prepared from cholestyramine-treated rabbits.[60] The preparation was fractionated into two subfractions by chromatography on an octylamine-Sepharose column. The two subfractions, cytochromes P-450 LM_4I and LM_4II, showed the same apparent molecular

Table 1
HYDROXYLASE ACTIVITIES IN PREPARATIONS OF LIVER MICROSOMAL CYTOCHROME P- 450 FROM UNTREATED AND DIFFERENTLY TREATED RABBITS

Fraction	M_r	Cytochrome P-450 (specific content)	Hydroxylation of Cholesterol	Hydroxylation of 5β-Cholestane-3α,7α-diol	Hydroxylation of 5β-Cholestane-3α,7α,12α-triol		Ref.
			7α	12α	25	26	
		(nmol/mg protein)	(pmol/nmol cytochrome P-450 × min)				
Cytochrome P-450 LM_2 (phenobarbital)	49,000	18.8	<1	<1	<1	<1	40
Cytochrome P-450 LM_{3b} (untreated)	52,000	19.7	<1	9	20	<1	55
Cytochrome P-450 LM_{3b} (cholestyramine)	52,000	15.0	<1	17	30	<1	55
Cytochrome P-450 LM_4 (untreated)	53,000	15.5	<1	165	160	<1	55,56
Cytochrome P-450 LM_4 (starved)	53,000	14.0	<1	481	162	<1	56
Cytochrome P-450 LM_4 (β-naphthoflavone)	53,000	15.0	<1	130	n.d.[a]	n.d.	56
Cytochrome P-450 LM_4 (β-naphthoflavone)	55,000	12.0—16.0	0	n.d.	n.d.	n.d.	38
Cytochrome P-450 LM_4 (phenobarbital)	53,000	19.0	<1	112	357	<1	40,56,57
Cytochrome P-450 LM_4 (phenobarbital)	55,000	12.0—16.0	0	n.d.	n.d.	n.d.	38
Cytochrome P-450 LM_4 (cholestyramine)	53,000	13.0	23	120	184	<1	55
Cytochrome P-450 LM_4 (cholestyramine)	55,000	6.0	38	n.d.	n.d.	n.d.	38
Cytochrome P-450 LM_4I (cholestyramine)	53,000	18.8	<1	160	120	<1	60,76
Cytochrome P-450 LM_4II (cholestyramine)	53,000	10.8	80	50	50	<1	60,76

[a] n.d. = not determined.

weight and spectral properties as the original cytochrome P-450 LM_4 fraction but differed in amino acid composition. The specific cytochrome P-450 content of cytochrome P-450 LM_4II was lower than that of cytochrome P-450 LM_4I and the original cytochrome P-450 LM_4 fraction, possibly due to heme losses during the preparation. Cytochrome P-450 LM_4I did not catalyze 7α-hydroxylation of cholesterol, but catalyzed efficient 12α-hydroxylation and 25-hydroxylation (Table 1). Cytochrome P-450 LM_4II catalyzed 7α-hydroxylation more efficiently and 12α- and 25-hydroxylation less efficiently than the original cytochrome P-450 LM_4 fraction.

B. Rat Liver Microsomal Cytochrome P-450

Several different isozymes of cytochrome P-450 have been purified from untreated and differently treated rats.[22,61-71] Some of these forms are apparently constitutive whereas others are induced by, for instance, phenobarbital, isosafrole, and 3-methylcholanthrene. However, it has not been possible, as in the case of rabbit liver cytochromes P-450, to establish a uniform nomenclature for rat liver cytochromes P-450 that will fit the results of various workers.

Multiple forms of cytochrome P-450 are induced by phenobarbital treatment.[24,61,62,72,73] The major phenobarbital-induced form has been purified by several laboratories and is called cytochrome P-450 B, cytochrome P-450 Fraction C, or cytochrome P-450_b.[17] Another phenobarbital-induced form, cytochrome P-450 D, has been purified by Guengerich.[61,62] The purified cytochrome P-450 species catalyze hydroxylations of various endogenous substrates such as fatty acids, prostaglandins, and steroids as well as of exogenous substrates such as drugs and carcinogens.

The involvement of cytochromes P-450 B, D, and Fraction C in bile acid biosynthesis has been studied by Hansson and Wikvall.[40,57,74] Table 2 summarizes the properties of some purified cytochrome P-450 species from rat liver microsomes. The three preparations from phenobarbital-treated rats were all apparently homogeneous on SDS-polyacrylamide gel electrophoresis and had a specific content of 14 to 18 nmol of cytochrome P-450 per milligram of protein. Judged by gel electrophoresis, these preparations had an apparent molecular weight of 50,000.

The catalytic properties towards various C_{24}- and C_{27}-steroids of the purified cytochrome P-450 fractions are shown in Table 2. Cytochromes P-450 B and Fraction C catalyzed 12α-, 25-, and 26-hydroxylations of C_{27}-steroids, but these hydroxylations were less efficient with the purified cytochromes P-450 than with microsomes. Cytochrome P-450 D showed very low activity towards the C_{27}-steroids, but catalyzed 7α-hydroxylation of taurodeoxycholic acid as efficiently as the microsomal fraction. None of these cytochromes P-450 catalyzed 7α-hydroxylation of cholesterol.

Recently, preparation of two cytochrome P-450 fractions called cytochrome P-$450_{7\alpha}$ and cytochrome P-450_{25} has been described.[75,76] The purification steps involved polyethylene glycol fractionation, octylamine-Sepharose-, hydroxylapatite-, and DEAE-Sepharose chromatography.

Cytochrome P-450_{25} prepared from untreated rats and cytochrome P-$450_{7\alpha}$ prepared from cholestyramine-treated rats each showed one major protein band on gel electrophoresis with an apparent molecular weight of 51,000 and 52,000, respectively. Cytochrome P-450_{25} had a specific content of 16 nmol of cytochrome P-450 per milligram of protein whereas the heme content in the cytochrome P-$450_{7\alpha}$ preparation was rather low, 3 nmol/mg of protein.

Cytochrome P-450_{25} did not catalyze 7α,12α-, and 26-hydroxylations, but catalyzed 25-hydroxylation with high efficiency. Cytochrome P-$450_{7\alpha}$ catalyzed 7α-hydroxylation of cholesterol 200 times more efficiently than the corresponding microsomes, showed very low 12α-hydroxylase activity, and no 25- and 26-hydroxylase activities.

Table 2
HYDROXYLASE ACTIVITIES IN PREPARATIONS OF LIVER MICROSOMAL CYTOCHROME P-450 FROM UNTREATED AND DIFFERENTLY TREATED RATS

Fraction	M_r	Cytochrome P-450 (specific content)	Hydroxylation of						Ref.
			Cholesterol 7α	Cholestanol 7α	5β-Cholestane-3α,7α-diol 12α	5β-Cholestane-3α,7α,12α-triol 25	5β-Cholestane-3α,7α,12α-triol 26	Taurodeoxycholic acid 7α	
		(nmol/mg protein)	(pmol/nmol cytochrome P-450 × min)						
Cytochrome P-450 B (phenobarbital)	50,000	18.0	<1	n.d.[a]	7	26	15	120	40
Cytochrome P-450 D (phenobarbital)	50,000	14.0	<1	n.d.	5	4	<1	540	40
Cytochrome P-450 Fraction C (phenobarbital)	50,000	18.3	<1	n.d.	5	46	15	70	57,74
Cytochrome $P\text{-}450_{25}$ (untreated)	51,000	15.6	<1	n.d.	<1	2,600	<1	n.d.	75
Cytochrome $P\text{-}450_{7\alpha}$ (cholestyramine)	52,000	3.0	7,500	2,500	50	<1	<1	<10	76

[a] n.d. = not determined.

C. Pig Liver Microsomal Cytochrome P-450

Multiple forms of cytochrome P-450 have been purified from phenobarbital-treated pigs by Ullrich and associates[77] and by Masters et al.[78] Some of these forms exhibit a high degree of substrate specificity towards drugs and steroids, but the catalytic properties towards cholesterol were not studied. Cholesterol 7α-hydroxylase activity has been studied in pig liver microsomes by Li and Kim.[79,80] It was shown that cholesterol 7α-hydroxylase activity was induced severalfold by feeding cholestyramine to the pigs, indicating that 7α-hydroxylation is the rate-limiting step for biosynthesis of bile acids also in the pig.

Three fractions of cytochrome P-450, cytochrome $P\text{-}450_x$, cytochrome $P\text{-}450_{6\alpha}$, and cytochrome $P\text{-}450_{7\alpha}$, have been purified by Boström from liver microsomes of untreated pigs (Table 3).[81] On gel electrophoresis these fractions each showed one major protein band. The specific cytochrome P-450 content in the fractions were 12, 4, and 6 nmol of cytochrome P-450 per milligram of protein, respectively. Cytochrome $P\text{-}450_x$ catalyzed 25-hydroxylation of C_{27}-steroids, but did not catalyze 7α-hydroxylation of cholesterol or 6α-hydroxylation of taurochenodeoxycholic acid. Cytochrome $P\text{-}450_{6\alpha}$ catalyzed 6α-hydroxylation of taurochenodeoxycholic acid, a reaction unique for pig liver, but did not catalyze 7α-hydroxylation of cholesterol or hydroxylation of the side chain. Cytochrome $P\text{-}450_{7\alpha}$ catalyzed an efficient 7α-hydroxylation of cholesterol, but showed no side chain hydroxylase activity or 6α-hydroxylase activity. It should be pointed out that pig liver microsomes do not catalyze 12α-hydroxylation of C_{27}-steroids.

It is noteworthy that in spite of the low specific content of cytochrome P-450 in cytochrome $P\text{-}450_{7\alpha}$ from pig liver, this fraction catalyzed cholesterol 7α-hydroxylation 40 times more efficiently than cytochrome P-450 LM_4II prepared from cholestyramine-treated rabbits (compare Table 1). It should also be borne in mind that cytochrome $P\text{-}450_{7\alpha}$ was prepared from untreated pigs.

D. Human Liver Microsomal Cytochrome P-450

Cytochrome P-450 has been purified from human liver microsomes in several laboratories and the results show the existence of multiple forms of cytochrome P-450 also in human liver.[17,19,82] Two independent reports on the solubilization and reconstitution of cholesterol 7α-hydroxylase activity from human liver microsomes were published in 1981.[41,83] Ahlsten et al.,[41] isolated three fractions of cytochrome P-450 using essentially the same procedure as in the isolation of rat liver cytochromes P-450. The fractions had a specific content of 5 to 9 nmol of cytochrome P-450 per milligram of protein. One fraction showed very low 7α-hydroxylase activity, low 25-hydroxylase activity, but efficient 12α-hydroxylase activity. The other fractions were able to catalyze cholesterol 7α-hydroxylation more efficiently than the microsomal fraction. These preparations also catalyzed 12α-hydroxylation and 25-hydroxylation of C_{27}-steroids. The human cholesterol 7α-hydroxylase system showed an absolute requirement for cytochrome P-450 and NADPH-cytochrome P-450 reductase. It was also possible to replace the human reductase with reductase prepared from rat liver. The conclusion from this study was that the 7α-hydroxylase systems in rat and human liver are closely related and show similar properties.

Erickson et al.[83] studied cholesterol 7α-hydroxylation in phospholipid-cholesterol vesicles containing a partially purified cytochrome P-450 fraction and rat liver NADPH-cytochrome P-450 reductase. The specific content of the cytochrome P-450 fraction was 7.6 nmol of cytochrome P-450 per milligram of protein and this fraction catalyzed 7α-hydroxylation more efficiently than microsomes. Catalytic properties of this preparation towards other substrates in bile acid biosynthesis were not studied.

Table 3
HYDROXYLASE ACTIVITIES IN PREPARATIONS OF LIVER MICROSOMAL CYTOCHROME P-450 FROM UNTREATED PIGS

Fraction	M_r	Cytochrome P-450 (specific content)	Hydroxylation of: Cholesterol 7α	5β-Cholestane-3α,7α-diol 25	5β-Cholestane-3α,7α-diol 26	Taurochenodeoxycholic acid 6α	Ref.
		(nmol/mg protein)	(pmol/nmol cytochrome P-450 × min)				
Cytochrome P-450$_x$	50,000	11.3	≤2	60	≤10	≤10	81
Cytochrome P-450$_{6\alpha}$	53,000	4.1	≤2	≤10	≤10	1,050	81
Cytochrome P-450$_{7\alpha}$	52,000	6.1	3,100	≤10	≤10	≤10	81

V. PROPERTIES OF CHOLESTEROL 7α-HYDROXYLASE

Using microsomal fractions it is difficult to study properties including substrate specificity of cholesterol 7α-hydroxylase. For instance, a substrate may be more or less available to the enzyme depending on the physical properties of the substrate and the enzyme environment. Further, microsomes may contain several 7α-hydroxylating systems specific for e.g., C_{19}-,C_{21}-,C_{24}-, and C_{27}-steroids, making it impossible to study substrate specificity of cholesterol 7α-hydroxylase with microsomal preparations. From studies with whole microsomes and partially purified cytochrome P-450 fractions from liver of rats treated with cholestyramine or phenobarbital it can be concluded that cholesterol and cholestanol are 7α-hydroxylated by the same enzyme, whereas 7α-hydroxylation of taurodeoxycholic acid is catalyzed by another cytochrome P-450 species.[6,39,40,84] These conclusions have been recently confirmed. In unpublished studies, cytochrome P-$450_{7\alpha}$ (compare Section IV.B and Table 2) has been shown to catalyze 7α-hydroxylation of both cholesterol and cholestanol.[85] The rate with cholesterol as substrate was about three times faster than with cholestanol. Cytochrome P-$450_{7\alpha}$ does not catalyze 7α-hydroxylation of taurodeoxycholic acid or testosterone. A highly purified cytochrome P-450 from rat liver microsomes, cytochrome P-450 D (compare Section IV.B and Table 2), catalyzes efficient 7α-hydroxylation of taurodeoxycholic acid, but has no capacity to 7α-hydroxylate cholesterol.[40]

Boström has published recently a study of the binding of cholesterol to different cytochromes P-450 from rabbit liver microsomes.[86] Cytochromes P-450 LM_2 and LM_{3b}, which do not catalyze 7α-hydroxylation of cholesterol, did not bind cholesterol. Cytochrome P-450 LM_4 from phenobarbital-treated rabbits and cytochromes P-450 LM_4I and LM_4II from cholestyramine-treated rabbits all bound cholesterol with roughly the same K_s value, 33 to 37 μM. It should be noted that of these species only cytochrome P-450 LM_4II shows significant 7α-hydroxylase activity. 7α-Hydroxycholesterol and 7-ketocholesterol inhibited competively the binding of cholesterol and inhibited by more than 50% the 7α-hydroxylation of cholesterol. Similar findings have been previously made by Boyd et al.,[3] van Cantfort,[87] and Schwartz and Margolis[88] in studies with whole microsomes. The implications of these findings in terms of regulation of bile acid biosynthesis can not be assessed at present.

Ahringer and Eneroth[89] and Boyd and associates[90] have found in studies with whole microsomes that the side chain of cholesterol is important in the interaction between substrate and enzyme. No such studies have been performed with purified cytochrome P-450 systems.

It was early shown that thiol compounds were necessary for optimal cholesterol 7α-hydroxylase activity and that sulfhydryl reagents were inhibitory.[91,92] The same findings have since been made with reconstituted systems. Addition of dithiothreitol, reduced glutathione, thioredoxin, and cysteine stimulated 7α-hydroxylation of cholesterol, whereas addition of glutathione disulfide inhibited the reaction.[93,94] Cholesterol 7α-hydroxylase was much more sensitive to sulfhydryl agents than other hydroxylases involved in bile acid biosynthesis.[93-95]

In this connection, a report concerning the effect of disulfiram is of interest. Andersson and Boström[96] found that this disulfide-containing drug, which induces increased levels of cholesterol in liver and serum, inhibited 7α-hydroxylation of cholesterol in vitro. The inhibition could be reversed by addition of dithiothreitol to the incubation mixture containing a purified fraction of cytochrome P-450.

Several reports concerning regulation of cholesterol 7α-hydroxylase activity by cytosolic proteins have appeared.[94,97-104] Some of these proteins appear to act on cholesterol 7α-hydroxylase through mechanisms involving phosphorylation-dephosphorylation, whereas others do not. A detailed discussion of these findings is found in Chapter 7.

Recently, a cytosolic protein able to stimulate reconstituted cholesterol 7α-hydroxylase activity has been purified to homogeneity from rat liver cytosol by Danielsson et al.[94] The protein required reduced glutathione or reduced thioredoxin for activity. The protein was specific for cholesterol 7α-hydroxylation and did not stimulate hydroxylations of other C_{27}-steroids. Although the stimulatory cytosolic protein showed glutathione *S*-transferase activity, purified glutathione *S*-transferases A, B, and C did not stimulate cholesterol 7α-hydroxylation. Addition of ATP and $MgCl_2$ or sodium fluoride inhibited cholesterol 7α-hydroxylation, but did not affect the stimulatory capacity of the cytosolic 7α-hydroxylase stimulator. The mechanism of action of the stimulatory protein can not yet be defined nor can the importance of this protein in regulation of cholesterol 7α-hydroxylase activity.

VI. SUMMARY

The cytochrome P-450, which in combination with NADPH-cytochrome P-450 reductase, catalyzes 7α-hydroxylation of cholesterol has been purified from pig, rabbit, and rat liver. Cytochrome P-$450_{7\alpha}$ is not identical with cytochromes P-450 induced by e.g., phenobarbital nor with some known constitutive cytochromes P-450. Cytochrome P-$450_{7\alpha}$ shows a high specificity for 7α-hydroxylation of cholesterol and cholestanol. The enzyme is dependent on reduced sulfhydryl groups for activity. The activity can be stimulated by a protein in rat liver cytosol in the presence of reduced glutathione or thioredoxin.

REFERENCES

1. **Wada, F., Hirata, K., Nakano, K., and Sakamato, Y.,** Participation of the microsomal electron transport system involving cytochrome P-450 in 7α-hydroxylation of cholesterol, *J. Biochem. (Tokyo),* 66, 699, 1969.
2. **Johansson, G.,** Oxidation of cholesterol, 3β-hydroxy-5-pregnen-20-one and 3β-hydroxy-5-androsten-17-one by rat liver microsomes, *Eur. J. Biochem.,* 21, 68, 1971.
3. **Boyd, G. S., Grimwade, A. M., and Lawson, M. E.,** Studies on rat liver microsomal cholesterol 7α-hydroxylase, *Eur. J. Biochem.,* 37, 334, 1973.
4. **Björkhem, I., Danielsson, H., and Wikvall, K.,** 7α-Hydroxylation of cholesterol by reconstituted systems from rat liver microsomes, *Biochem. Biophys. Res. Commun.,* 61, 934, 1974.
5. **Danielsson, H., Einarsson, K., and Johansson, G.,** Effect of biliary drainage on individual reactions in the conversion of cholesterol to taurocholic acid, *Eur. J. Biochem.,* 2, 44, 1967.
6. **Shefer, S., Hauser, S., and Mosbach, E. H.,** 7α-Hydroxylation of cholestanol by rat liver microsomes, *J. Lipid Res.,* 9, 328, 1968.
7. **Mitropoulos, K. A. and Balasubramaniam, S.,** Cholesterol 7α-hydroxylase in rat liver microsomal preparations, *Biochem. J.,* 128, 1, 1972.
8. **Gielen, J., Van Cantfort, J., Robaye, B., and Renson, J.,** Rythme circadien de la cholesterol-7α-hydroxylase chez la rat, *C.R. Acad. Sci. Paris,* 269, 731, 1969.
9. **Sanghvi, A., Grassi, E., Bartman, C., Lester, R., Kienle, M., and Galli, G.,** Measurement of cholesterol 7α-hydroxylase activity with selected ion monitoring, *J. Lipid Res.,* 22, 720, 1981.
10. **Sato, R. and Omura, T.,** *Cytochrome P-450,* Academic Press, New York, 1978.
11. **Wikvall, K.,** Hydroxylations in biosynthesis of bile acids. Isolation of a cytochrome P-450 from rabbit liver mitochondria catalyzing 26-hydroxylation of C_{27}-steroids, *J. Biol. Chem.,* 259, 3800, 1984.
12. **Burke, D. M. and Orrenius, S.,** Isolation and comparison of endoplasmic reticulum membranes and their mixed function oxidase activities from mammalian extrahepatic tissues, in *International Encyclopedia of Pharmacology and Therapeutics,* Sect. 108, *Hepatic Cytochrome P-450 Monooxygenase System,* Schenkman, J. B. and Kupfer, D., Eds., Pergamon Press, Oxford, 1982, chap. 2.
13. **Lu, A. Y. H. and West, S. B.,** Reconstituted mammalian mixed-function oxidases: requirements, specificities and other properties, in *International Encyclopedia of Pharmacology and Therapeutics,* Sect. 108, *Hepatic Cytochrome P-450 Monooxygenase System,* Schenkman, J. B. and Kupfer, D., Eds., Pergamon Press, Oxford, 1982, chap. 20.
14. **Conney, A. H.,** Induction of microsomal enzymes by foreign chemicals and carcinogenesis by polycyclic aromatic hydrocarbons: G.H.A. Clowes Memorial Lecture, *Cancer Res.,* 42, 4875, 1982.

15. **Coon, M. J. and Koop, D. R.,** P-450 Oxygenases in lipid transformation, in *The Enzymes,* Boyer, P. D., Ed., Academic Press, New York, 1983, chap. 19.
16. **Orrenius, S. and Ernster, L.,** Phenobarbital-induced synthesis of the oxidative demethylating enzymes of rat liver microsomes, *Biochem. Biophys. Res. Commun.,* 16, 60, 1964.
17. **Lu, A. Y. H. and West, S. B.,** Multiplicity of mammalian microsomal cytochromes P-450, *Pharmacol. Rev.,* 31, 277, 1980.
18. **Lu, A. Y. H. and Coon, M. J.,** Role of hemoprotein P-450 in fatty acid ω-hydroxylation in a soluble enzyme system from liver microsomes, *J. Biol. Chem.,* 243, 1331, 1968.
19. **Guengerich, F. P.,** Isolation and purification of cytochrome P-450, and the existence of multiple forms, in *International Encyclopedia of Pharmacology and Therapeutics,* Sect. 108, *Hepatic Cytochrome P-450 Monooxygenase System,* Schenkman, J. B. and Kupfer, D., Eds., Pergamon Press, Oxford, 1982, chap. 19.
20. **van der Hoeven, T. A., Haugen, D. A., and Coon, M. J.,** Cytochrome P-450 purified to apparent homogeneity from phenobarbital-induced rabbit liver microsomes: catalytic activity and other properties, *Biochem. Biophys. Res. Commun.,* 60, 569, 1974.
21. **Imai, Y. and Sato, R.,** A gel-electrophoretically homogeneous preparation of cytochrome P-450 from liver microsomes of phenobarbital-pretreated rabbits, *Biochem. Biophys. Res. Commun.,* 60, 8, 1974.
22. **Ryan, D., Lu, A. Y. H., Kawalek, J., West, S. B., and Levin, W.,** Highly purified cytochrome P-448 and P-450 from rat liver microsomes, *Biochem. Biophys. Res. Commun.,* 64, 1134, 1975.
23. **Johnson, E. F. and Muller-Eberhard, U.,** Purification of the major cytochrome P-450 of liver microsomes from rabbits treated with 2,3,7,8-tetrachlorodibenzo-p-dioxin (TCDD), *Biochem. Biophys. Res. Commun.,* 76, 652, 1977.
24. **Vlasuk, G. P., Ghrayeb, J., Ryan, D. E., Reik, L., Thomas, P. E., Levin, W., and Walz, F. G., Jr.,** Multiplicity, strain differences, and topology of phenobarbital-induced cytochromes P-450 in rat liver microsomes, *Biochemistry,* 21, 789, 1982.
25. **Koop, D. R., Persson, A. V., and Coon, M. J.,** Properties of electrophoretically homogeneous constitutive forms of liver microsomal cytochrome P-450, *J. Biol. Chem.,* 256, 10704, 1981.
26. **Dean, W. L. and Coon, M. J.,** Immunochemical studies on two electrophoretically homogeneous forms of rabbit liver microsomal cytochrome P-450: P-450 LM_2 and P-450 LM_4, *J. Biol. Chem.,* 252, 3255, 1977
27. **White, R. E. and Coon, M. J.,** Oxygen activation by cytochrome P-450, *Annu. Rev. Biochem.,* 49, 315, 1980.
28. **Peterson, J. A.,** The ligands of cytochrome P-450: identification and role in the activation of dioxygen, in *Biochemical and Clinical Aspects of Oxygen,* Caughey, W. S., Ed., Academic Press, New York, 1979, 227.
29. **Hildebrandt, A. and Estabrook, R. W.,** Evidence for the participation of cytochrome b_5 in hepatic microsomal mixed-function oxidation reactions, *Arch. Biochem. Biophys.,* 143, 66, 1971.
30. **Sugiyama, T., Miki, N., and Yamanu, T.,** NADH- and NADPH-dependent reconstituted p-nitroanisole O-demethylation system containing cytochrome P-450 with high affinity for cytochrome b_5, *J. Biochem. (Tokyo),* 87, 1457, 1980.
31. **Ingelman-Sundberg, M. and Johansson, I.,** Cytochrome b_5 as electron donor to rabbit liver cytochrome P-450_{LM2} in reconstituted phospholipid vesicles, *Biochem. Biophys. Res. Commun.,* 97, 582, 1980.
32. **Björkhem, I., Danielsson, H., and Wikvall, K.,** 7α-Hydroxylation of cholesterol by reconstituted systems from rat liver microsomes, *Biochem. Biophys. Res. Commun.,* 61, 934, 1974.
33. **Lu, A. Y. H., Kuntzman, R., West, S., Jacobson, M., and Conney, A. H.,** Reconstituted liver microsomal enzyme system that hydroxylates drugs, other foreign compounds, and endogenous substrates. II. Role of the cytochrome P-450 and P-448 fractions in drug and steroid hydroxylations, *J. Biol. Chem.,* 247, 1727, 1972.
34. **Hattersley, N. G. and Boyd, G. S.,** Solubilization of rat liver microsomal cholesterol 7α-hydroxylase, *FEBS Lett.,* 53, 80, 1975.
35. **Ozasa, S. and Boyd, G. S.,** Cholesterol 7α-hydroxylase of rat liver: Studies on the solubilisation, resolution and reconstitution of the enzyme complex, *Eur. J. Biochem.,* 119, 263, 1981.
36. **Rush, W. R., Gibson, G. G., and Parke, D. V.,** The purification and reconstitution of cytochrome P-450-dependent cholesterol 7α-hydroxylase activity from rat liver microsomal fractions, *Biochem. Soc. Trans.,* 8, 102, 1980.
37. **Nimmannit, S. and Porter, J. W.,** Cholesterol 7α-hydroxylase: Solubilization and determination of enzyme activity, *Arch. Biochem. Biophys.,* 201, 533, 1980.
38. **Chiang, J. Y. L., Malmer, M., and Hutterer, F.,** A form of rabbit liver cytochrome P-450 that catalyzes the 7α-hydroxylation of cholesterol, *Biochim. Biophys. Acta,* 750, 291, 1983.
39. **Hansson, R. and Wikvall, K.,** Properties of reconstituted cholesterol 7α-hydroxylase system from rat and rabbit liver microsomes, *Eur. J. Biochem.,* 93, 419, 1979.

40. **Hansson, R. and Wikvall, K.**, Hydroxylations in biosynthesis and metabolism of bile acids: catalytic properties of different forms of cytochrome P-450, *J. Biol. Chem.*, 255, 1643, 1980.
41. **Ahlsten, M., Hansson, R., Holmberg, I., Ljungdahl, I., and Wikvall, K.**, Reconstitution of cholesterol 7α-hydroxylase system from human liver microsomes, *Biochem. Med.*, 26, 307, 1981.
42. **Danielsson, H. and Wikvall, K.**, Evidence for a specific cytochrome P-450 with short half-life catalyzing 7α-hydroxylation of cholesterol, *Biochem. Biophys. Res. Commun.*, 103, 46, 1981.
43. **Haugen, D. A. and Coon, M. J.**, Properties of electrophoretically homogeneous phenobarbital-inducible and β-naphthoflavone-inducible forms of liver microsomal cytochrome P-450, *J. Biol. Chem.*, 251, 7929, 1976.
44. **Johnson, E. F., Zounes, M. C., and Muller-Eberhard, U.**, Characterization of three forms of rabbit microsomal cytochrome P-450 by peptide mapping utilizing limited proteolysis in sodium dodecyl sulfate and analysis by gel electrophoresis, *Arch. Biochem. Biophys.*, 192, 282, 1979.
45. **Koop, D. R. and Coon, M. J.**, Purification and properties of P-450 LM_{3b}, a constitutive form of cytochrome P-450, from rabbit liver microsomes, *Biochem. Biophys. Res. Commun.*, 91, 1075, 1979.
46. **Johnson, E. F.**, Isolation and characterisation of a constitutive form of rabbit liver microsomal cytochrome P-450, *J. Biol. Chem.*, 255, 304, 1980.
47. **Dieter, H. H., Muller-Eberhard, U., and Johnson, E. F.**, Identification of rabbit microsomal cytochrome P-450 isozyme, form 1, as a hepatic progesterone 21-hydroxylase, *Biochem. Biophys. Res. Commun.*, 105, 515, 1982.
48. **Koop, D. R., Morgan, E. T., Tarr, G. E., and Coon, M. J.**, Purification and characterization of a unique isozyme of cytochrome P-450 from liver microsomes of ethanol-treated rabbits, *J. Biol. Chem.*, 257, 8472, 1982.
49. **Chiang, J. Y. L., DiLella, A. G., and Steggles, A. W.**, Effect of inducers and aging on rabbit liver microsomal drug-metabolizing enzymes, *Mol. Pharmacol.*, 23, 244, 1983.
50. **Bonfils, D., Dalet, C., Dalet-Beluche, I., and Maurel, P.**, Cytochrome P-450 isozyme LM_{3b} from rabbit liver microsomes: induction by triacetylolendomycin, purification, and characterization, *J. Biol. Chem.*, 258, 5358, 1983.
51. **Delaforge, M., Koop, D. R., and Coon, M. J.**, Role of isosafrole as complexing agent and inducer of P-450 LM_4 in rabbit liver microsomes, *Biochem. Biophys. Res. Commun.*, 108, 59, 1982.
52. **Vermilion, J. L. and Coon, M. J.**, Highly purified detergent-solubilized NADPH-cytochrome P-450 reductase from phenobarbital-induced rat liver microsomes, *Biochem. Biophys. Res. Commun.*, 60, 1315, 1974.
53. **Dignam, J. D. and Strobel, H. W.**, Preparation of homogeneous NADPH-cytochrome P-450 reductase from rat liver, *Biochem. Biophys. Res. Commun.*, 63, 845, 1975.
54. **Yasukochi, Y. and Masters, B. S. S.**, Some properties of a detergent-solubilized NADPH-cytochrome c (cytochrome P-450) reductase purified by biospecific affinity chromatography, *J. Biol. Chem.*, 251, 5337, 1976.
55. **Boström, H., Hansson, R., Jönsson, K.-H., and Wikvall, K.**, Cytochromes P-450 LM_{3b} and LM_4 in biosynthesis of bile acids, *Eur. J. Biochem.*, 120, 29, 1981.
56. **Hansson, R. and Wikvall, K.**, Hydroxylations in biosynthesis of bile acids: cytochrome P-450 LM_4 and 12α-hydroxylation of 5β-cholestane-3α,7α-diol, *Eur. J. Biochem.*, 125, 423, 1982.
57. **Hansson, R., Holmberg, I., and Wikvall, K.**, 25-Hydroxylation of vitamin D_3 and side chain hydroxylations of 5β-cholestane-3α, 7α,12α-triol by purified rabbit and rat liver microsomal cytochromes P-450, *J. Biol. Chem.*, 256, 4345, 1981.
58. **Haugen, D. A., Armes, L. G., Yasunobu, K. T., and Coon, M. J.**, Amino-terminal sequence of phenobarbital-inducible cytochrome P-450 from rabbit liver microsomes. Similarity to hydrophobic amino-terminal segments of proproteins, *Biochem. Biophys. Res. Commun.*, 77, 967, 1977.
59. **Fujita, V. S., Black, S. D., Tarr, G. E., Koop, D. R., and Coon, M. J.**, On the amino acid sequence of cytochrome P-450 isozyme 4 from rabbit liver microsomes, *Proc. Natl. Acad. Sci. U.S.A.*, 21, 4260, 1984.
60. **Boström, H. and Wikvall, K.**, Hydroxylations in biosynthesis of bile acids: Isolation of subfractions with different substrate specificity from cytochrome P-450 LM_4, *J. Biol. Chem.*, 257, 11755, 1982.
61. **Guengerich, F. P.**, Separation and purification of multiple forms of microsomal cytochrome P-450: activities of different forms of cytochrome P-450 towards several compounds of environmental interest, *J. Biol. Chem.*, 252, 3970, 1977.
62. **Guengerich, F. P.**, Separation and purification of multiple forms of microsomal cytochrome P-450: partial characterization of three apparently homogeneous cytochromes P-450 prepared from livers of phenobarbital and 3-methylcholanthrene-treated rats, *J. Biol. Chem.*, 253, 7931, 1978.
63. **West, S. B., Huang, M.-T., Miwa, G. T., and Lu, A. Y. H.**, A simple and rapid procedure for the purification of phenobarbital-inducible cytochrome P-450 from rat liver microsomes, *Arch. Biochem. Biophys.*, 193, 42, 1979.

64. **Ryan, D. E., Thomas, P. E., Korzeniowski, D., and Levin, W.,** Separation and characterization of highly purified forms of liver microsomal cytochrome P-450 from rats treated with polychlorinated biphenyls, phenobarbital and 3-methylcholanthrene, *J. Biol. Chem.*, 254, 1365, 1979.
65. **Agosin, M., Morello, A., White, R., Repetto, Y., and Pedemonte, J.,** Multiple forms of noninduced rat liver cytochrome P-450: metabolism of 1-(4′-ethylphenoxy)-3,7-dimethyl-6,7-expoxy-trans-2-octene by reconstituted preparations, *J. Biol. Chem.*, 254, 9915, 1979.
66. **Ryan, D. E., Thomas, P. E., and Levin, W.,** Hepatic microsomal cytochrome P-450 from rats treated with isosafrole: purification and characterization of four enzymic forms, *J. Biol. Chem.*, 255, 7941, 1980.
67. **Elshourbagy, N. A. and Guzelian, P. S.,** Separation, purification, and characterization of a novel form of hepatic cytochrome P-450 from rats treated with pregnenolone-16α-carbonitrile, *J. Biol. Chem.*, 255, 1279, 1980.
68. **Cheng, K.-C. and Schenkman, J. B.,** Purification and characterization of two constitutive forms of rat liver microsomal cytochrome P-450, *J. Biol. Chem.*, 257, 2378, 1982.
69. **Guengerich, F. P., Dannan, G. A., Wright, S. T., Martin, M. V., and Kaminsky, L. S.,** Purification and characterization of liver microsomal cytochromes P-450: electrophoretic, spectral, catalytic, and immunochemical properties and inducibility of eight isozymes isolated from rats treated with phenobarbital or β-naphthoflavone, *Biochemistry*, 21, 6019, 1982.
70. **Gibson, G. G., Orton, T. C., and Tamburini, P. P.,** Cytochrome P-450 induction by clofibrate: purification and properties of a hepatic cytochrome P-450 relatively specific for the 12- and 11-hydroxylation of dodecanoic acid (lauric acid), *Biochem. J.*, 203, 161, 1982.
71. **Ryan, D. E., Iida, S., Wood, A. W., Thomas, P. E., Lieber, C. S., and Levin, W.,** Characterization of three highly purified cytochromes P-450 from hepatic microsomes of adult male rats, *J. Biol. Chem.*, 259, 1239, 1984.
72. **Rampersaud, A. and Walz, F. G.,** At least six forms of extremely homologous cytochromes P-450 in rat liver are encoded at two closely linked genetic loci, *Proc. Natl. Acad. Sci. U.S.A.*, 80, 6542, 1983.
73. **Waxman, D. J. and Walsh, C.,** Phenobarbital-induced rat liver cytochrome P-450, *J. Biol. Chem.*, 257, 10446, 1982.
74. **Wikvall, K., Danielsson, H., and Hansson, R.,** Cytochrome P-450 in biosynthesis and metabolism of bile acids, in *Biochemistry, Biophysics and Regulation of Cytochrome P-450*, Gustafsson, J.-Å., Carlstedt-Duke, J., Mode, A., and Rafter, J., Eds., Elsevier/North-Holland, Amsterdam, 1980, 499.
75. **Andersson, S., Holmberg, I., and Wikvall, K.,** 25-Hydroxylation of C_{27}-steroids and vitamin D_3 by a constitutive cytochrome P-450 from rat liver microsomes, *J. Biol. Chem.*, 258, 6777, 1983.
76. **Andersson, S., Boström, H., Danielsson, H., and Wikvall, K.,** Purification from rabbit and rat liver of cytochromes P-450 involved in bile acid biosynthesis, *Methods Enzymol.*, 111B, 364, 1985.
77. **Tsuji, H., Muta, E., and Ullrich, V.,** Separation and purification of liver microsomal monooxygenases from induced and untreated pigs, *Hoppe-Seyler's Z. Physiol. Chem.*, 361, 681, 1980.
78. **Masters, B. S. S., Yasukochi, Y., Okita, R. T., Parkhill, L. K., Taniguchi, H., and Dees, J. H.,** Laurate hydroxylation and drug metabolism in pig liver and kidney microsomes and in reconstituted systems from pig liver and kidney, in *Microsomes, Drug Oxidations and Chemical Carcinogenesis*, Coon, M. J., Conney, A. H., Estabrook, R. W., Gelboin, H. V., Gillette, J. R., and O'Brien, P. J., Eds., Academic Press, New York, 1980, 709.
79. **Li, J. R. and Kim, D. N.,** The effect of bile acids on "solubilized" cholesterol 7α-hydroxylase activity in swine, *Steroids*, 33, 317, 1979.
80. **Li, J. R. and Kim, D. N.,** Effects of superoxide dismutase on cholesterol 7α-hydroxylation in swine, *Steroids*, 35, 459, 1980.
81. **Boström, H.,** Hydroxylations in biosynthesis of bile acids. Multiple forms of cytochrome P-450 in biosynthesis of bile acids in pig liver, submitted for publication.
82. **Wang, P. P., Beaune, P., Kaminsky, L. S., Dannan, G. A., Kadlubar, F. F., Larrey, D., and Guengerich, F. P.,** Purification and characterization of six cytochrome P-450 isozymes from human liver microsomes, *Biochemistry*, 22, 5375, 1983.
83. **Erickson, S. K. and Bösterling, B.,** Cholesterol 7α-hydroxylase from human liver:partial purification and reconstitution into defined phospholipid-cholesterol vesicles, *J. Lipid Res.*, 22, 872, 1981.
84. **Björkhem, I. and Danielsson, H.,** Hydroxylations in biosynthesis and metabolism of bile acids, *Molec. Cell. Biochem.*, 4, 79, 1974.
85. **Andersson, S.,** unpublished data, 1985.
86. **Boström, H.,** Binding of cholesterol to cytochromes P-450 from rabbit liver microsomes, *J. Biol. Chem.*, 258, 15091, 1983.
87. **van Cantfort, J.,** The *in vitro* formation and the strong inhibitory action of 7-keto-cholesterol on cholesterol 7α-hydroxylase activity, *Life Sci.*, 11, 773, 1972.
88. **Schwartz, M. A. and Margolis, S.,** Effect of drugs and sterols on cholesterol 7α-hydroxylase activity in rat liver microsomes, *J. Lipid Res.*, 24, 28, 1983.

89. **Ahringer, L. and Eneroth, P.,** Studies on the formation of C_7-oxygenated cholesterol and β-sitosterol metabolites in cell-free preparations of rat liver, *J. Lipid Res.*, 14, 563, 1973.
90. **Boyd, G. S., Brown, M. J. G., Hattersley, N. G., Lawson, M. E., and Suckling, K. E.,** Rat liver cholesterol 7α-hydroxylase, in *Advances in Bile Acid Research,* Vol. III, Matern, S., Hackenschmidt, J., Back, P., and Gerok, W., Eds., F. K. Schattauer Verlag, Stuttgart, 1975, 45.
91. **Scholan, N. A. and Boyd, G. S.,** The cholesterol 7α-hydroxylase enzyme system, *Hoppe-Seyler's Z. Physiol. Chem.*, 349, 1628, 1968.
92. **van Cantfort, J. and Gielen, J.,** Cholesterol 7α-hydroxylase. 2. Biochemical properties and participation of endogenous cholesterol in the assay *in vitro, Eur. J. Biochem.*, 55, 33, 1975.
93. **Kalles, I. and Wikvall, K.,** Role of sulfhydryl groups in catalytic activity of purified cholesterol 7α-hydroxylase system from rabbit and rat liver microsomes, *Biochem. Biophys. Res. Commun.*, 100, 1361, 1981.
94. **Danielsson, H., Kalles, I., and Wikvall, K.,** Regulation of hydroxylations in biosynthesis of bile acids: isolation of a protein from rat liver cytosol stimulating reconstituted cholesterol 7α-hydroxylase activity, *J. Biol. Chem.*, 259, 4258, 1984.
95. **Danielsson, H., Kalles, I., Lidström, B., Lundell, K., and Wikvall, K.,** Regulation of hydroxylations in biosynthesis of bile acids: modulation of reconstituted 12α-hydroxylase activity by protein fractions from rabbit liver cytosol and microsomes, *Biochem. Biophys. Res. Commun.*, 113, 212, 1983.
96. **Andersson, S. and Boström, H.,** Effect of disulfiram on rat liver cholesterol 7α-hydroxylase, *Biochem. Pharmacol.*, 33, 2930, 1984.
97. **Danielsson, H., Kalles, I., and Wikvall, K.,** Modulation of reconstituted cholesterol 7α-hydroxylase activity by protein fractions from rat liver cytosol, *Biochem. Biophys. Res. Commun.*, 97, 1459, 1980.
98. **Spence, J. T. and Gaylor, J. L.,** Investigation of regulation of microsomal hydroxymethylglutaryl coenzyme A reductase and methyl sterol oxidase of cholesterol biosynthesis, *J. Biol. Chem.*, 252, 5852, 1977.
99. **Craig, I. F., Boyd, G. S., McLeod, R. J., and Suckling, K. E.,** Partial characterization of a protein from rat liver 100,000 g supernatant which stimulates cholesterol 7α-hydroxylase, *Biochem. Soc. Trans.*, 7, 967, 1979.
100. **Kwok, C. T., Burnett, W., and Hardie, I. R.,** Regulation of rat liver microsomal cholesterol 7α-hydroxylase: presence of a cytosolic activator, *J. Lipid Res.*, 22, 570, 1981.
101. **Kwok, C. T., Burnett, W., and Hardie, I. R.,** Regulation of rat liver microsomal cholesterol 7α-hydroxylase: reversible inactivation by ATP + Mg^{2+} and a cytosolic activator, *J. Lipid Res.*, 22, 580, 1981.
102. **Kwok, C. T., Pillay, S. P., and Hardie, I. R.,** Regulation of rat liver microsomal cholesterol 7α-hydroxylase by MgATP: effect of pH, *Biochem. Biophys. Res. Commun.*, 116, 966, 1983.
103. **Sanghvi, A., Grassi, E., Warty, V., Diven, W., Wight, C., and Lester, R.,** Reversible activation-inactivation of cholesterol 7α-hydroxylase possibly due to phosphorylation-dephosphorylation, *Biochem. Biophys. Res. Commun.*, 103, 886, 1981.
104. **Goodwin, C. D., Cooper, B. W., and Margolis, S.,** Rat liver cholesterol 7α-hydroxylase: modulation of enzyme activity by changes in phosphorylation state, *J. Biol. Chem.*, 257, 4469, 1982.

Chapter 6

REQUIREMENTS FOR CYTOCHROME P-450: EVIDENCE FROM METABOLIC STUDIES AND THE RESPONSE TO XENOBIOTICS

W. R. Rush and R. Fears

TABLE OF CONTENTS

I. INTRODUCTION

The discovery of cholesterol 7α-hydroxylase activity (EC 1.14.13.17) in liver homogenates by Danielsson and Einarsson in 1964 was closely followed by reports that the enzyme resided in the microsomal fraction of the liver homogenate,[1-3] a location which it shares with the hemoprotein cytochrome P-450, discovered some years earlier.[4,5] Cytochrome P-450 was eventually established as the terminal oxidase in a hepatic enzyme system responsible for the biotransformation of drugs and many endogenous compounds.[6-8] This hepatic enzyme system catalyzes mainly oxidation reactions and exhibits true mixed-function oxidase properties requiring oxygen and reduced cofactor, usually reduced nicotinamide adenine dinucleotide phosphate (NADPH), for full activity. It was noted that, in addition to sharing the endoplasmic reticulum site with the cytochrome P-450-dependent oxygenases, cholesterol 7α-hydroxylase was also a monooxygenase enzyme. These common features have prompted numerous investigators to consider the interrelationship between the two enzyme systems and, in particular, the possible requirement by 7α-hydroxylation for cytochrome P-450. These investigations will be reviewed in the present chapter as three categories: characterization, modulation, and purification studies (see Chapter 5 for related aspects of enzyme characterization).

II. EVIDENCE FOR A CYTOCHROME P-450 REQUIREMENT FROM CHARACTERIZATION STUDIES

Cholesterol 7α-hydroxylase, like the cytochrome P-450-dependent monooxygenases, behaves as a true mixed-function oxidase requiring oxygen and NADPH for the expression of complete activity.[9,10] It can be inhibited by carbon monoxide,[11,12] a feature typical of cytochrome P-450-dependent reactions and attributable to reversible binding of the carbon monoxide to the oxygen-binding site of the hemoprotein. Reactions of this type are further characterized by reversal of the inhibition after irradiation of the microsomal fractions with monochromatic light at 450 nm and the inhibition of cholesterol 7α-hydroxylase by an atmosphere enriched with carbon monoxide is relieved in this manner.[6,13]

Another component of the hepatic microsomal cytochrome P-450-dependent oxygenase system is a flavoprotein, NADPH-cytochrome P-450 reductase, also named NADPH-cytochrome c reductase in recognition of the cytochrome c reduction assay used to measure its activity. This flavoprotein catalyzes the transfer of electrons from NADPH to cytochrome P-450 and early evidence for the involvement in the 7α-hydroxylation of cholesterol was adduced from the observation that an antibody raised against NADPH-cytochrome c reductase inhibited the hydroxylation reaction.[14] NADPH normally acts as the donor of the two electrons required to facilitate cytochrome P-450-dependent oxygenation and if the cholesterol 7α-hydroxylation reaction proceeds by the same general mechanism that has been proposed for the other cytochrome P-450-dependent hydroxylations then the sequence of reactions will be as outlined in Figure 1.[15] The nature of the activated oxygen inserted into cholesterol in this sequence of reactions is not totally clear although a role for the superoxide anion in the activation of this oxygen has been recently suggested. Investigations with rat liver microsomal fractions conducted by us (Table 1) and liver microsomes from swine and rat in other laboratories corroborate the finding that removal of superoxide from these tissues by the use of superoxide dismutase enzyme decreases the measurable cholesterol 7α-hydroxylase activity.[16,17]

The possibility that another reduced cofactor, reduced nicotinamide adenine dinucleotide (NADH), might also act as a donor for the second electron through another reductase enzyme, NADH-cytochrome b_5 reductase, has been raised.[15,19] Experiments that support this suggestion indicate that the addition of NADH to microsomal fractions fortified with a suboptimal

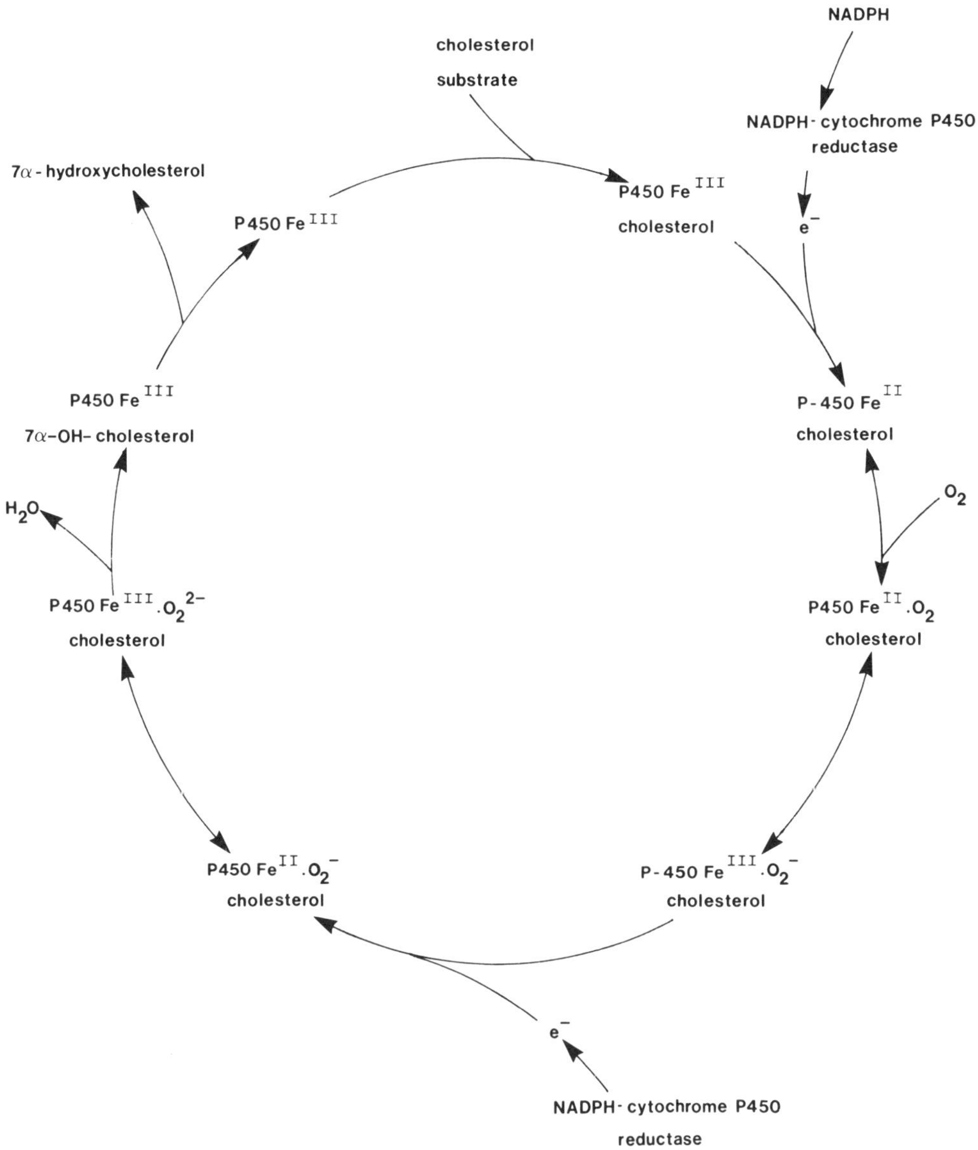

FIGURE 1. Possible mechanism for cholesterol hydroxylation based on the general sequence of cytochrome P-450-dependent reactions.

level of NADPH increases oxygenation of the substrate to a greater extent than can be accounted for by metabolism in the presence of the NADPH alone.[20]

Experiments on cholesterol 7α-hydroxylase conducted by us using Wistar rats (Table 2) and results by other workers using Sprague-Dawley rats show that the rate of the reaction supported by a suboptimal level of NADPH can be stimulated by NADH whereas the rate supported by an optimal level of NADPH cannot be further increased.[21] Our conclusions contradict those made by another group of workers using Wistar rats;[22] one explanation that has been proposed for the discrepancy is a variable contribution by the reaction in the metabolic sequence immediately following 7α-hydroxylation: NAD-dependent oxidation to yield 7α-hydroxy-4-cholesten-3-one.[22] Whether or not NADH has a physiological role, there is no doubt that NADPH is the principal cofactor so that, together with the other characteristic

Table 1
THE EFFECT OF SUPEROXIDE DISMUTASE ON IN VITRO CHOLESTEROL 7α-HYDROXYLASE ACTIVITY IN RAT LIVER MICROSOMAL FRACTIONS

Superoxide dismutase (units · mℓ$^{-1}$)	Cholesterol 7α-hydroxylase[a] (pmol · min^{-1} · g liver^{-1})
—	6.10 ± 0.13[b]
2.0	5.50 ± 0.15[c]
20.0	5.57 ± 0.20[c]

[a] Hepatic microsomal activity measured by the procedure previously reported.[18]
[b] Values are mean ± standard deviation for three determinations.
[c] Values significantly lower than control activity measured in the absence of added superoxide dismutase ($p<0.05$).

Table 2
THE EFFECT OF NADH ON CHOLESTEROL 7α-HYDROXYLASE ACTIVITY IN VITRO

Cofactor	Cholesterol 7α-hydroxylase[a] (pmol · min^{-1} · g liver^{-1})
—	0.3 ± 0.1[b]
6 m*M* NADPH (optimal)	5.8 ± 0.8
2 m*M* NADPH (suboptimal)	2.7 ± 0.4
6 m*M* NADH	1.6 ± 0.2[c]
6 m*M* NADPH + 6 m*M* NADH	5.6 ± 0.2
2 m*M* NADPH + 6 m*M* NADH	4.0 ± 0.5[d]

[a] Hepatic activity was measured as described previously.[18]
[b] Values are mean ± standard deviation for three determinations.
[c] Significantly different from assay containing 6 m*M* NADPH ($p<0.01$).
[d] Significantly different from assay containing 2 m*M* NADPH ($p<0.05$).

features described, there is strong evidence that cholesterol 7α-hydroxylase contains cytochrome P-450 as a prosthetic group.

III. MODULATION OF CHOLESTEROL 7α-HYDROXYLASE ACTIVITY

A. Arthritis

The occurrence of arthritis in experimental animals, such as that induced in male rats by a single subcutaneous injection of adjuvant into the foot pad, has been linked with decreased liver drug-metabolizing enzyme activity and depletion of the hepatic cytochrome P-450 content.[23-25] Atkin and co-workers (1972) used this adjuvant-induced model to evaluate the relationship between hepatic cholesterol 7α-hydroxylase activity,[26] microsomal drug monooxygenation and cytochrome P-450 content. They found that although the hepatic microsomal cytochrome P-450 concentration and the rate of the *N*-demethylation of aminopyrine were lowered in the arthritic animals, there was no change in cholesterol 7α-hydroxylase activity from control levels. The result was not interpreted to indicate an absence of a requirement for cytochrome P-450 by cholesterol 7α-hydroxylase, but rather that the induction of arthritis had decreased the level of a form of cytochrome P-450 involved solely in the metabolism of drugs without modulating the level of the cholesterol 7α-hydroxylating fraction. This conclusion was one of the earliest to claim the existence of a specific form of the hemoprotein for cholesterol 7α-hydroxylation in liver microsomes. In a more recent study the age-related effect of genetically linked arthritis in mice on liver microsomal mixed-function oxidase components, drug monooxygenases, and cholesterol 7α-hydroxylase activity was followed.[18] Once more, the incidence of arthritis was associated with lower hepatic drug monooxygenase activity, exemplified by biphenyl 4-hydroxylase, and cytochrome P-450 content, for although the cytochrome P-450 concentration increased with the age in both the obese/arthritic strain (STR/1N) and control strains (Figure 2) the control increases were significantly greater. Cholesterol 7α-hydroxylase activity also increased with the age but there was no correlation with change in cytochrome P-450 levels so that those results on a genetically determined variant of arthritis substantiate the earlier findings in experimental arthritis.[26]

B. Physiological Controls

A chronic dietary deficiency of ascorbic acid in guinea pigs can produce lower hepatic microsomal mixed-function oxidase activity and cytochrome P-450 content along with depression of cholesterol 7α-hydroxylase activity.[27-29] In contrast, excessive vitamin C intake in guinea pigs is not accompanied by any change in hepatic drug monooxygenase activity except for a small increase in cytochrome P-450 level, while cholesterol 7α-hydroxylase activity is apparently depressed.[30] Taken together, these studies indicate a poor correlation between cholesterol 7α-hydroxylase activity and both changes in hepatic microsomal cytochrome P-450 and in the drug-metabolizing monooxygenase activities, but the disparity might reflect the existence of a minor cholesterol-metabolizing form of cytochrome P-450 which is particularly sensitive to variations in dietary ascorbate.[30]

Treatment of castrated male rats with estradiol provides another example where divergent effects on cholesterol 7α-hydroxylase and hepatic microsomal drug metabolizing activity are elicited (see Chapter 9 for further discussion of the effects of estrogens).[31] A compensatory increase in cholesterol hydroxylating isozyme(s) and decrease in drug metabolizing isozyme(s) would explain the respective induction and depression of these enzyme activities provided by the estradiol and account for the unchanged cytochrome P-450 content observed.

Thyroid hormone administration, thyroidectomy, and portacaval anastomosis in rats are all further examples of metabolic states where changes in cholesterol 7α-hydroxylase activity are not accompanied by related variations in total cytochrome P-450 level or drug monooxygenase activity.[32] Furthermore, the elevation in cholesterol 7α-hydroxylase activity noted after interruption of the enterohepatic recirculation of bile acids and during the dark period of the diurnal cycle is not associated with changes in total cytochrome P-450 or NADPH-cytochrome c reductase and drug monooxygenase activities.[33-35] Cytochrome P-450 involvement in the 7α-hydroxylation of cholesterol can be inferred from experiments using cholesterol-enriched rat diets where short-term supplementation (3 to 5 days) enhances cholesterol 7α-hydroxylase activity while prolonged supplementation (4 to 5 weeks) supports an induction of the hepatic cytochrome P-450-dependent monooxygenases although the potential influence of changes in membrane fluidity must also be considered.[36-37] The well-recognized stimulation of cholesterol 7α-hydroxylase by the bile acid sequestrant cholestyramine (for further discussion, see Chapter 11) is not usually accompanied by any change in microsomal total cytochrome P-450 content although the activity of the flavoprotein reductase component is raised.[22,38-40]

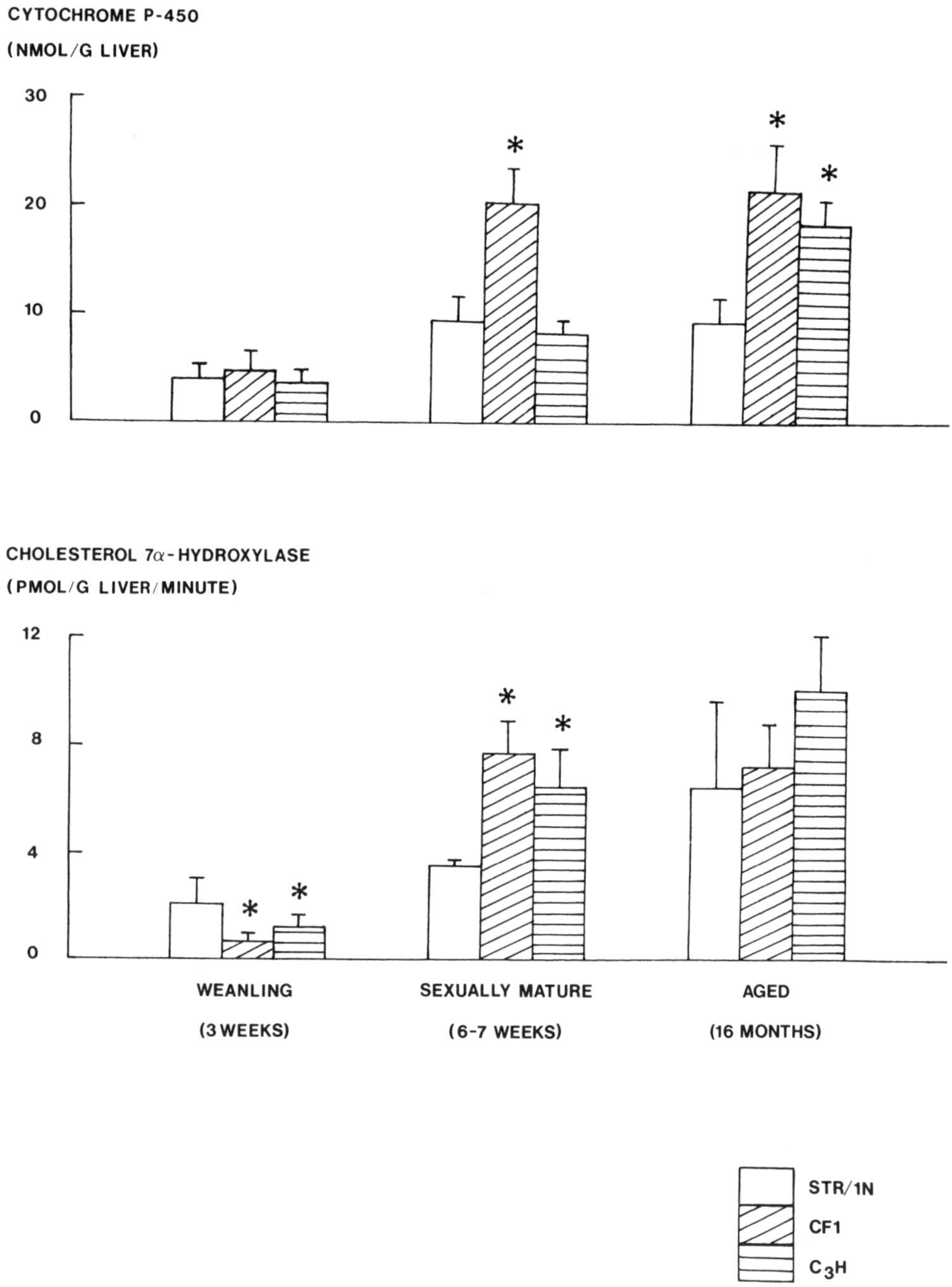

FIGURE 2. Development of cholesterol 7α-hydroxylase activity and cytochrome P-450 concentration in three strains of mice. * Indicates a significant difference from STR/1N mice ($p < 0.01$).

C. Xenobiotics

The ability of a variety of compounds to inhibit and then induce the oxidative metabolism of substrates by hepatic microsomal fractions has been used by several investigators to provide evidence of a cytochrome P-450 involvement. For example, Remmer and Merker (1965) observed a parallel rise and fall of microsomal cytochrome P-450 with drug metabolism following the administration and withdrawal of phenobarbitone.[41] Ernster and Orrenius

(1965) noted concomitant increases in hepatic microsomal oxidative demethylation,[42] cytochrome P-450, and the flavoprotein reductase component in response to phenobarbitone or 3-methylcholanthrene. The use of chemicals that bind to the substrate site of cytochrome P-450 such as carbon monoxide, SKF-525A, piperonyl butoxide, metyrapone, and the alkylimidazoles have been associated with hindrance of the oxidative metabolism of many xenobiotics in vitro.[6,43-46]

Contradictory reports as to the effect of phenobarbitone treatment on rat liver cholesterol 7α-hydroxylase activity have appeared in which a consistent induction of microsomal cytochrome P-450 content has been associated either with no stimulation of the 7α-hydroxylase,[22,38,47] or with enhancement.[11,47-50] Inconsistencies have also been documented for 3-methylcholanthrene (cytochrome P-448) type induction in the rat but this agent either has no effect on,[33,40] or depletes,[14,22,51] 7α-hydroxylase activity. Induction of drug-metabolizing activity by the insecticide DDT has also been associated with depression of cholesterol 7α-hydroxylase activity,[51] and experiments using rats treated with the catatoxic steroid pregnenolone 16α-carbonitrile, yet another type of inducer of the hepatic drug monooxygenation system,[52,53] have repeatedly produced a decrease in the activity of cholesterol 7α-hydroxylase.[54-56] The treatment of rats with the hypolipidemic agent clofibrate induced hepatic cytochrome P-450 and drug metabolizing activity, but failed to produce concomitant increases in cholesterol 7α-hydroxylase activity.[51,57,58]

Evidence for a cytochrome P-450 requirement by cholesterol 7α-hydroxylase obtained from studies with inducers of hepatic cytochrome P-450-dependent monooxygenation is, therefore, at best, equivocal. However, experimental details have often varied considerably and some attempts at rationalization have been made. For example, Shefer and co-workers (1972) in a study of the inductive effect of phenobarbitone in Wistar and Sprague-Dawley rats noted that cholesterol 7α-hydroxylase activity was increased in Wistar rats, but not in Sprague-Dawley rats and they proposed that the disparity was the result of a strain difference in the responsiveness of this enzyme.[49] If this were the explanation, then the results for cholesterol 7α-hydroxylase induction by phenobarbitone treatment in Sprague-Dawley rats and lack of any response in Wistar rats become difficult to rationalize.[11,22,38] In an attempt to resolve this problem, Mellon et al.[22] proposed the additional existence of intrastrain differences in responsiveness, but although such variations may contribute to the demonstrated incongruities alternative, more fundamental explanations should not be overlooked. One common variation is in the dosage regime employed. For example, Shefer et al.[49] used a 5 day intraperitoneal treatment with phenobarbitone at 100 mg · kg^{-1}, while Wada et al.[11] gave phenobarbitone in the diet at 0.1% w/w for 7 days and Boyd et al.[38] used subcutaneous administration for 4 days at 100 mg · kg^{-1}. The difference noted between the results of Shefer et al.[49] and Mellon et al.[22] for the phenobarbitone effect might merely reflect the dose levels of 100 mg · kg^{-1} used by the former and 80 mg · kg^{-1} by the latter. The discrepancy between the 3-methylcholanthrene effect noted by Mellon et al.[22] and the absence of an effect shown by Brown and Boyd (1974) could be in the 5 and 3 day dosage regime used,[40] respectively. Another area of potential variation lies with the procedure used to measure cholesterol 7α-hydroxylase activity (see Chapter 4 for a discussion of methodology). Techniques that employ isotope incorporation to measure the conversion of radiolabeled cholesterol to 7α-hydroxycholesterol are not considered to be as accurate or as reliable as techniques that measure the mass of 7α-hydroxycholesterol formed.[32] This variability in precision may account for the differences observed among the early work, all of which was performed by isotope incorporation, and between this work and the more recent studies which have used more refined techniques such as dual-isotope labeling and tritium exchange.[22,23] Another important factor is the tissue fraction used for the measurement of activities; some studies have assayed hepatic postmitochondrial supernatant while others have used microsomal fractions. The work of Mason and Boyd (1978) with pregnenolone

16α-carbonitrile illustrates the variation that can be obtained using the two tissue fractions in that the depression of cholesterol 7α-hydroxylase activity could only be detected in the microsomal fraction.[56]

Few attempts to characterize a cytochrome P-450 requirement using compounds that actively inhibit reactions dependent on the hemoprotein have been made although those that have suggest a role for cytochrome P-450. Schwartz and Margolis (1983) demonstrate that SKF-525A and metyrapone,[59] two classical inhibitors of hepatic drug metabolism,[43,45] inhibit microsomal cholesterol 7α-hydroxylase. Ethylmorphine, a xenobiotic metabolized by cytochrome P-450-dependent monooxygenases,[60] can competitively inhibit cholesterol 7α-hydroxylase activity in vitro and, likewise, cholesterol can inhibit the metabolism of ethylmorphine.[35] A similarity between substrate or other binding sites for cholesterol 7α-hydroxylase and cytochrome P-450-dependent monooxygenases, therefore, seems plausible and this observation may be taken to imply cytochrome P-450 involvement in the hydroxylation step.

We consider that some, at least, of the inconsistencies have arisen in this area because there have been few direct comparisons of the different classes of drug inducer. In an attempt to clarify the position we made a comprehensive comparison of the effects of selected xenobiotics as inducers of cytochrome P-450 (on dosing to rats) and as inhibitors (the acute effect measured in vitro) together with the response to supplementary dietary cholesterol and cholestyramine. The structures of the agents used are shown in Figure 3.

Male Wistar rats (180 to 200 g) were used in these studies with the exception of one experiment when male Sprague-Dawley rats of similar weight were used. To induce the liver microsomal enzymes phenobarbitone sodium (80 mg $\cdot$ kg^{-1}) in isotonic saline, 3-methylcholanthrene (20 mg $\cdot$ kg^{-1}) in corn oil, or safrole (150 mg $\cdot$ kg^{-1}) in corn oil were given intraperitoneally once daily for 3 days. Pregnenolone 16α-carbonitrile (20 mg $\cdot$ kg^{-1}) was dosed orally twice daily for 3 days as an aqueous microcrystalline suspension containing 1 drop of Tween 80 dℓ^{-1}. SKF-525A (50 mg $\cdot$ kg^{-1}) in isotonic saline was administered once daily for 2 days intraperitoneally. Cholesterol and cholestyramine were given in the diet at a level of 1 and 4% w/w, respectively, for 5 days. All rats were dosed between 0830 and 0930 hr and killed 24 hr after the last dose with the exception of the pregnenolone 16α-carbonitrile-treated rats which were killed 15 hr after the last dose. Control rats received drug vehicle only. Ethylmorphine and aminopyrine *N*-demethylase activities were measured by the procedure of Holtzman and co-workers with the formaldehyde liberated assayed by the method of Nash.[60,61] Methods for the isolation of hepatic subcellular fractions, determination of microsomal cytochrome P-450 and protein concentrations, and assay of cholesterol 7α-hydroxylase and biphenyl 2- and 4-hydroxylase activities have appeared elsewhere.[18]

Rats treated with phenobarbitone exhibited enhanced hepatic mixed-function oxidase activity characterized by two- to threefold increases in ethylmorphine *N*-demethylase and biphenyl 4-hydroxylase activities along with significant increases in the total cytochrome P-450 and NADPH-cytochrome c reductase components, and in microsomal protein (Table 3). However, no change in cholesterol 7α-hydroxylase activity could be detected. 3-Methylcholanthrene administration to rats also produced increases in the hepatic mixed-function oxidases, although biphenyl 4-hydroxylase and ethylmorphine *N*-demethylase activities, and cytochrome P-450 (P-448) concentration were not so highly induced as with phenobarbitone (Table 3). No increases in NADPH-cytochrome c reductase activity or microsomal protein were seen and cholesterol 7α-hydroxylase activity was significantly decreased.

Safrole treatment produced significant increases in the activities of biphenyl 2-hydroxylase, ethylmorphine *N*-demethylase, and NADPH-cytochrome c reductase with more moderate increases in microsomal cytochrome P-450 (P-448) and protein contents (Table 3). Pregnenolone 16α-carbonitrile administration also induced significant increases in ethylmorphine *N*-demethylase and biphenyl 2-hydroxylase and more minor increases in NADPH-cytochrome

A. DRUGS

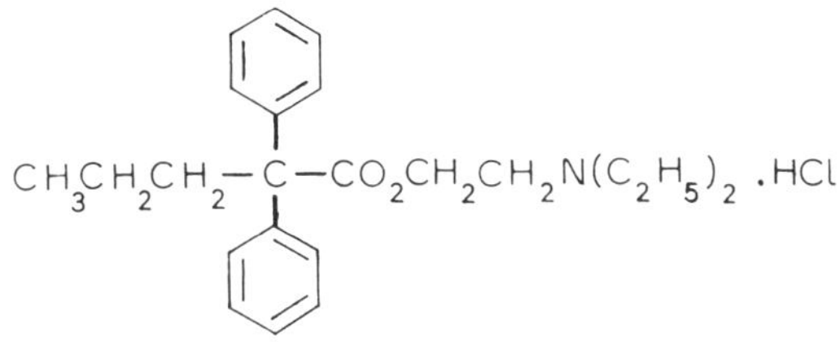

SKF-525A

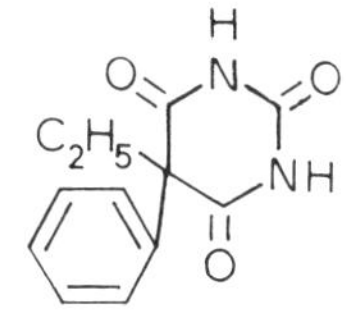

PHENOBARBITONE

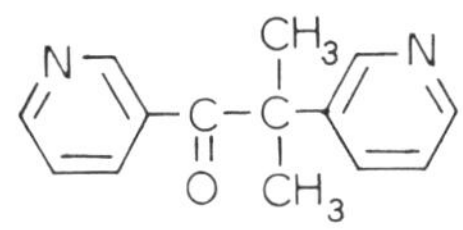

METYRAPONE

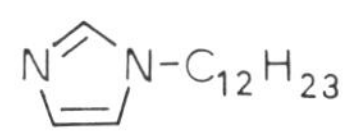

DODECYLIMIDAZOLE

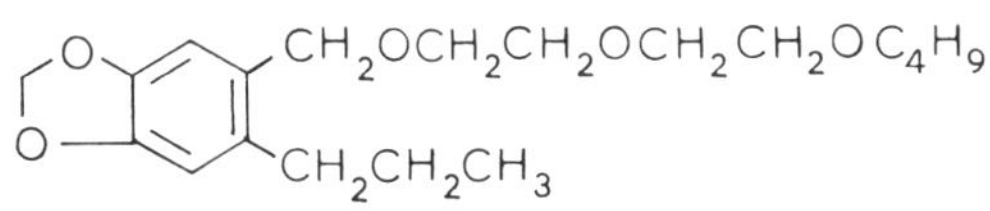

PIPERONYL BUTOXIDE

B. STEROID

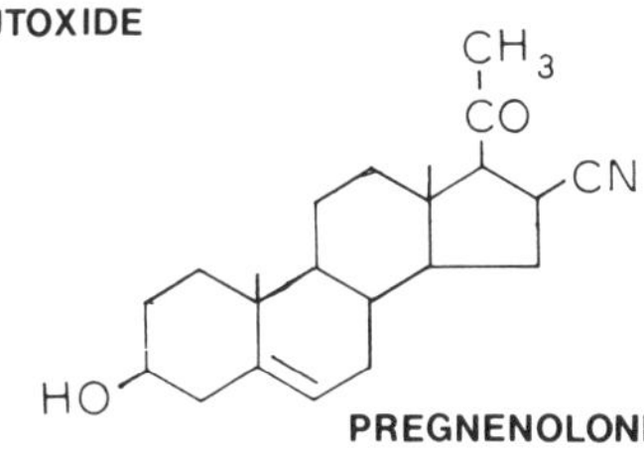

PREGNENOLONE 16α-CARBONITRILE

C. CARCINOGENIC POLYCYCLIC HYDROCARBONS

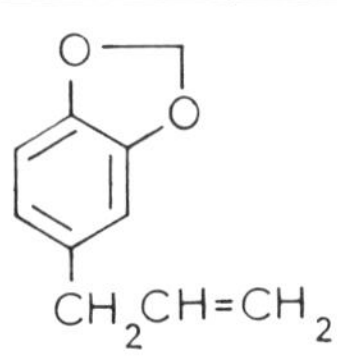

SAFROLE

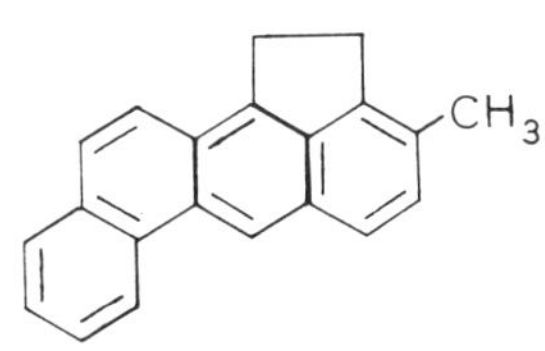

3-METHYLCHOLANTHRENE

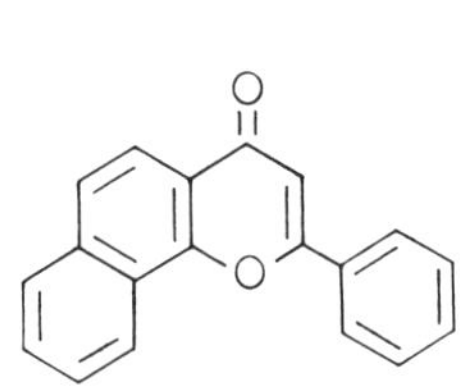

α-NAPHTHOFLAVONE

FIGURE 3. Chemical structures of the agents used in vitro and in vivo to explore the relationship between cholesterol 7α-hydroxylase and the hepatic concentration of cytochrome P-450. The classification is according to Parke.[87]

Table 3
EFFECTS OF HEPATIC ENZYME-INDUCING AGENTS ON WISTAR RAT LIVER CHOLESTEROL 7α-HYDROXYLASE AND OTHER MICROSOMAL MIXED-FUNCTION OXIDASE ENZYMES

Measurement	Enzyme activities as percent control values following pretreatment with inducer:						
	Cholesterol	Cholestyramine	SKF-525A	Phenobarbitone	3-Methylcholanthrene	Pregnenolone 16α-carbonitrile	Safrole
Microsomal protein	110 ± 5[ab]	120 ± 10[b]	110 ± 5[b]	120 ± 5[b]	100 ± 5	90 ± 5	120 ± 5
Cytochrome P-450	130 ± 20[b]	160 ± 20	170 ± 10[b]	280 ± 40[b]	210 ± 20[b]	130 ± 30	130 ± 20[b]
NADPH-cytochrome c reductase	110 ± 20	120 ± 10[b]	100 ± 10	160 ± 20[b]	90 ± 10	110 ± 20[b]	150 ± 20[b]
Cholesterol 7α-hydroxylase	130 ± 10[b]	330 ± 40[b]	230 ± 30[b]	80 ± 10	65 ± 20[b]	90 ± 20	120 ± 30
Ethylmorphine *N*-demethylase	125 ± 30	100 ± 10	100 ± 5	260 ± 30[b]	120 ± 20[b]	130 ± 10[b]	150 ± 20[b]
Biphenyl 4-hydroxylase	110 ± 10	110 ± 10	100 ± 10	230 ± 20[b]	190 ± 30[b]	110 ± 30	120 ± 30
Biphenyl 2-hydroxylase	60 ± 20	100 ± 20	350 ± 50[b]	100 ± 50	2000 ± 100[b]	130 ± 30[b]	320 ± 30[b]

[a] Mean values ± standard deviation for at least four rats. Control values were: microsomal protein, 23 ± 1 mg · g liver^{-1}; cytochrome P-450, 20 ± 3 nmol · g liver^{-1}, NADPH-cytochrome c reductase, 0.96 ± 0.13 μmol · min^{-1} · g liver^{-1}; cholesterol 7α-hydroxylase, 19 ± 4 pmol · min^{-1} · g liver^{-1}; ethylmorphine *N*-demethylase, 0.19 ± 0.03 μmol · min^{-1} · g liver^{-1}, biphenyl 4-hydroxylase, 36 ± 4 nmol · min^{-1} · g liver^{-1}; biphenyl 2-hydroxylase, 1.01 ± 0.23 nmol · min^{-1} · g liver^{-1}.

[b] $p < 0.05$.

Table 4
EFFECT OF FERRICYANIDE ON MIXED-FUNCTION OXIDASE ACTIVITIES IN LIVER MICROSOMAL FRACTIONS PREPARED FROM SPRAGUE-DAWLEY RATS TREATED CHRONICALLY WITH SKF-525A

Mixed-function oxidase component	Treatment:	Control microsomes		SKF-525A microsomes[a]	
		None	$Fe^{3+}(CN)_6$	None	$Fe^{3+}(CN)_6$
Cytochrome P-450 (nmol · mg protein^{-1})		0.88 ± 0.10[b]	0.92 ± 0.13	1.61 ± 0.16[c]	2.26 ± 0.23[cd]
Aminopyrine *N*-demethylase (nmol · min^{-1} · mg protein^{-1})		0.73 ± 0.52	0.91 ± 0.42	1.29 ± 0.50	1.90 ± 0.63[c]
Cholesterol 7α-hydroxylase (pmol · min^{-1} · mg protein^{-1})		0.76 ± 0.08	1.08 ± 0.25	1.89 ± 0.25[c]	2.31 ± 0.42[c]

[a] SKF-525A (50 mg · kg^{-1}) in isotonic NaCl was administered once daily, intraperitoneally, for two days. Control rats received NaCl. Rats were killed 24 hr after the last dose and ferricyanide treatment was performed.[62]
[b] Mean values ± standard deviation for six rats.
[c] Significantly different ($p<0.05$) from respective control.
[d] Significantly different ($p<0.05$) from untreated fraction.

c reductase and cytochrome P-450. Neither safrole nor pregnenolone 16α-carbonitrile elicited any detectable change in cholesterol 7α-hydroxylase activity (Table 3).

Rats treated with SKF-525A exhibited significantly increased cholesterol 7α-hydroxylase activity which was associated with significant increases in microsomal cytochrome P-450 and biphenyl 2-hydroxylase activity (Table 3). No changes in ethylmorphine *N*-demethylase, biphenyl 4-hydroxylase, or NADPH-cytochrome c reductase activities were observed. The cytochrome P-450 induced by SKF-525A had a typical absorbance maximum at 450 nm for the reduced-carbon monoxide adduct and the SDS-polyacrylamide gel electrophoretic pattern showed that a protein of molecular weight 52,000 in the molecular weight range characteristic of the cytochromes P-450 had been induced.

The induction of cytochrome P-450 by SKF-525A has been associated with formation of a metabolically inactive ligand complex between a proportion of the hemoprotein and SKF-525A or one of its metabolites.[62] This complex can be dissociated and biological activity restored by oxidation with ferricyanide. Ferricyanide treatment of liver microsomal fractions from our SKF-525A-treated rats in a second experiment illustrated this effect indicating that a proportion of the induced cytochrome was not measured by the reduced carbon monoxide-difference spectrum (Table 4). Release of the complexed fraction produced only a marginal further elevation of cholesterol 7α-hydroxylase activity (20%), but it did appreciably increase aminopyrine *N*-demethylase activity (45%) in the microsomal fractions.

The inclusion of cholesterol in the diet of rats induced small but nonetheless significant increases in cholesterol 7α-hydroxylase activity and total cytochrome P-450 (Table 3). A more marked enhancement of cholesterol 7α-hydroxylase activity was observed following addition of cholestyramine to the diet, and was also accompanied by significant increases in total cytochrome P-450, by contrast with the results of others,[40] and in NADPH-cytochrome c reductase and microsomal protein (Table 3).

The acute expression of inhibition of drug monooxygenase activity by the addition of α-naphthoflavone, SKF-525A, piperonyl butoxide, metyrapone, or dodecylimidazole to rat liver microsomal fractions in vitro was associated with inhibition of both cholesterol 7α- and biphenyl 4-hydroxylase activities, often to quite similar degrees (Table 5) although metyrapone did not significantly inhibit the biphenyl 4-hydroxylase.

Table 5
THE EFFECTS OF CYTOCHROME P-450 INHIBITORS ON IN VITRO CHOLESTEROL 7α-HYDROXYLASE AND BIPHENYL 4-HYDROXYLASE ACTIVITIES IN RAT LIVER MICROSOMAL FRACTIONS

	% Inhibition of control activity[b]	
Inhibitor[a]	Cholesterol 7α-hydroxylase[c]	Biphenyl 4-hydroxylase[c]
α-Naphthoflavone	21 ± 1[de]	10 ± 4
SKF-525A	50 ± 3[e]	22 ± 6[e]
Piperonyl butoxide	23 ± 3[e]	58 ± 3[e]
Metyrapone	21 ± 3[e]	17 ± 6
Dodecylimidazole	43 ± 1[e]	66 ± 0[e]

[a] Each inhibitor added to a final concentration of 100 μ*M*.
[b] Control cholesterol 7α-hydroxylase activity 10.2 ± 0.1 pmol · min^{-1} · g $liver^{-1}$ and biphenyl 4-hydroxylase activity 16.9 ± 0.9 nmol · min^{-1} · g $liver^{-1}$.
[c] Measured by procedures previously documented.[18]
[d] Values are mean ± standard deviation for three determinations.
[e] Value significantly lower than control activity ($p<0.01$).

It would seem fair to conclude that our results (Tables 3 to 5) reflect and illustrate the diversity of responses that can be obtained with these standard compounds. The absence of an effect on cholesterol 7α-hydroxylase by phenobarbitone agrees with the results of Boyd et al.[38] and Mellon et al.[22] in Wistar rats, but not with the results of Wada et al.[11] in the Sprague-Dawley rat strain and by Shefer et al.[49] in the Wistar rat strain. An absence of a pregnenolone 16α-carbonitrile effect in our study also disagrees with the observations of Einarsson and Gustafsson[54] and Mason and Boyd who report a decrease in cholesterol 7α-hydroxylase activity after repeated administration of this steroid to rats.[56] However, the decrease in cholesterol 7α-hydroxylase activity produced by 3-methylcholanthrene, an inducer of cytochrome P-448,[63] resembled the findings of Wada et al.[14] and Mellon et al.[22] The conclusions of Mellon et al.[22] that cytochrome P-448 is less efficient than cytochrome P-450 as the terminal oxidase for the 7α-hydroxylation of cholesterol, or that some component(s) of this enzyme system are adversely affected by 3-methylcholanthrene treatment appear reasonable explanations. However, safrole, a compound similar to 3-methylcholanthrene in its pattern of induction and formation of cytochrome P-448,[64,65] showed no such lowering of cholesterol 7α-hydroxylase activity in our study, which would favor the second explanation, proposed by Mellon et al.,[22] that 3-methylcholanthrene specifically inhibits cholesterol 7α-hydroxylase. Thus, although the 7α position of cholesterol is sterically hindered similar to the "bay region" positions of the polycyclic aromatic hydrocarbons, which require cytochrome P-448 for oxygenation,[66] this form of the hepatic microsomal hemoprotein does not appear to be responsible for the 7α-hydroxylation of cholesterol.

Perhaps the most interesting of our results was the effect of SKF-525A, which induced a doubling in cholesterol 7α-hydroxylase activity accompanied by a similar increase in microsomal cytochrome P-450 content (the response to SKF-525A is discussed in further detail in Chapter 9). Using ferricyanide treatment to dissociate the complexed proportion of the cytochrome P-450, we showed (Table 4) that the uncomplexed cytochrome P-450 fraction induced by SKF-525A supported the increase in cholesterol 7α-hydroxylase activity while the complexed fraction, when dissociated, exhibited greater potential to hydroxylate aminopyrine. We suggest that this is further evidence for the proposition that among the different forms of hepatic microsomal cytochrome P-450 there are distinct forms with preferential ability to hydroxylate cholesterol or xenobiotics. Moreover, in our experiments, the stimulatory effects of dietary cholesterol and cholestyramine on cholesterol 7α-hydroxylase were also associated with small increases in total cytochrome P-450. In both cases the fraction of cytochrome P-450 induced did not demonstrate enhanced drug-metabolizing activity, but appeared more selective for cholesterol 7α-hydroxylation and it is possible that these increases in concentration of cytochrome P-450 are an adaptive response to the demand for enhanced bile acid synthesis.

To summarize our experiments: there is evidence for a specific fraction of cytochrome P-450, inducible by SKF-525A, cholestyramine, and perhaps by dietary cholesterol, that can support the 7α-hydroxylation step. When the total pool of cytochrome P-450 is bound and inhibited in vitro, then cholesterol 7α-hydroxylase activity is impaired. Whether or not other types of monooxygenase inducer also affect cholesterol 7α-hydroxylase when administered chronically depends on which point in the inhibition-induction sequence is being studied and how general the effect is on the various isozymes of cytochrome P-450. This will vary, *inter alia,* according to dose and route of administration and strain sensitivity.

IV. EVIDENCE FOR A REQUIREMENT FOR CYTOCHROME P-450 FROM PURIFICATION AND RECONSTITUTION STUDIES

Procedures to resolve the components of the liver microsomal cytochrome P-450-dependent mixed-function oxidase systems have been available for some time.[67-71] The abundance of data plausibly suggesting a cytochrome P-450 involvement in cholesterol 7α-hydroxylase activity, but an absence of absolute proof made this enzyme an obvious candidate for such studies. This subject is discussed in greater detail in Chapter 5.

Partial purification of cholesterol 7α-hydroxylase was first reported by Björkhem and coworkers from the liver microsomes of rats treated with dietary cholestyramine to maximize the hydroxylase activity.[72] The procedure they adopted involved cholate solubilization of the microsomal fractions followed by fractionation with polyethylene glycol and ammonium sulfate. However, only a low cholesterol 7α-hydroxylase activity was obtained in their final fractions and the extent of purification was small. Despite these shortcomings, it was still possible to demonstrate in a reconstituted system, containing the partially purified fractions of cytochrome P-450 and NADPH-cytochrome P-450 reductase in the presence of NADPH and a synthetic phospholipid, that the rate of cholesterol 7α-hydroxylation was proportional both to the concentration of the hemoprotein and to the flavoprotein reductase fractions. In subsequent years, other workers using more efficient detergent solubilization of the enzyme from microsomal fractions and more sophisticated chromatographic techniques were able to confirm that cytochrome P-450 and NADPH-cytochrome P-450 reductase were essential components of the cholesterol 7α-hydroxylase reaction.[73-75]

For example, when the components were purified according to the scheme shown in Figure 4 the reconstituted system exhibited an absolute requirement for cytochrome P-450 and NADPH-cytochrome P-450 reductase and a partial requirement for phospholipid (Table 6). The complete reconstituted system also supported *N*-demethylation of ethylmorphine (12.3 nmol product per minute per nmol cytochrome P-450). It is interesting to note (Table 6) that the 7α-hydroxylation rate in the reconstituted system could be enhanced by increasing the concentration of cholesterol. We suggest that this effect is mediated either by an additional allosteric action on the enzyme or by a modification of the fluidity of the phospholipids present.

The existence of multiple forms of cytochrome P-450 in liver microsomes with differing abilities to catalyze monooxygenation reactions and with differing responsiveness to induction by various agents has resulted in the development of the concept of substrate-specific cytochromes P-450.[69,76-78] The possibility that a cholesterol-specific cytochrome P-450 is present in liver microsomes has been considered by a number of workers and is supported by some of the results described in Section III.C.

FIGURE 4. Purification of cytochrome P-450 and NADPH-cytochrome c reductase.

In some of their earlier work Hansson and Wikvall (1979 and 1980) were able to identify a form of cytochrome P-450[73,79] among those isolated from the liver microsomes of cholestyramine-treated rats or rabbits, that preferentially catalyzed the 7α-hydroxylation of cholesterol. However, this fraction of the total hemoprotein complement was not devoid of other drug monooxygenase or steroid hydroxylase activities. Cytochrome P-450 LM_4, the major isozyme of cytochrome P-450 present in control rabbit liver microsomes, was subsequently shown to be the fraction from the liver microsomes of cholestyramine-treated rabbits in which these activities resided.[80] In this respect, it was particularly interesting to find that cytochrome P-450 LM_4 purified from phenobarbitone-treated, control, or starved rabbits did not support cholesterol 7α-hydroxylase activity although it could reconstitute other steroid hydroxylation reactions. A similar disparity was observed with the liver mi-

Table 6
THE RECONSTITUTION OF CHOLESTEROL 7α-HYDROXYLASE ACTIVITY

Assay system	Cholesterol concentration (μM)	Cholesterol 7α-hydroxylase (pmol · min^{-1} · nmol cytochrome P-450^{-1})
Microsomal fraction	4.5	1.8[a]
Complete reconstituted	4.5	0.7
Complete reconstituted	45	10.4
Omitting cytochrome P-450 (normally 4 nM)	45	0
Omitting reductase (normally 6 U)	45	0.2
Omitting phospholipid (normally 60 μM)	45	5.3

[a] Mean values for two assays performed according to the procedure described previously.[18]

crosomal cytochrome P-450 fractions from phenobarbitone-treated rats, none of which catalyzed cholesterol 7α-hydroxylation but did catalyze other steroid hydroxylations. Convincing evidence for a specific hemoprotein has been provided by Danielsson and Wikvall who administered radioactive δ-aminolevulinic acid to cholestyramine-induced rats,[81] prior to killing, in order to label the newly synthesized cytochrome P-450 pool. In this study they identified a form of cytochrome P-450 which selectively supported cholesterol 7α-hydroxylase activity. This function was localized in a minor fraction of the cytochrome P-450 pool with low specific radioactivity and was separated from other steroid hydroxylases that were isolated in fractions of higher specific radioactivity. The occurrence of cholesterol 7α-hydroxylase-supporting activity in a fraction of low specific radioactivity was taken to indicate that the reaction was catalyzed by a cytochrome P-450 with a short half-life.

Continued attempts to characterize the cholesterol 7α-hydroxylase cytochrome P-450 have used the cytochrome P-450 LM_4 fraction from cholestyramine-pretreated rabbits. Boström and Wikvall further resolved the cytochrome P-450 LM_4 fraction into one which catalyzed steroid and drug monooxygenation reactions, but which was devoid of cholesterol 7α-hydroxylase activity and another which efficiently catalyzed the latter reaction, but had a lower activity towards the other substrates.[82] The forms were electrophoretically and spectrally identical, and gave electrophoretically similar peptide fragments using limited proteolysis. But they differed significantly in their total amino acid composition. Cytochrome P-450 LM_4 from cholestyramine-treated rabbits also exhibited identical electrophoretic and immunochemical characteristics to that from phenobarbitone-treated rabbits although minor structural differences were evident by two-dimensional iodinated tryptic peptide mapping.[83] The cytochrome P-450 LM_4 from cholestyramine-treated rabbits has now been further purified into fractions I and II.[84] Both subfractions and the cytochrome P-450 LM_4 subfraction from phenobarbitone-treated rabbits can bind cholesterol, but only fraction II catalyzes the 7α-hydroxylation (fractions of cytochrome P-450 LM_2 and LM_{3b} cannot bind cholesterol). Thus, to summarize, cytochrome P-450 LM_4 is the important isozyme for the binding and 7α-hydroxylation of cholesterol, but it is now apparent that this subfraction is heterogenous. We have yet to establish which moiety is induced by SKF-525A.

V. CONCLUSIONS

The evidence discussed in this review suggests convincingly that cholesterol 7α-hydroxylase exhibits the characteristic features of a cytochrome P-450-dependent monooxygenase. In general, attempts to establish the cytochrome P-450 requirement of the enzyme using drug inducers have been only moderately successful although much of the uncertainty could be accounted for if the isozyme was present normally in only low abundance. The advent of the purification and reconstitution studies has provided the definitive proof of the requirement for cytochrome P-450: the precise characterization of the specific subfraction is now in progress.

It has been suggested that the dependence of cholesterol 7α-hydroxylation on cytochrome P-450 may have clinical consequences for man. The genetic polymorphism of drug metabolic oxidation is a well-established phenomenon and if these "poor metabolizers" were also less efficient at catabolizing cholesterol — so that circulating levels increased — then they might be predisposed to atherosclerosis and coronary heart disease (see Chapter 9 and 11 for a discussion of other aspects of this association). In a selected rat strain the relative impairment of debrisoquine hydroxylation and phenacetic-*O*-deethylation is inherited as a simple Mendelian recessive trait and such rats (females) are hypercholesterolemic.[85] However, a recent investigation of poor metabolizer subjects[86] controlled for age and weight, disclosed no difference from the control levels of cholesterol, triglycerides, and high-density lipoprotein.

This lack of apparent association in man is consistent with the existence of a distinct form of cytochrome P-450 for cholesterol catabolism and, when considered teleologically, would seem to be generally beneficial. Cholesterol 7α-hydroxylase is a highly regulated enzyme in a metabolic sequence of physiological importance and thus differs from most other microsomal cytochrome P-450-dependent reactions. There would seem to be no metabolic purpose for a fluctuation in the rate of bile acid synthesis dependent on the variable exposure to xenobiotics.

ACKNOWLEDGMENTS

We thank Miss P. Walker for her assistance in the performance of some of these studies. We are also indebted to Professor D. V. Parke, Dr. J. Green, Dr. H. Ferres, and Dr. G. G. Gibson for their interest and advice, and to the late Professor G. S. Boyd for his contribution to this field of investigation.

REFERENCES

1. **Danielsson, H. and Einarsson, K.,** The enzymic formation of 7α-hydroxycholesterol from cholesterol in rat liver homogenates, *Acta Chem. Scand.*, 18, 831, 1964.
2. **Mitton, J. R. and Boyd, G. S.,** The hydroxylation of cholesterol by liver microsomes, *Biochem. J.*, 96, 60P, 1965.
3. **Danielsson, H. and Einarsson, K.,** On the conversion of cholesterol to 7α, 12α-dihydroxycholest-4-en-3-one, *J. Biol. Chem.*, 241, 1449, 1966.
4. **Klingenberg, M.,** Pigments of rat liver microsomes, *Arch. Biochem. Biophys.*, 75, 376, 1958.
5. **Klingenberg, M.,** Pigments of rat liver microsomes, *Arch. Biochem. Biophys.*, 75, 376, 1958.
5. **Omura, T. and Sato, R.,** A new cytochrome in liver microsomes, *J. Biol. Chem.*, 237, 1375, 1962.
6. **Cooper, D. Y., Levin, S., Narasimhulu, S., Rosenthal, O., and Estabrook, R. W.,** Photochemical action spectrum of the terminal oxidase of mixed-function oxidase systems, *Science*, 147, 400, 1965.
7. **Conney, A. H.,** Pharmacological implications of microsomal enzyme induction, *Pharmacol. Rev.*, 19, 317, 1967.
8. **Gillette, J. R., Davis, D. C., and Sasame, H. A.,** Cytochrome P-450 and its role in drug metabolism, *Annu. Rev. Pharmacol.*, 12, 57, 1972.
9. **Shefer, S., Hauser, S., and Mosbach, E. H.,** 7α-Hydroxylation of cholestanol by liver microsomes, *J. Lipid Res.*, 9, 328, 1968.
10. **Mitton, J. R., Scholan, N. A., and Boyd, G. S.,** The oxidation of cholesterol in rat liver sub-cellular particles; the cholesterol 7α-hydroxylase enzyme system, *Eur. J. Biochem.*, 20, 569, 1971.
11. **Wada, F., Hirata, K., Nakao, K., and Sakamoto, Y.,** Participation of P-450 in 7α-hydroxylation of cholesterol, *J. Biochem. (Tokyo)*, 64, 415, 1968.

12. **Mitropoulos, K. A. and Balasubramaniam, S.,** Cholesterol 7α-hydroxylase in rat liver microsomal preparations, *Biochem. J.*, 128, 1, 1972.
13. **Boyd, G. S., Grimwade, A. M., and Lawson, M. E.,** Studies on rat-liver microsomal cholesterol 7α-hydroxylase, *Eur. J. Biochem.*, 37, 334, 1973.
14. **Wada, F., Hirata, K., Nakao, K., and Sakamoto, Y.,** Participation of the microsomal electron transport system involving cytochrome P-450 in 7α-hydroxylation of cholesterol, *J. Biochem. (Tokyo)*, 66, 699, 1969.
15. **Estabrook, R. W., Hildebrandt, A. G., Baron, J., Netter, K. J., and Leibman, K.,** A new spectral intermediate associated with cytochrome P-450 function in liver microsomes, *Biochem. Biophys. Res. Commun.*, 42, 132, 1971.
16. **Li, J. R. and Kim, D. N.,** Effects of superoxide dismutase on cholesterol 7α-hydroxylation in swine, *Steroids*, 35, 459, 1980.
17. **Baillie, J., Crankshaw, D. L., Hanson, R. F., and Holtzman, J. L.,** The effect of superoxide dismutase and catalase on cholesterol 7-α-hydroxylase of rat hepatic microsomes, *Fed. Proc.*, 42, 1776, 1983.
18. **Rush, W. R., Fears, R., Green, J. and Parke, D. V.,** Liver microsomal mixed-function oxidase activity (MFO) and cholesterol metabolism in the obese arthritic STR/1N strain of mouse, *Comp. Biochem. Physiol.*, 69B, 493, 1981.
19. **Hildebrandt, A. and Estabrook, R. W.,** Evidence for the participation of cytochrome b_5 in hepatic microsomal mixed-function oxidation reactions, *Arch. Biochem. Biophys.*, 143, 66, 1971.
20. **Cohen, B. S. and Estabrook, R. W.,** Cooperative interactions between reduced diphosphopyridine nucleotide and reduced triphosphopyridine nucleotide linked reactions, *Arch. Biochem. Biophys.*, 143, 54, 1971.
21. **Björkhem, I. and Danielsson, H.,** Heterogeneity of hepatic mixed function oxidases, *Biochem. Biophys. Res. Commun.*, 51, 766, 1973.
22. **Mellon, W. S., Witiak, D. T., and Feller, D. R.,** Cholesterol 7α-hydroxylase. Evidence for a distinct hepatic microsomal enzyme system, *Biochem. Pharmacol.*, 27, 1055, 1978.
23. **Morton, D. M. and Chatfield, D. H.,** Effects of adjuvant-induced arthritis on the liver metabolism of drugs in rats, *Biochem. Pharmacol.*, 19, 473, 1970.
24. **Cawthorne, M. A., Palmer, E. D., and Green, J.,** Adjuvant-induced arthritis and drug-metabolizing enzymes, *Biochem. Pharmacol.*, 25, 2683, 1976.
25. **Mathur, P. P., Carr, G. S., and Smyth, R. D.,** Sex-related differences in the impairment of hepatic drug-metabolizing enzymes in rats with adjuvant disease, *Biochem. Pharmacol.*, 27, 1379, 1978.
26. **Atkin, S. D., Palmer, E. D., English, P. D., Morgan, B., Cawthorne, M. A., and Green, J.,** The role of cytochrome P-450 in cholesterol biogenesis and catabolism, *Biochem. J.*, 128, 237, 1972.
27. **Zannoni, V. G.,** Ascorbic acid and liver microsomal drug metabolism, *Acta Vitaminol. Enzymol.*, 31, 17, 1977.
28. **Sikic, B. I., Mimnaugh, E. G., Litterst, C. L., and Gram, T. E.,** The effects of ascorbic acid deficiency and repletion on pulmonary, renal and hepatic drug metabolism in the guinea pig, *Arch. Biochem. Biophys.*, 179, 663, 1977.
29. **Björkhem, I. and Kallner, A.,** Hepatic 7α-hydroxylation of cholesterol in ascorbate-deficient and ascorbate-supplemented guinea pigs, *J. Lipid Res.*, 17, 360, 1976.
30. **Peterson, F. J., Holloway, D. E., Duquette, P. H., and Rivers, J. M.,** Dietary ascorbic acid and hepatic mixed function oxidase activity in the guinea pig, *Biochem. Pharmacol.*, 32, 91, 1983.
31. **Deliconstantinos, G. H., Livanou, T., and Margioris, A. N.,** Cholesterol-7α-hydroxylase and drug oxidation in castrated rats after estradiol treatment, *IRCS Libr. Compend.*, 8, 136, 1980.
32. **Myant, N. B. and Mitropoulos, K. A.,** Cholesterol 7α-hydroxylase, *J. Lipid Res.*, 18, 135, 1977.
33. **Gielen, J. E., Van Cantfort, J., and Kremers, P.,** Genetic and hormonal regulation of steroid hydroxylases and drug metabolizing enzymes in rat liver, *Arch. Toxicol.*, 36, 255, 1976.
34. **Mitropoulos, K. A., Balasubramaniam, S., Gibbons, G. F., and Reeves, B. E. A.,** Diurnal variation in the activity of cholesterol 7α-hydroxylase in livers of fed and fasted rats, *FEBS Lett.*, 27, 203, 1972.
35. **Balasubramaniam, S. and Mitropoulos, K. A.,** The role of cytochrome P-450 in the 7α-hydroxylation of cholesterol, *Biochem. Soc. Trans.*, 3, 964, 1975.
36. **Lang, M. and Hänninen, O.,** Alterations in the activities of drug metabolising enzymes in rat liver microsomes by dietary and in vitro incorporated cholesterol, in *Microsomes and Drug Oxidations*, Ullrich, V., Roots, I., and Hildebrandt, A., Eds., Pergamon Press, Oxford, 1977, 597.
37. **Hietanen, E., Hänninen, O., Laitinen, M., and Lang, M.,** Dietary cholesterol-induced enhancement of hepatic biotransformation rate in male rats, *Pharmacology*, 17, 163, 1978.
38. **Boyd, G. S., Scholan, N. A., and Mitton, J. R.,** Factors influencing cholesterol 7α-hydroxylase activity in the rat liver, *Adv. Exp. Med. Biol.*, 4, 443, 1969.
39. **Johansson, G.,** Effect of cholestyramine and diet on hydroxylations in the biosynthesis and metabolism of bile acids, *Eur. J. Biochem.*, 17, 292, 1970.

40. **Brown, M. J. G. and Boyd, G. S.,** The specificity of the rat-liver cholesterol 7α-hydroxylase, *Eur. J. Biochem.*, 44, 37, 1974.
41. **Remmer, H. and Merker, H. J.,** Effect of drugs on the formation of smooth endoplasmic reticulum and drug-metabolizing enzymes, *Ann. N.Y. Acad. Sci.*, 123, 79, 1965.
42. **Ernster, L. and Orrenius, S.,** Substrate-induced synthesis of the hydroxylating enzyme system of liver microsomes, *Fed. Proc.*, 24, 1190, 1965.
43. **Stitzel, R. E., Anders, M. W., and Mannering, G. J.,** Inhibition of drug metabolism. III. Inhibition of hexobarbital metabolism in the intact rat and in the isolated perfused liver by 2-diethylaminoethyl-2,2-diphenyl valerate hydrochloride (SKF-525A) and its N-deethylated derivatives, *Mol. Pharmacol.*, 2, 335, 1966.
44. **Anders, M. W.,** Inhibition of microsomal drug metabolism by methylenedioxybenzenes, *Biochem. Pharmacol.*, 17, 2367, 1968.
45. **Leibman, K. C.,** Effects of metyrapone on liver microsomal drug oxidations, *Mol. Pharmacol.*, 5, 1, 1969.
46. **Palmer, E. D. and Cawthorne, M. A.,** Effects of 1-alkylimidazoles on hepatic drug-metabolizing enzyme activity, *Xenobiotica*, 4, 209, 1974.
47. **Cohen, B. I., Raicht, R. F., Nicolau, G., and Mosbach, E. H.,** Effects of phenobarbital upon bile acid synthesis in two strains of rats, *Lipids*, 10, 168, 1975.
48. **Einarsson, K. and Johansson, G.,** Effect of phenobarbital on the conversion of cholesterol to taurocholic acid, *Eur. J. Biochem.*, 6, 293, 1968.
49. **Shefer, S., Hauser, S., and Mosbach, E. H.,** Stimulation of cholesterol 7α-hydroxylase by phenobarbital in two strains of rats, *J. Lipid Res.*, 13, 69, 1972.
50. **Brandau, K. and Keup, U.,** Studies on the determination and induction of cholesterol 7α-hydroxylase, *Arzneim-Forsch.*, 26, 1837, 1976.
51. **Appleton, B. S., Shriver, C. N., Arnrich, L., and Hathcock, J. N.,** Effects of 3-methylcholanthrene and DDT on cholesterol 7α-hydroxylase in rats, *Drug-Nutr. Interact.*, 1, 15, 1981.
52. **Solymoss, B., Werringloer, J., and Toth, S.,** Influence of pregnenolone-16α-carbonitrile on hepatic mixed-function oxygenases, *Steroids*, 17, 427, 1971.
53. **Japundzic, I., Japundzic, M., and Szabo, S.,** Induction of ethylmorphine demethylation and aniline p-hydroxylation by various catatoxic steroids in regenerating rat liver, *Eur. J. Biochem.*, 43, 79, 1974.
54. **Einarsson, K. and Gustafsson, J. A.,** Effect of a potent catatoxic steroid, 16α-cyanopregnenolone, on microsomal metabolism of steroid hormones, sterols and bile acids in rats, *Eur. J. Biochem.*, 32, 197, 1973.
55. **Honohan, T. and Parkinson, T. M.,** Enhancement of cholesterol turnover in rats by a catatoxic steroid (pregnenolone 16α-carbonitrile) and a bile acid sequestrant (colestipol-HCl), *Biochem. Pharmacol.*, 24, 899, 1975.
56. **Mason, J. I. and Boyd, G. S.,** The suppressive effect of the catatoxic steroid, pregnenolone-16α-carbonitrile, on liver microsomal cholesterol-7α-hydroxylase, *Steroids*, 31, 849, 1978.
57. **Angelin, B., Björkem, I., and Einarsson, K.,** Effects of clofibrate on some microsomal hydroxylations involved in the formation and metabolism of bile acids in rat liver, *Biochem. J.*, 156, 445, 1976.
58. **Mellon, W. S., Goldberg, A. P., Witiak, D. T., and Feller, D. R.,** Differential effect of clofibrate on hepatic drug oxidation and cholesterol 7α-hydroxylation, *Biochem. Pharmacol.*, 25, 2403, 1976.
59. **Schwartz, M. A. and Margolis, S.,** Effects of drugs and sterols on cholesterol 7α-hydroxylase activity in rat liver microsomes, *J. Lipid Res.*, 24, 28, 1983.
60. **Holtzman, J. L., Gram, T. E., Gigon, P. L., and Gillette, J. R.,** The distribution of the components of mixed-function oxidase between the rough and the smooth endoplasmic reticulum of liver cells, *Biochem. J.*, 110, 407, 1968.
61. **Nash, T.,** The colorimetric estimation of formaldehyde by means of the Hantzsch reaction, *Biochem. J.*, 55, 416, 1953.
62. **Buening, M. K. and Franklin, M. R.,** SKF-525A inhibition, induction, and 452-nm complex formation, *Drug Metab. Dispos.*, 4, 244, 1976.
63. **Shoeman, D. W., Chaplin, M. D., and Mannering, G. J.,** Induction of drug metabolism. III. Further evidence for the formation of a new P-450 haemoprotein after treatment of rats with 3-methylcholanthrene, *Mol. Pharmacol.*, 5, 412, 1969.
64. **Lake, B. G., Hopkins, R., Chakraborty, J., Bridges, J. W., and Parke, D. V.,** Influence of some hepatic enzyme inducers and inhibitors on extrahepatic drug metabolism, *Drug. Metab. Dispos.*, 1, 342, 1973.
65. **Elcombe, C. R., Bridges, J. W., Gray, T. J. B., Nimmo-Smith, R. H., and Netter, K. J.,** Interaction of safrole with rat hepatic microsomes, *Biochem. Pharmacol.*, 24, 1427, 1975.
66. **Parke, D. V. and Ioannides, C.,** Role of mixed-function oxidases in the formation of biological reactive intermediates, *Adv. Exp. Med. Biol.*, 136A, 23, 1982.

67. **Lu, A. Y. H., Kuntzman, R., West, S., Jacobson, M., and Conney, A. H.,** Reconstituted liver microsomal enzyme system that hydroxylates drugs, other foreign compounds, and endogenous substrates, *J. Biol. Chem.,* 247, 1727, 1972.
68. **Imai, Y. and Sato, R.,** Gel-electrophoretically homogeneous preparation of cytochrome P-450 from liver microsomes of phenobarbital-pretreated rabbits, *Biochem. Biophys. Res. Commun.,* 60, 8, 1974.
69. **Haugen, D. A., Van Der Hoeven, T. A., and Coon, M. J.,** Purified liver microsomal cytochrome P-450. Separation and characterisation of multiple forms, *J. Biol. Chem.,* 250, 3567, 1975.
70. **Dignam, J. D. and Strobel, H. W.,** Preparation of homogeneous NADPH-cytochrome P-450 reductase from rat liver, *Biochem. Biophys. Res. Commun.,* 63, 845, 1975.
71. **Yasukochi, Y. and Masters, B. S. S.,** Some properties of a detergent-solubilised NADPH-cytochrome c (cytochrome P-450) reductase purified by biospecific affinity chromatography, *J. Biol. Chem.,* 251, 5337, 1976.
72. **Björkhem, I., Danielsson, H., and Wikvall, K.,** 7α-Hydroxylation of cholesterol by reconstituted systems from rat liver microsomes, *Biochem. Biophys. Res. Commun.,* 61, 934, 1974.
73. **Hansson, R. and Wikvall, K.,** Properties of reconstituted cholesterol 7α-hydroxylase system from rat and rabbit liver microsomes, *Eur. J. Biochem.,* 93, 419, 1979.
74. **Rush, W. R., Gibson, G. G., and Parke, D. V.,** The purification and reconstitution of cytochrome P-450-dependent cholesterol 7α-hydroxylase activity from rat liver microsomal fractions, *Biochem. Soc. Trans.,* 8, 102, 1980.
75. **Boyd, G. S., Hattersley, N. G., Ozasa, S., and Suckling, K. E.,** Cholesterol 7α-hydroxylase in rat liver microsomes-characteristics, solubilization, resolution and substrate specificity, in *Microsomes, Drug Oxidations and Chemical Carcinogenesis,* Vol. 2, Coon, M. J., Conney, A. H., and Estabrook, R. W., Eds., Academic Press, New York, 1980, 699.
76. **Guengerich, F. P.,** Separation and purification of multiple forms of microsomal cytochrome P-450, *J. Biol. Chem.,* 252, 3970, 1977.
77. **Lu, A. Y. H. and West, S. B.,** Reconstituted mammalian mixed-function oxidases: requirements, specificities and other properties, *Pharmacol. Ther.,* Part A, 2, 337, 1978.
78. **Ryan, D. E., Thomas, P. E., Korzeniowski, D., and Levin, W.,** Separation and characterization of highly purified forms of liver microsomal cytochrome P-450 from rats treated with polychlorinated biphenyls, phenobarbital, and 3-methylcholanthrene, *J. Biol. Chem.,* 254, 1365, 1979.
79. **Hansson, R. and Wikvall, K.,** Hydroxylations in biosynthesis and metabolism of bile acids, *J. Biol. Chem.,* 255, 1643, 1980.
80. **Wikvall, K., Danielsson, H., and Hansson, R.,** Cytochrome P-450 in biosynthesis and metabolism of bile acids, *Dev. Biochem.,* 13, 499, 1980.
81. **Danielsson, H. and Wikvall, K.,** Evidence for a specific cytochrome P-450 with short half-life catalyzing 7α-hydroxylation of cholesterol, *Biochem. Biophys. Res. Commun.,* 103, 46, 1981.
82. **Boström, H. and Wikvall, K.,** Hydroxylations in biosynthesis of bile acids, *J. Biol. Chem.,* 257, 11755, 1982.
83. **Chiang, J. Y. L., Malmer, M., and Hutterer, F.,** A form of rabbit liver cytochrome P-450 that catalyzes the 7α-hydroxylation of cholesterol, *Biochim. Biophys. Acta,* 750, 291, 1983.
84. **Boström, H.,** Binding of cholesterol to cytochrome P-450 from rabbit liver microsomes, *J. Biol. Chem.,* 258, 15091, 1983.
85. **Nadir, H. H., Al-Dabbagh, S. G., and Idle, J. R.,** Elevated serum cholesterol in drug-oxidation-deficient rats, *Biochem. Pharmacol.,* 31, 1665, 1982.
86. **Dayer, P., Suenram, A., Pometta, D., and Fabre, J.,** Apparent lack of influence of the genetic drug oxidation status on plasma cholesterol concentration in man, *Pharmacol. Res. Commun.,* 16, 129, 1984.
87. **Parke, D. V.,** *Enzyme Induction,* Plenum Press, London, 1975, 207.

Control Mechanisms

Chapter 7

MOLECULAR CONTROL OF ACTIVITY BY REVERSIBLE PHOSPHORYLATION

C. T. Kwok, S. P. Pillay, and I. R. Hardie

TABLE OF CONTENTS

I. INTRODUCTION

Cholesterol is essential for the metabolic and physiological functions of the body such as the biosynthesis of bile acids and vitamin D, steroid hormone production, the structural support in plasma and intracellular membranes that serve as a barrier to foreign matter, and as partitions for the various subcellular functions. The accumulation of unesterified cholesterol intracellularly, however, is potentially deleterious to cell function.[74] In the liver cell and also in the nonhepatic cell this accumulation is prevented by a coordinate regulation of possibly four key enzymes involved in the intermediate metabolism of cholesterol, namely, 3-hydroxy-3-methylglutaryl-coenzyme A (HMG-CoA) reductase [EC 1.1.1.34], cholesterol 7α-hydroxylase [EC 1.14.13.7], acyl-coenzyme A: cholesterol acyltransferase [ACAT, EC 2.3.1.26], and cholesterol ester-hydrolase [EC 3.1.1.13]. The activities of these enzymes are controlled by the circadian rhythm, having a diurnal high and a diurnal low, by diet, drugs, and hormones.[9,25,47,57,62]

The consequence of abnormal regulation of these enzymes results in atherosclerosis, a disease linked to altered regulation of cholesterol biosynthesis[29] and esterification of cholesterol.[40,73] Another consequence of abnormal cholesterol metabolism is cholesterol gallstones, in which patients have an increased activity of HMG-CoA reductase and the activity of cholesterol 7α-hydroxylase is invariably depressed by 36 to 47%, compared with controls without gallstones.[14,17,35,48,64] These observations indicate that cholesterol 7α-hydroxylase plays a very important role in the metabolism of cholesterol, since bile acids constitute the main metabolic route of cholesterol degradation and elimination from the body. The understanding of the molecular regulation of this enzyme is therefore fundamental to the overall understanding of how cholesterol homeostasis in the body is maintained (see Chapter 11).

Cholesterol 7α-hydroxylase [EC 1.14.13.7] is a cytochrome P-450-linked mixed-function monooxygenase[10,69,85] requiring NADPH-cytochrome-C reductase and cytochrome P-450 for activity.[1,63] It catalyzes the 7α-hydroxylation of cholesterol — a committal step in the biosynthesis of bile acids from cholesterol.[20,72] The activity of cholesterol 7α-hydroxylase is responsive to various dietary, drug, and hormonal manipulations.[38,57,71,80] These aspects are reviewed separately in this monograph. There are, however, some inconsistencies as to the mode of control of the catalytic activity of cholesterol 7α-hydroxylase. The studies of Mitropoulos et al.[49,53] and Gielen et al.[27] indicated that long-term regulation of the enzyme activity was mediated via protein synthesis and supply of substrate; but the amount of protein synthesized and the substrate pool-size showed no direct relationship to the changes in the enzyme activity under different experimental manipulations. Two questions thus arise: Is there a specific cytochrome P-450 involved in the activity of cholesterol 7α-hydroxylase; and is it regulated by a mechanism involving phosphorylation-dephosphorylation similar to that of HMG-CoA reductase?[59] Such a mechanism could explain the sharp rise in the enzyme activity without a concomitant increase in either the amount of enzyme or the substrate concentration. The background leading to the proposal of the phosphorylation mechanism will be briefly reviewed. The physiological significance of the phosphorylation mechanism and its relationship to the control of other systems including HMG-CoA reductase, ACAT, and cholesterol ester-hydrolase will be described with respect to their metabolic importance in the maintenance of cholesterol homeostasis in the liver cells.

II. CONTROL OF CHOLESTEROL 7α-HYDROXYLASE VIA ENZYME CONCENTRATION

Cholesterol 7α-hydroxylase activity has a short half-life of 2 to 4 hr.[24,61] with a diurnal high at midnight and a diurnal low at noon.[19,27,50] The diurnal high of the enzyme activity is associated with synthesis of new enzyme. The concentration of the enzyme, therefore, was seen to play a regulatory role in the catalytic activity of cholesterol 7α-hydroxylase in long-term experiments involving cholesterol feeding,[50,71] biliary drainage by bile fistula, and cholestyramine administration[20,50,56] and thyroid hormone administration[3,77,85] (see Myant and Mitropoulos[57] for full review). These manipulations increased the rate of total bile acid synthesis and the activity of cholesterol 7α-hydroxylase. There was, however, no direct proportional increase in the total concentration of cytochrome P-450[2,13,15,76,85] nor in the activity of NADPH-cytochrome-C reductase.[12] Instead, the concentration of total cytochrome P-450 in liver microsomes was decreased by thyroxine and cholestyramine treatments.[33,52] In a feeding experiment, rats fed trilinolein and triolein had lower cholesterol 7α-hydroxylase activity than those fed tripalmitin and trierucin, but the concentration of cytochrome P-450 in the former group was increased while in the latter group decreased.[10] This data suggests that the regulation of cholesterol 7α-hydroxylase activity was not directly linked to that of cytochrome P-450 and that perhaps the enzyme was a separate entity from the cytochrome P-450 system.

Recent studies, however, showed that a specific cytochrome P-450 was involved in the 7α-hydroxylation of cholesterol.[11,15,23,33,34,63] The presence of such a specific cytochrome P-450 is significant and would therefore explain the paradox of the inverse relationship of the content of cytochrome P-450 and the catalytic activity of cholesterol 7α-hydroxylase since it constitutes only a minor fraction of the total cytochrome P-450 concentration[23] and according to Myant and Mitropoulos[57] would contribute no detectable change in the microsomal concentration of cytochrome P-450 when the specific fraction was induced or depressed. However, evidence for such a direct relationship is yet to be established (see Chapters 5 and 6).

III. REGULATION OF CHOLESTEROL 7α-HYDROXYLASE ACTIVITY BY CYTOSOLIC FRACTIONS

Cytosolic fractions obtained by cellular fractionation of livers obtained from rats with bile fistula and treated by cholestyramine, stimulated 7α-hydroxylation of cholesterol to the same extent as that from control rats.[20,41] Similar results were also obtained with cytosol from fasted rats[86] indicating that long-term regulation of cholesterol 7α-hydroxylase activity is not mediated via modulating the activity of the cytosolic fraction.

IV. CONTROL OF CHOLESTEROL 7α-HYDROXYLASE ACTIVITY VIA SUBSTRATE CONCENTRATION

The stimulation of cholesterol 7α-hydroxylase activity by cholestyramine treatment was due to an increase in the Vmax from 2.15 ± 0.60 in the control on normal diet to 3.31 ± 0.43 nmol. 20 $min^{-1} \cdot mg^{-1}$ protein in the experimental groups, while the apparent Km remained at 226 μ*M*. Similarly, in the same experiment,[41] the Vmax was increased from 1.42 ± 0.35 at the diurnal low to 3.31 ± 0.43 nmol · 20 $min^{-1} \cdot mg^{-1}$ protein at the diurnal high of cholesterol 7α-hydroxylase activity. This suggests that the increased enzyme activity was linked to the increased supply of cholesterol during the acrophase of the circadian rhythm since HMG-CoA reductase activity was similarly increased.[50,55] This was also supported by the observation that less exogenous cholesterol was required to saturate the enzyme during the acrophase compared with the nadir (100 μ*M* against 160 μ*M*).[41] However, no significant change in the concentration of nonesterified or esterified cholesterol in the microsomal fraction was found.[51,55] Other investigations into the regulatory mechanism involving the direct modulation of gene expression by bile acids showed that gene expression was not directly involved in the regulation of bile acid synthesis.[78]

V. REGULATION OF CHOLESTEROL 7α-HYDROXYLASE ACTIVITY BY REVERSIBLE PHOSPHORYLATION

There is good evidence that cholesterol 7α-hydroxylase and HMG-CoA reductase activities are similarly controlled by the circadian rhythm,[19,70] cholestyramine and cholic acid feeding, dietary fat, ethanol and phenobarbital administration.[10,37,45,46,68] Intracellularly, both enzymes are located in the endoplasmic reticulum with low RNA density.[55] The findings that HMG-CoA reductase is modulated by phosphorylation,[6,7,31,36,59] a regulatory mechanism now truly and well established for HMG-CoA reductase,[8,28,39] also suggest that cholesterol 7α-hydroxylase could be similarly regulated. Both enzymes, present in considerable amount in the inactive forms, were activated by cytosolic factors, washing of the microsomal preparation, and by a 15 to 20 min preincubation, which in the case of HMG-CoA reductase was associated with desensitization of ATP + Mg^{2+} inactivation.[6,31,41,59,60,75] The modulation of cholesterol 7α-hydroxylase by MgATP, however, differs in many respects from that of HMG-CoA reductase.

Evidence available to date for the reversible phosphorylation of cholesterol 7α-hydroxylase has been obtained from rat liver microsomes and is based on the measurement of the enzyme catalytic activity using crude or partially purified cytosolic fractions, rabbit and bovine protein kinases, and *E. coli* alkaline phosphatases.[29,41-43,66]

A. Effect of MgATP Concentration and pH

The modulation of cholesterol 7α-hydroxylase activity was dependent on the MgATP concentration and pH of the incubation medium. ATP concentrations up to 1 m*M* stimulated cholesterol 7α-hydroxylase activity, with 0.5 m*M* MgATP[29,43] providing maximum stimulation at both acidic and alkaline pH, while 2 and 3 m*M* MgATP inhibited 7α-hydroxylation of cholesterol.[21,43] The inactivation was significantly greater at pH 6.7 (26%) than at pH 7.4 (12%) in the presence of 3 m*M* MgATP.[43] Sanghvi et al.,[66] however, reported that 2 m*M* MgATP activated cholesterol 7α-hydroxylase when the enzyme was assayed in an incubation medium containing imidazole buffer as against the phosphate buffer used in our studies[41-43] and that of Danielsson et al.[21] The imidazole buffer was itself inhibitory to the 7α-hydroxylation of cholesterol[66] suggesting a similar effect to that of Tris-buffer, which was also shown to be inhibitory, by increasing the formation of 7α-ketocholesterol, a potent inhibitor of cholesterol 7α-hydroxylase ($pKi = 7 \times 10^{-6}$ *M*).[81,83,84] The mode of action of MgATP in this study is therefore not clear; the activation may have arisen from MgATP preventing the formation of 7α-ketocholesterol and thereby relieving the inhibition by the imidazole-buffer per se.

B. Effect of NaF and Metallic Bivalent Cations

NaF, when incubated alone or together with MgATP, inhibited HMG-CoA reductase activity,[59] but stimulated cholesterol 7α-hydroxylase activity.[30,65] maximum activation being obtained with 10 m*M* NaF.[41] When 50 m*M* NaF was incubated in the presence of 2 or 3 m*M* MgATP, the inactivation of 7α-hydroxylation was prevented.[21,41,42] The results are summarized in Table 1. The nature of NaF interacting with the enzyme is not clear. NaF appeared to exert its activating effect via inhibiting the dissociation of the -phosphodiester bond of ATP[42] or, as indicated by Sanghvi et al.[66] and Goodwin et al.,[30] by inhibiting dephosphorylation catalyzed by phosphoprotein phosphatase. It may exert an independent ionic effect, as seen with the bivalent cations such as Hg^{2+}, Pb^{2+}, Cu^{2+}, and Zn^{2+}, which were highly inhibitory.[83] Other metallic bivalent cations such as Ca^{2+}, Mn^{2+}, and Mg^{2+} were also inhibitory;[42,66] the latter was required in the ATP-mediated inactivation of cholesterol 7α-hydroxylase.[42,43]

Table 1
EFFECT OF NaF ON MODULATION OF CHOLESTEROL 7α-HYDROXYLASE BY MgATP

Addition	7α-Hydroxycholesterol formed (nmol · 20 min^{-1} · mg^{-1} protein)	p Value
Control (4)	1.09 ± 0.18	
ATP + Mg^{2+} (5)	0.62 ± 0.07	<0.02[a]
ATP + Mg^{2+} + NaF (5)	0.88 ± 0.10	<0.05[b]

Note: The activity of cholesterol 7α-hydroxylase was assayed in washed microsomal fraction from rat liver as previously described[41] without prior preincubation of the microsomes. Added cofactors are 3 mMATP, 8 m*M* Mg^{2+}, 50 m*M* NaF as indicated; heat-treated cytosolic fraction (40 to 60 μg protein) was included in each assay. The results are means ± SEM of the indicated number of independent experiments in parenthesis. Significance of the results was assessed by Student's paired *t*-test (unpublished data).

[a] Compared with control without ATP added.
[b] Compared with ATP + Mg^{2+} without NaF added.

C. Mechanism of MgATP Reaction with Cholesterol 7α-Hydroxylase

The cytosolic fractions prepared from rat liver contained an activator and an inactivator which modulate cholesterol 7α-hydroxylase activity.[18,21,41,75] Danielsson et al.[21,22] showed that the activities of the inactivator and activator were independent of Mg^{2+} and ATP. The identity of the inactivator is yet to be characterized and identified. Sanghvi et al.[65] failed to demonstrate the correspondence of the activity of this cytosolic inactivator to that of the cytosolic phosphoprotein phosphatase. We thought that the effect of MgATP was mediated via a protein kinase-catalyzed phosphorylation reaction.[42]

Consistent with the phosphorylation mechanism, protein kinases prepared from bovine heart (cyclic AMP-independent) and rabbit muscle (cyclic AMP-dependent) were shown to enhance the activation of cholesterol 7α-hydroxylase by 0.5 m*M* MgATP at both pH 7.4 and pH 6.9, while greater effect was exerted in the latter pH. In addition, these kinases increased the inactivation of cholesterol 7α-hydroxylase by 2.0 and 3.0 m*M* MgATP at pH 6.9; in this reaction the cyclic AMP-independent protein kinase appears to have greater effect than the cyclic AMP-dependent protein kinase (Figure 1).[43]

The inactivation reaction was reversible, depending on the reaction time, i.e., when the activator from the cytosolic fraction was introduced: complete reactivation was achieved when the crude cytosol was added following an assay carried out after a 10 min preincubation with 3.0 m*M* MgATP; while only partial reactivation was obtained when the inactivation was challenged after 60 min preincubation (Figure 2).[42] In a different approach, the reversibility of the phosphorylation mechanism was demonstrated according to Sanghvi et al.[6] and Goodwin et al.[30] who incubated "dephosphorylated" microsomes, prepared by preincubation with *E. coli* alkaline phosphatase with MgATP, cytosolic fraction, or cyclic AMP-dependent protein kinase from rabbit muscle. However, the significance arising from the use of *E. coli* alkaline phosphatase, a nonspecific dephosphorylating enzyme, should be considered in relation to a recent report of Sanghvi et al.,[65] who showed that a purified cytosolic phosphoprotein phosphatase which activated HMG-CoA reductase had comparatively very little effect on cholesterol 7α-hydroxylase. This is an important observation suggesting either the specificity of phosphoprotein phosphatase-catalyzed dephosphorylation or, less likely, that cholesterol 7α-hydroxylase is not modulated by phosphoprotein phosphatase. Thus, the significance placed on the dephosphorylation experiments using *E. coli* alkaline phosphatase is uncertain.

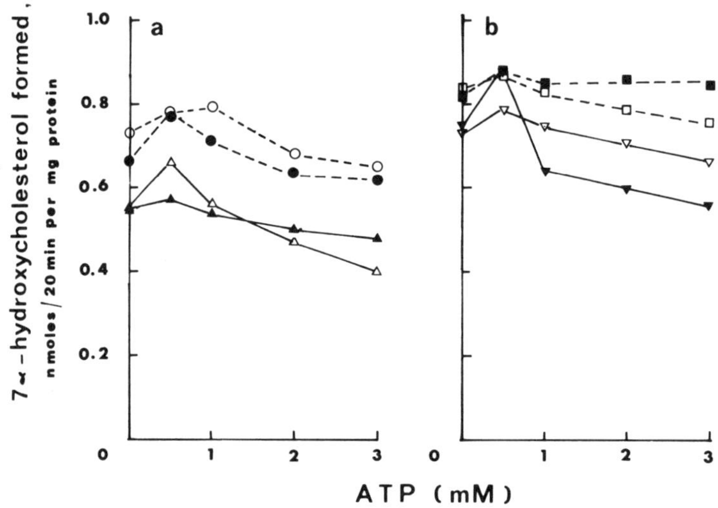

FIGURE 1. Effect of rabbit protein (cAMP-dependent) and bovine protein kinase (cAMP-independent) on MgATP modulation of cholesterol 7α-hydroxylase. Rat liver microsomes were pre-incubated in the presence of MgATP as previously described.[43] Figure 1a shows assays in the presence of rabbit protein kinase (PK) and 40 μM cAMP: (i) pH 6.9; 20 μg PK, ▲-▲; 50 μg PK, △-△; (ii) pH 7.4: 20 μg PK, ●-●; 50 μg PK, ○-○. Figure 1b shows assays in the presence of bovine PK: (i) pH 6.9: 2 units PK, ▽-▽; 6 units PK, ▼-▼; (ii) pH 7.4: 2 units PK, □-□; 6 units PK, ■-■. (From Kwok, C. T., Pillay, S. P., and Hardie, I. R., *Biochem. Biophys. Res. Commun.*, 116, 966, 1983. With permission).

Taken together, these studies so far indicate that cholesterol 7α-hydroxylase is regulated by a mechanism different from that of HMG-CoA reductase. The MgATP may play a dual role in the activation or inactivation of cholesterol 7α-hydroxylase; although the nature of the activator and inactivator has yet to be defined.

D. Physiological Significance of the MgATP Modulation of Cholesterol 7α-Hydroxylase Activity

The physiological significance of the regulatory role of MgATP was investigated in conditions of acute energy consumption and of rapid enzyme turnover as represented by fasting (Figure 3) and the nadir of the diurnal rhythm (Figure 4), respectively. The enzyme activity was greatly reduced in the fasted rats (1/7) and at the nadir (1/2) compared with that of the fed rat at the acrophase of enzyme activity. The response of cholesterol 7α-hydroxylase from fasted rats differed both qualitatively and quantitatively from that of control rats fed *ad libitum*. Activation of enzyme activity in the control by preincubation (Figure 3a) was not observed in the fasted rats. This change of enzyme catalytic efficiency was not related to substrate dilution as the total cholesterol content in the microsomes was similar in both groups: 0.11 ± 0.02 and 0.11 ± 0.05 nmol · mg^{-1} protein (from six separate determinations) for control and fasted rats, respectively. Neither was this change of enzyme activity due to change in the molar phospholipid cholesterol ratio of the microsomal membrane, which was again similar for the respective groups (2.72 ± 0.38 against 2.33 ± 0.30).

As already mentioned earlier, the modulation of cholesterol 7α-hydroxylase by MgATP in the fed rats was concentration-dependent with 0.5 m*M* MgATP providing maximum activation, a 22% increase in cholesterol 7α-hydroxylase activity over the O m*M* ATP ($p < 0.01$) (Figure 3b); 3 m*M* MgATP reduced the enzyme activity 16% as compared with

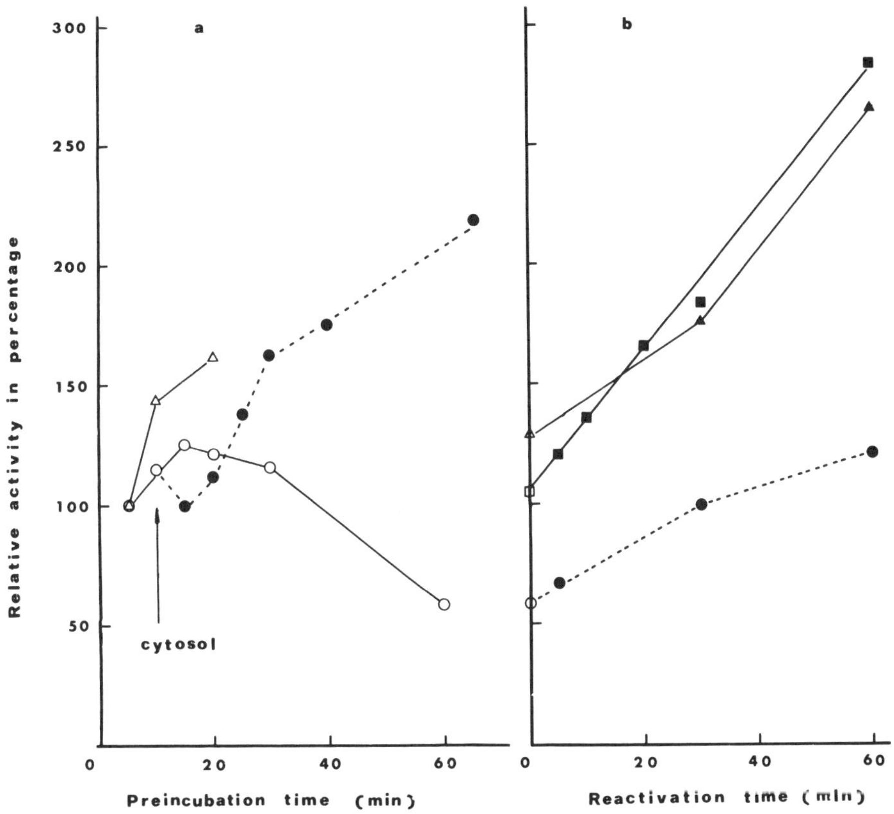

FIGURE 2. The inactivation and reactivation of cholesterol 7α-hydroxylase. Preincubation was carried out with (○,●) or without 3 m*M* MgATP (△,▲) for the indicated time intervals. The open symbols represent the hydroxylase activity remaining at the end of the various preincubation time, while in Figure b they represent that of 60 min preincubation. The closed symbols represent enzyme activities measured at 5, 10, 15, 20, 30, and 60 min intervals following the introduction of crude cytosol (300 μg protein) to the assays after a 10 and 60 min preincubation as shown in Figures a and b, respectively. The effect of 50 m*M* NaF was indicated by (□ and ■). The ordinate represents the percentage of cholesterol 7α-hydroxylase activity relative to that of the 5 min preincubation. (From Kwok, C. T., Burnett, W. and Hardie, I. R., *J. Lipid Res.*, 22, 580, 1981. With permission.)

the control without ATP ($p < 0.05$). In the fasted rats no stimulation of cholesterol 7α-hydroxylase activity was seen with any of the MgATP concentration tested but 3 m*M* Mg ATP decreased the enzyme activity 15% compared with its zero ATP control ($p < 0.01$).

At the nadir of the diurnal rhythm, the modulation of cholesterol 7α-hydroxylase by MgATP was similar to that observed during fasting. In this study cholesterol 7α-hydroxylase activity at the nadir was compared with that from the acrophase at pH 7.4 and pH 6.7, the latter pH was shown to be more effective in the inactivation of cholesterol 7α-hydroxylase.[43] The dual effects of MgATP at 0.5 and 3.0 m*M* were observed as previously described for enzyme activity at the acrophase at both pH 7.4 and pH 6.7 (Figure 4). At the nadir, however, no activation or inactivation was caused by 0.5 or 3.0 m*M* MgATP, respectively, at pH 7.4 (Figure 4a); similarly at pH 6.7 (Figure 4b) no activation was observed with all MgATP concentrations tested except an inactivation occurred at 3.0 m*M* MgATP (21%) and was significant compared with its 0 m*M* ATP control ($p < 0.05$).

This data showed that the activating effect of MgATP was reduced in these physiological states; but the inactivating effect of high MgATP concentrations (3.0 m*M* MgATP) was retained, suggesting that the enzyme was fully activated under these conditions, hence not

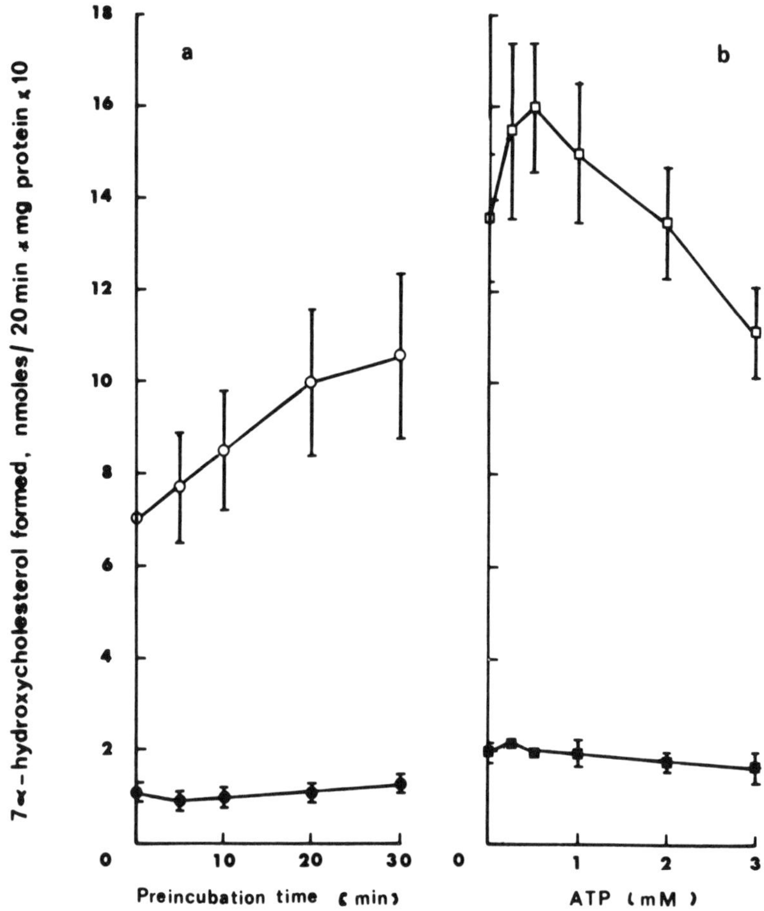

FIGURE 3. Effect of fasting and MgATP on the modulation of cholesterol 7α-hydroxylase activity. Activities of cholesterol 7α-hydroxylase in microsomal preparations obtained from fed (open-symbols) and fasted (48 hr) (closed symbols) were compared in relation to: (a) preincubation time and (b) MgATP concentration. In the latter Mg^{2+} and ATP were maintained at a molar ratio of 2:1. Heat-treated cytosol (60 to 80 μg protein) was added together with the NADPH-regenerating system to start the assay as previously described for the inactivation assay.[43] The values are means ± SEM of five to six independent experiments (unpublished data).

reacting to the activating effect of low MgATP concentrations. The retention of the response to the inactivating effect of 3.0 m*M* MgATP lends further support for the dual role of MgATP.

Our study[41] showed that cholesterol 7α-hydroxylase existed predominantly in the inactive form at the acrophase of the diurnal rhythm. It was estimated to be 58 to 88% of the native enzyme activity becoming activated during the microsomal preparation by washing, preincubation and the addition of the cytosolic factors, indicating a large reserve of the inactive form of the enzyme. The physiological significance appears to be to provide a steady supply of active enzyme when new enzyme synthesis is greatly reduced as indicated by the enzyme activity during fasting and at the nadir of the diurnal rhythm. The dual roles of MgATP can thus be summarized by the following metabolic sequence:

$$\text{Protein synthesis} \rightarrow \underset{\text{(nonphosphorylated)}}{\text{inactive}} \underset{}{\overset{\text{MgATP}}{\rightleftharpoons}} \underset{\text{(phosphorylated)}}{\text{active}} \overset{\text{MgATP}}{\rightleftharpoons} \underset{\text{(phosphorylated)}}{\text{deactivated}}$$

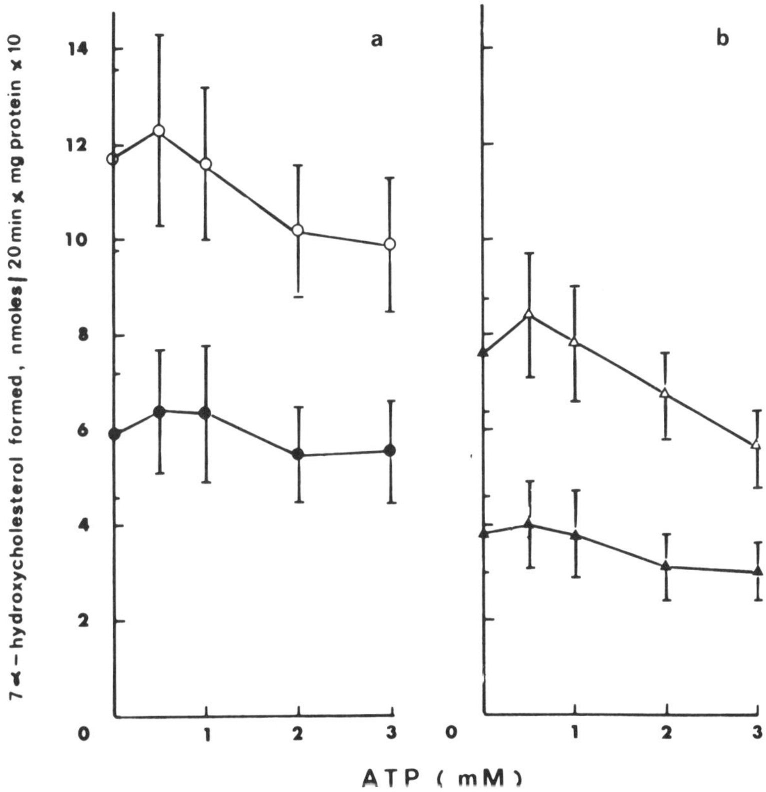

FIGURE 4. Modulation of cholesterol 7α-hydroxylase activity by MgATP at the acrophase and the nadir of the diurnal rhythm. Microsomes prepared from rat livers at the acrophase (open symbols) and the nadir (closed symbols) of the diurnal rhythm were preincubated with various MgATP concentrations which were maintained at a relative molar concentration of 2:1 (Figure a) at pH 7.4 (○-○, ●-●) and Figure b at pH 6.7 (△-△, ▲-▲). The assay was initiated after a 10 min pre-incubation by adding crude cytosol (270 to 300 μg protein) from the respective preparations and the NADPH-regenerating system as previously described.[43] Each point represents mean ± SEM of four independent experiments (unpublished data).

The newly formed *in situ* enzyme was inactive and became activated in contact with low concentration of MgATP; further phosphorylation in the presence of high concentration of MgATP, however, induced the ultimate degradation of the enzyme. Evidence to support these reactions is derived from our observation[42] that the inactivation reaction of cholesterol 7α-hydroxylase was reversible, depending on the reactivation time and the presence of NaF (Figure 2). As discussed earlier, the inactivation was fully reversible after a 10 min preincubation with 3.0 m*M* MgATP, but was only partially reversible after 60 min preincubation unless NaF was included, indicating some irreversible deactivation, due perhaps to enzyme degradation.

Further support for this hypothesis is drawn from a study of the half-life of cholesterol 7α-hydroxylase by Gielen et al.[27] They showed that cholesterol 7α-hydroxylase had a half-life of 2 hr if assessed by the method of spontaneous decrease of the enzyme activity; the half-life went up to 4 hr if the enzyme activity was measured following cyclohexamide-directed decay. In accord with our hypothesis, this difference was seen to be the result of the involvement of an active mechanism in the control of the enzyme degradation. Unlike the spontaneous decay study the presence of cyclohexamide blocks new protein synthesis, thereby altering the preexisting equilibrium of the different enzymic forms. The establishment

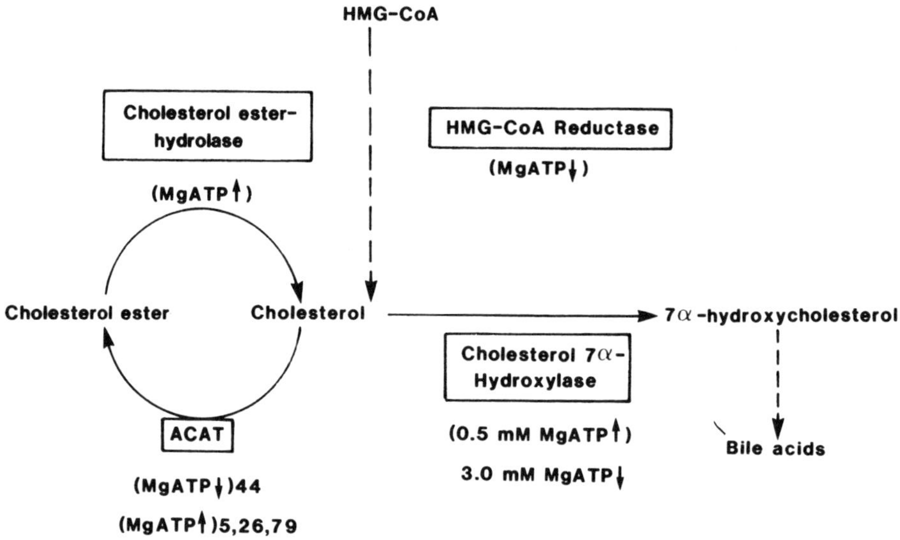

FIGURE 5. Schematic presentation of coordinate regulation of the four key enzymes involved in cholesterol homeostasis by phosphorylation: activation indicated by (↑) and inactivation by (↓).

of a new steady state in the amount of the active enzymes is achieved by increased conversion to the active form, with a concomitant reduction in the conversion to the deactivated form. This shift in enzyme activity is in accord with the apparent increased conversion to the active form of cholesterol 7α-hydroxylase during fasting and at the nadir of the circadian rhythm when enzyme synthesis was greatly reduced.[57] Thus, the shift increased the pool of the active form of cholesterol 7α-hydroxylase leading to an apparently longer half-life.

E. Coordinate Regulation of HMG-CoA Reductase, Cholesterol 7α-Hydroxylase, ACAT and Cholesterol Ester-Hydrolase

Metabolic studies based on dietary and hormonal manipulation indicate that in rat liver the activities of HMG-CoA reductase cholesterol 7α-hydroxylase and ACAT are coordinately controlled.[4,10,37,51,53,54,82] The findings that the activity of these enzymes is being modulated by MgATP further supports the coordinate nature of the regulation of cholesterol metabolism in the liver, although evidence for a similar molecular mechanism is lacking for cholesterol ester-hydrolase in this organ. The schematic representation of this coordinate regulatory mechanism by phosphorylation is shown in Figure 5. The reader will find this scheme differs from that proposed by Scallen and Sanghvi.[67]

HMG-CoA reductase, which catalyzes the limiting step in cholesterol biosynthesis, was inactivated by phosphorylation catalyzed by protein kinase and was reactivated by phosphoprotein phosphatase (Figure 5).[8,28,36] Cholesterol 7α-hydroxylase was activated by low concentration of MgATP, a reaction appeared to be mediated by protein kinase, while high MgATP concentrations inactivated the enzyme.[43] ACAT, catalyzed the esterification of free cholesterol, but was inactivated by MgATP as shown by our study,[44] contrary to the findings of Basheeruddin et al.,[5] Suckling et al.,[79] and Gavey et al.[26] This discrepancy, we believe, is linked to the omission or adequate addition of bovine serum albumin in their studies, resulting in marked inhibition of ACAT activity by the free acids released during the reaction catalyzed by acyl-CoA hydrolase. Mechanistically, activation of ACAT by phosphorylation is not tenable because cholesterol ester-hydrolase, which catalyzes the formation of nonesterified cholesterol, was also activated by phosphorylation in tissues so far studied such as bovine adrenal gland,[58] bovine corpus luteum,[16] and rabbit arterial tissues.[32] The molecular mechanism involved in the control of cholesterol ester-hydrolase in the liver is unknown. Nevertheless, a concomitant activation of ACAT and cholesterol ester-hydrolase would generate a futile cycle, in which a recycling of esterified and nonesterified cholesterol would be produced by these enzymes. According to our data the phosphorylation of ACAT would complement that of cholesterol ester-hydrolase.

In summary, the data reviewed to date strongly support the theory that cholesterol 7α-hydroxylase is regulated by phosphorylation mechanism. Phosphorylation by MgATP both activates and inactivates cholesterol 7α-hydroxylase depending on pH and MgATP concentrations. This dual role of MgATP provides a large reserve of the inactive cholesterol 7α-hydroxylase and the ultimate degradation of the enzyme. Thus, the short-term cholesterol homeostasis in the body is delicately balanced and maintained by a coordinate regulation of the key enzymes in the synthesis and utilization of cholesterol by phosphorylation.

REFERENCES

1. **Ahlsten, M., Hansson, R., Holmberg, I., Ljungdahl, I., and Wikvall, K.,** Reconstitution of cholesterol 7α-hydroxylase system from human liver microsomes, *Biochem. Med.*, 26, 307, 1981.
2. **Balasubramaniam, S. and Mitropoulos, K. A.,** The role of cytochrome P-450 in the 7α-hydroxylation of cholesterol, *Biochem. Soc. Trans.*, 3, 194, 1975.
3. **Balasubramaniam, S., Mitropoulos, K. A., and Myant, N. B.,** Hormonal control of the activities of cholesterol 7α-hydroxylase and hydroxymethyl- glutaryl-CoA reductase in rats, in *Advances in Bile Acid Research, III, Bile Acid Meeting,* Matern, S., Hackenschmidt, J., Back, P., and Geroks, W., Eds., F.K. Schattauer Verlag, Stuttgart, 1975, 61.
4. **Balasubramaniam, S., Mitropoulous, K. A., and Venkatesan, S.,** Rat liver acyl-CoA: cholesterol acyltransferase, *Eur. J. Biochem.*, 90, 377, 1978.
5. **Basheeruddin, K., Rawstorne, S., and Higgins, M. J. P.,** Reversible activation of rat liver acyl-CoA: cholesterol acyltransferase *in vitro, Biochem. Soc. Trans.*, 10, 390, 1982.
6. **Beg, Z. H., Allmann, D. W., and Gibson, D. M.,** Modulation of 3-hydroxy-3-methylglutaryl coenzyme A reductase activity with cAMP and with protein fractions of rat liver cytosol, *Biochem. Biophys. Res. Commun.*, 54, 1362, 1973.
7. **Beg, Z. H., Stonik, J. A., and Brewer, H. B., Jr.,** 3-Hydroxy-3-methylglutaryl coenzyme A reductase: regulation of enzyme activity by phosphorylation and dephosphorylation, *Proc. Natl. Acad. Sci. U.S.A.*, 75, 3678, 1978.
8. **Beg, Z. H., Stonik, J. A., and Brewer, H. B., Jr.,** Characterization and regulation of reductase kinase, a protein kinase that modulates the enzyme activity of 3-hydroxy-3-methylglutaryl-coenzyme A reductase, *Pro. Natl. Acad. Sci. U.S.A.*, 76, 4375, 1979.
9. **Beins, D. M., Vining, R., and Balasubramaniam, S.,** Regulation of neutral cholesterol esterase and acyl-CoA: cholesterol acyltransferase in the rat adrenal gland, *Biochem. J.*, 202, 631, 1982.
10. **Bjorkhem, I., Blomestrand, R., and Svensson, L.,** Effect of different dietary triglycerides on 7α-Hydroxylation of cholesterol and other mixed function oxidations, *J. Lipid Res.*, 19, 359, 1978.
11. **Bjorkhem, I., Danielsson, H., and Wikvall, K.,** 7α-hydroxylation of cholesterol by reconstituted systems from rat liver microsomes, *Biochem. Biophys. Res. Commun.*, 61, 934, 1974.
12. **Boyd, G. S., Brown, M. J. G., Hattersby, N. G., Lawson, M. E., and Suckling, K. E.,** Rat liver cholsterol 7α-hydroxylase, in *Advances in Bile Acid Research,* Vol. III, Matern, S., Hackenschmidt, J., Back, P., and Geroks, W., Eds., F.K. Schattauer Verlag, Stuttgart, 1975, 45.
13. **Boyd, G. S., Scholan, N. A., and Mitton, J. R.,** Factors influencing cholesterol 7α-hydroxylase activity in rat liver, *Adv. Exp. Med. Biol.*, 4, 443, 1969.
14. **Carulli, N., Deleon, M. P., Zironi, F., Pineti, A., Smerieri, R., Lori, R., and Loria, P.,** Hepatic cholesterol and bile acid metabolism in subjects with gallstones: comparative effects of short term feeding of chenodeoxycholic and ursodeoxycholic acid, *J. Lipid Res.*, 21, 35, 1980.
15. **Chiang, J. Y. L., Malmer, M., and Hutterer, F.,** A form of rabbit liver cytochrome P-450 that catalyzes the 7α-hydroxylation of cholesterol, *Biochim. Biophys. Acta,* 750, 291, 1983.

16. **Cook, K. G., Colban, R. J., Snee, J., and Yeoman, S. J.,** Cytosolic cholesterol ester hydrolase from bovine corpus luteum. Its purification, identification and relationship to hormone-sensitive lipase, *Biochem. Biophys. Acta,* 752, 46, 1983.
17. **Coyne, M. J., Bonorris, G. E., Goldstein, L. I., and Schoenfield, L. J.,** Effect of chenodeoxycholic acid and phenobarbital on the rate-limiting enzymes of hepatic cholesterol and bile acid synthesis in patients with gallstones, *J. Lab. Clin. Med.,* 87, 281, 1976.
18. **Craig, I. F., Boyd, G. S., McLeod, R. J., and Suckling, K. E.,** Partial characterization of a protein from rat liver 100 000g supernatant which stimulates cholesterol 7α-hydroxylase, *Biochem. Soc. Trans.,* 7, 967, 1979.
19. **Danielsson, H.,** Relationship between diurnal variations in biosynthesis of cholesterol and bile acids, *Steroids,* 20, 63, 1972.
20. **Danielsson, H., Einarsson, K., and Johansson, G.,** Effect of biliary drainage on individual reactions in the conversion of cholesterol to taurocholic acid bile acids and steroid, *Eur. J. Biochem.,* 2, 44, 1967.
21. **Danielsson, H., Kalles, I., and Wikvall, K.,** Modulation of reconstituted cholesterol 7α-hydroxylase activity by protein fractions from rat liver cytosol, *Biochem. Biophys. Res. Commun.,* 97, 1459, 1980.
22. **Danielsson, H., Kalles, I., and Wikvall, K.,** Regulation of hydroxylations in biosynthesis of bile acids. Isolation of a protein from rat liver cytosol stimulating reconstituted cholesterol 7α-hydroxylase activity, *J. Biol. Chem.,* 259, 4258, 1984.
23. **Danielsson, H. and Wikvall, K.,** A specific cytochrome P-450 with short half-life catalyzing 7α-hydroxylation of cholesterol, *Biochem. Biophys. Res. Commun.,* 103, 46, 1981.
24. **Einarsson, K. and Johansson, G.,** Bile acids and steroids 206. Effect of actinomycin D and puromycin on the conversion of cholesterol into bile acids in bile-fistula rats, *FEBS Lett.,* 1, 219, 1968.
25. **Erickson, S. K., Shrewsbury, M. A., Brooks, C., and Myer, D. J.,** Rat liver acyl-coenzyme A: cholesterol acyltransferase: its regulation *in vivo* and some of its property *in vitro, J. Lipid Res.,* 21, 930, 1980.
26. **Gavey, K. L., Trujillo, D. L., and Scallen, T. J.,** Evidence for phosphorylation/dephosphorylation of rat liver acyl-CoA: cholesterol acyltransferase, *Proc. Natl. Acad. Sci. U.S.A.,* 80, 2171, 1983.
27. **Gielen, J., Van Cantfort, J., Robaye, B., and Renson, J.,** Rat liver cholesterol 7α-hydroxylase. 3. New results about its circadian rhythm, *Eur. J. Biochem.,* 55, 41, 1975.
28. **Gil, G., Stiges, M., Bove, J., and Hegardt, F. G.,** Phosphorylation-dephosphorylation of rat liver 3-hydroxy-3-methylglutaryl coenzyme A reductase associated with changes in activity, *FEBS Lett.,* 110, 195, 1980.
29. **Goldstein, J. L. and Brown, M. S.,** Familial hypercholesterolemia: Pathogenesis of a receptor disease, *Johns Hopkins Med. J.,* 143, 8, 1978.
30. **Goodwin, C. D., Cooper, B. W., and Margolis, S.,** Rat liver cholesterol 7α-hydroxylase. Modulation of enzyme activity by changes in phosphorylation state, *J. Biol. Chem.,* 257, 4469, 1982.
31. **Goodwin, C. D. and Margolis, S.,** Cyclic AMP-sensitive activation of hepatic sterol synthesis and 3-hydroxy-3-methylglutaryl coenzyme A reductase, *J. Lipid Res.,* 19, 747, 1978.
32. **Hijjar, D. P., Minick, C. R., and Fowler, S.,** Arterial neutral cholesteryl esterase. A hormone-sensitive enzyme distinct from lysosomal cholesteryl esterase, *J. Biol. Chem.,* 258, 192, 1983.
33. **Hansson, R. and Wikvall, K.,** Properties of reconstituted cholesterol 7α-hydroxylase system from rat and rabbit microsomes, *Eur. J. Biochem.,* 93, 419, 1979.
34. **Hansson, R. and Wikvall, K.,** Hydroxylations in biosynthesis and metabolism of bile acids. Catalytic properties of different forms of cytochrome P-450, *J. Biol. Chem.,* 255, 1643, 1980.
35. **Hepher, G. W. and Quardfordt, S. H.,** Kinetics of cholesterol and bile acids in patients with cholesterol cholelithiasis, *Gastroenterology,* 69, 318, 1975.
36. **Ingebritsen, T. S., Lee, H. S., Parker, R. A., and Gibson, D. M.,** Reversible modulation of the activities of both liver microsomal hydroxymethyl glutaryl coenzyme A reductase and its inactivating enzyme. Evidence for regulation by phosphorylation-dephosphorylation, *Biochem. Biophys. Res. Commun.,* 81, 1268, 1978.
37. **Jenke, H-S., Lowel, M., and Berndt, J.,** Effect of dietary lipids on the activity of 3-hydroxy-3-methylglutaryl-CoA reductase in rat liver, *Hoppe-Seylers Z. Physiol. Chem.,* 363, 725, 1982.
38. **Johansson, G.,** Effect of cholestyramine and diet on hydroxylations in the biosynthesis and metabolism of bile acids, *Eur. J. Biochem.,* 17, 292, 1970.
39. **Keith, M. L., Rodwell, V. W., Rogers, D. H., and Rudney, H.,** *In vitro* phosphorylation of 3-hydroxy-3-methylglutaryl coenzyme A reductase: Analysis of ^{32}P-labelled inactivated enzyme, *Biochem. Biophys. Res. Commun.,* 90, 969, 1979.
40. **Kritchevsky, D. and Kothari, H. V.,** Arterial enzymes of cholesteryl ester metabolism, *Adv. Lipid Res.,* 16, 221, 1978.
41. **Kwok, C. T., Burnett, W., and Hardie, I. R.,** Regulation of rat liver microsomal cholesterol 7α-hydroxylase: presence of a cytosolic activator, *J. Lipid Res.,* 22, 570, 1981a.

42. **Kwok, C. T., Burnett, W., and Hardie, I. R.,** Regulation of rat liver microsomal cholesterol 7α-hydroxylase: reversible inactivation by ATP + Mg^{2+} and a cytosolic activator, *J. Lipid Res.*, 22, 580, 1981b.
43. **Kwok, C. T., Pillay, S. P., and Hardie, I. R.,** Regulation of rat liver microsomal cholesterol 7α-hydroxylase by MgATP. Effect of pH, *Biochem. Biophys. Res. Commun.*, 116, 966, 1983.
44. **Kwok, C. T., Smith, J. L., Pillay, S. P., and Hardie, I. R.,** Inactivation of acyl-CoA: cholesterol acyltransferase by ATP + Mg^{2+} in rat liver microsomes, *Biochem. Int.*, 8, 271, 1984.
45. **Lakshmanan, M. R. and Veech, R. L.,** Short- and long-term effects of ethanol administration *in vivo* on rat liver HMG-CoA reductase and cholesterol 7α-hydroxylase activities, *J. Lipid Res.*, 18, 325, 1977.
46. **Lehrer, G., Panini, S. R., Rogers, D. H., and Rudney, H.,** Modulation of rat liver 3-hydroxy-3-methylglutaryl coenzyme A reductase by lipid inhibitors, substrates and cytosolic factors, *J. Biol. Chem.*, 256, 5612, 1981.
47. **Lichtenstein, A. H. and Brecher, P.,** Properties of acyl-CoA: cholesterol acyltransferase in rat liver microsome. Topological localization and effects of detergents, albumin and polar steroids, *J. Biol. Chem.*, 255, 9098, 1980.
48. **Maton, P. N., Ellis, H. J., Higgins, M. J. P., and Dowling, R. H.,** Hepatic HMG-CoA reductase in human cholelithiasis: effects of chenodeoxycholic and ursodeoxycholic acids, *Eur. J. Clin. Invest.*, 10, 325, 1980.
49. **Mitropoulos, K. A., Balasubramaniam, S., Gibbons, G. F., and Reeves, B. E. A.,** Diurnal variation in the activity of cholesterol 7α-hydroxylase in the livers of fed and fasted rats, *FEBS Lett.*, 27, 203, 1972.
50. **Mitropoulos, K. A., Balasubramaniam, S., and Myant, N. B.,** The effect of interruption of the enterohepatic circulations of bile acids and cholesterol feeding on cholesterol 7α-hydroxylase in relation to the diurnal rhythm in its activity, *Biochem. Biophys. Acta,* 326, 428, 1973.
51. **Mitropoulos, K. A., Balasubramaniam, S., Venkatesan, S., and Reeves, B. E. A.,** On the mechanism for the regulation of 3-hydroxy-3-methylglutaryl-coenzyme A reductase, of cholesterol 7α-hydroxylase and of acyl-coenzyme A. Cholesterol acyltransferase by free cholesterol, *Biochem. Biophys. Acta,* 530, 99, 1978b.
52. **Mitropoulos, K. A., Suzuki, M., Myant, N. B., and Danielsson, H.,** Effects of thyroidectomy and thyroxine treatment on the activity of 12α-hydroxylase and of some components on microsomal electron transfer chains in rat liver, *FEBS Lett.*, 1, 13, 1968.
53. **Mitropoulos, K. A., Venkatesan, S., and Balasubramaniam, S.,** Compartmentation and supply of cholesterol: two important factors in the co-ordinate regulation of hydroxymethylglutaryl-CoA reductase and cholesterol 7α-hydroxylase, *Biochem. Soc. Trans.*, 6, 878, 1978c.
54. **Mitropoulos, K. A., Venkatesan, S., and Balasubramaniam, S.,** On the mechanism of regulation of hepatic 3-hydroxy-3-methylglutaryl coenzyme A reductase and of acyl-coenzyme: cholesterol acyltransferase by dietary fat, *Biochem. Biophys. Acta,* 619, 247, 1980.
55. **Mitropoulos, K. A., Venkatessan, S., Balasubramaniam, S., and Peters, T. J.,** The submicrosomal localization of 3-hydroxy-3-methylglutary coenzyme A reductase, cholesterol 7α-hydroxylase and cholesterol in rat liver, *Eur. J. Biochem.*, 82, 419, 1978a.
56. **Myant, N. B. and Eder, H. A.,** The effect of biliary drainage upon the synthesis of cholesterol in the liver, *J. Lipid Res.*, 2, 363, 1961.
57. **Myant, N. B. and Mitropoulos, K. A.,** Cholesterol 7α-hydroxylase, *J. Lipid Res.*, 18, 135, 1977.
58. **Naghshineh, S., Treadwell, C. R., Gallo, L. L., and Vahouny, G. V.,** Protein kinase-mediated phosphorylation of a purified sterol ester hydrolase from bovine adrenal cortex, *J. Lipid Res.*, 19, 561, 1978.
59. **Nordstrom, J. L., Rodwell, V. W., and Mitschelen, J. J.,** Interconversion of active and inactive forms of rat liver hydroxymethylglutaryl-CoA reductase, *J. Biol. Chem.*, 252, 8924, 1977.
60. **Rajan, G., Menon, A. S., and Ramasarma, T.,** Desensitization of ATP-Mg inhibition of 3-hydroxy-3-methylglutaryl CoA reductase by heat treatment of microsomes, *Biochem. Biophys. Res. Commun.*, 84, 544, 1978.
61. **Renson, J., Van Cantfort, J., Robaye, B., and Gielen, J.,** Measure de la demi-vic de la cholesterol 7α-hydroxylase, *Arch. Int. Physiol.*, 77, 972, 1969.
62. **Rodwell, V. W., Nordstrom, J. L., and Mitschelen, J. J.,** Regulation of HMG-CoA reductase, *Adv. Lipid Res.*, 14, 1, 1976.
63. **Rush, W. R., Gibson, G. G., and Parke, D. V.,** The purification and reconstitution of cytochrome P-450 dependent cholesterol 7α-hydroxylase activity from rat liver microsomal fractions, *Biochem. Soc. Trans.*, 8, 102, 1980.
64. **Salen, G., Nicolau, G., Shefer, S., and Mosbach, E. H.,** Hepatic cholesterol metabolism in patients with gallstones, *Gastroenterology,* 69, 676, 1975.
65. **Sanghvi, A., Grassi, E., and Diven, W.,** Loss of cholesterol 7α-hydroxylase activity *in vitro* in the presence of bivalent metal ions and by dialysis of rat liver microsomes, *Proc. Natl. Acad. Sci. U.S.A.*, 80, 2175, 1983.

66. **Sanghvi, A., Grassi, E., Warty, V., Diven, W., Wight, C., and Lester, R.,** Reversible activation — inactivation of cholesterol 7α-hydroxylase possibly due to phosphorylation-dephosphorylation, *Biochem. Biophys. Res. Commun.*, 103, 886, 1981.
67. **Scallen, T. J. and Sanghvi, A.,** Regulation of three key enzymes in cholesterol metabolism by phosphorylation/dephosphorylation, *Proc. Natl. Acad. Sci. U.S.A.*, 80, 2477, 1983.
68. **Schoenfield, L. J., Bonorris, G. G., and Ganze, P.,** Induced alterations in the rate-limiting enzymes of hepatic cholesterol and bile acid synthesis in the hamster, *J. Lab. Clin. Med.*, 82, 858, 1973.
69. **Scholan, N. A. and Boyd, G. S.,** The cholesterol 7α-hydroxylase enzyme system, *Hoppe-Seyler's Z. Physiol. Chem.*, 349, 1628, 1968.
70. **Shapiro, D. J. and Rodwell, V. W.,** Diurnal variation and cholesterol regulation of hepatic HMG-CoA reductase activity, *Biochem. Biophys. Res. Commun.*, 37, 867, 1969.
71. **Shefer, S., Hauser, S., Berkersky, I., and Mosbach, E. H.,** Feedback regulation of bile acid biosynthesis in the rat, *J. Lipid Res.*, 10, 646, 1969.
72. **Shefer, S., Hauser, S., Berkersky, I., and Mosbach, E. H.,** Biochemical site of regulation of bile acid biosynthesis in the rat, *J. Lipid. Res.*, 11, 404, 1970.
73. **Small, D. M.,** Cellular mechanisms for lipid deposition in atherosclerosis, *N. Engl. J. Med.*, 297, 873, 1977.
74. **Spector, A. A., Mathur, S. N., and Kaduce, T. L.,** Role of acyl-coenzyme A: cholesterol O-actyltransferase in cholesterol metabolism, *Prog. Lipid Res.*, 18, 31, 1979.
75. **Spence, J. T. and Gaylor, J. L.,** Investigation of regulation of microsomal hydroxymethylglutaryl coenzyme A reductase and methyl sterol oxidase of cholesterol biosynthesis, *J. Biol. Chem.*, 252, 5852, 1977.
76. **Story, J. A., Tepper, S. A., and Kritchevsky, D.,** Influence of thyroid state on cholesterol hydroxylation and absorption in the rat, *Biochem. Med.*, 10, 214, 1974.
77. **Strand, O.,** Effect of D- and L-triidothyronine and of propylthiouracil on the production of bile acids in the rat, *J. Lipid Res.*, 4, 305, 1963.
78. **Strange, R. C., Beckett, G. J., and Percey-Robb, I. W.,** Nuclear and cytosolic distribution of conjugated cholic acid and radiolabelled glycocholic acid in rat liver, *Biochem. J.*, 178, 71, 1979.
79. **Suckling, K. E., Stange, E. F., and Dietschy, J. M.,** Dual modulation of hepatic and intestinal acyl-CoA: cholesterol acyltransferase activity by (de-)phosphorylation and substrate supply *in vitro, FEBS Lett.*, 151, 111, 1983.
80. **Takeuchi, N., Ito, M., and Yamamura, Y.,** Regulation of cholesterol synthesis in rat liver, *Atherosclerosis,* 20, 481, 1974.
81. **Van Cantfort, J.,** The *in vitro* formation and the strong inhibitory action of 7-keto-cholesterol on cholesterol-7α-hydroxylase activity, *Life Sci.*, 11, 773, 1972.
82. **Van Cantfort, J.,** Controle par les glucocortico-steroides de l'activite circadienne de la cholesterol-7α-hydroxylase, *Biochemie,* 55, 1171, 1973.
83. **Van Cantfort, J. and Gielen, J.,** Cholesterol 7α-hydroxylase. 2. Biochemical properties and participation of endogenous cholesterol in the assay *in vitro, Eur. J. Biochem.*, 55, 33, 1975.
84. **Van Cantfort, J., Renson, J., and Gielen, J.,** Rat liver cholesterol 7α-hydroxylase. 1. Development of a new assay based on the enzymic exchange of the tritium located on the 7α-position of the substrate, *Eur. J. Biochem.*, 55, 23, 1975.
85. **Wada, F., Hirata, K., Nakao, K., and Sakamoto, Y.,** Participation of the microsomal electron transport system involving cytochrome P-450 in 7α-hydroxylation of cholesterol, *J. Biochem. (Tokyo),* 66, 699, 1969.
86. **Kwok, C. T., Pillay, S. P., and Hardie, I. R.,** unpublished data.

Chapter 8

REGULATION BY ASCORBIC ACID AND OTHER NUTRIENTS

Emil Ginter

TABLE OF CONTENTS

I. INTRODUCTION

This chapter presents data on the specific effects of some nutrients on the activity of cholesterol 7α-hydroxylase. The survey does not include the data achieved in animals with reduced food consumption in which a marked fall in the activity and the lowering of the amplitude of the diurnal rhythm in the activity of 7α-hydroxylase was observed.[1] Administration of glucose to fasting animals sharply increases 7α-hydroxylase activity.[2] In such case, however, it is a nonspecific consequence of the increased hepatic cholesterologenesis resulting in the increased supply of newly synthesized cholesterol, which is an optimum substrate for 7α-hydroxylase. This chapter does not include the studies describing the effect of cholesterol feeding on 7α-hydroxylase, as the effects of the substrate supply on the 7α-hydroxylase activity are discussed elsewhere in the volume.

II. ASCORBIC ACID

Vitamin C is the first nutrient the role of which in the 7α-hydroxylase regulation was proved. In many older studies describing metabolic disorders in scorbutic guinea pigs and monkeys disorders in the cholesterol metabolism were repeatedly observed, including reduced cholesterol transformation into bile acids.[3] However, scurvy is a very complicated pathological situation accompanied with refusal of food and decrease in body weight. Thus, the doubts as to the specific consequence of ascorbic acid deficiency are justified. The effects of ascorbic acid on cholesterol and bile acid metabolism have been most effectively isolated from other variables through the use of a model of marginal (= latent, subclinical) vitamin C deficiency. In this model, guinea pigs are fed just enough ascorbic acid to maintain normal food consumption and to avoid secondary complications of scurvy (anorexia, weight loss, inanition, hemorrhagic phenomena), yet tissue ascorbate levels are similar to those observed in scurvy.[4,5] A great number of authors has found hypercholesterolemia and/or accumulation of cholesterol in the liver of the vitamin C-deficient guinea pigs.[4,6-19] The attempts to explain this phenomenon by cholesterol redistribution in different cholesterol pools, by increased absorption of dietary cholesterol, or by increased synthesis of endogenous cholesterol failed.[4,17,20] It is very probable that the main reason for the cholesterol accumulation in vitamin C-deficient animals in diminished conversion of cholesterol into bile acids. Three laboratories have found 30 to 50% reduction in bile acid synthesis in marginally deficient guinea pigs by a variety of techniques including $^{14}CO_2$ production after administration of (26-^{14}C)cholesterol,[6,21] fecal bile acid output and bile acid pool size,[22] and bile acid kinetics following (24-^{14}C)chenodeoxycholic acid administration.[12,16]

Ginter et al.[4,23] demonstrated identical catabolism of (26-^{14}C) 7α-hydroxycholesterol in vitamin C-supplemented and vitamin C-deficient guinea pigs, suggesting that the impairment of bile acid synthesis occurs at the level of 7α-hydroxylase. Three laboratories measured 7α-hydroxylase directly in hepatic microsomes isolated from vitamin C-deficient guinea pigs and found depression of enzyme activity in deficient animals.[9,12,22] In sub-scorbutic animals with lowered body weight 7α-hydroxylation of endogenous microsomal cholesterol was very markedly reduced to less than 10% of control values.[9] On the other hand, hydroxylation of laurate as well as 25- and 26-hydroxylation of 5β-cholestane-3α,7α-diol were not significantly affected. Partially purified cytochrome P-450 isolated from sub-scorbutic hepatal microsomes had much lower capacity to 7α-hydroxylate (4-^{14}C)cholesterol than a corresponding system isolated from the liver of normal animals. Björkhem and Kallner suggested that vitamin C deficiency affected in their experiments the synthesis or breakdown of the cholesterol 7α-hydroxylation system, in particular the cytochrome P-450 component.[9] Harris et al.[22] found that even short-term vitamin C deficiency without a decrease in the body weight resulted in a marked decrease in the 7α-hydroxylase activity in the hepatic microsomes of guinea pigs. After 7 days of vitamin C-free regimen the activity of 7α-hydroxylase was reduced by 26% and after 10 days it had fallen by 46% when compared to control animals. Holloway and Rivers found in marginally deficient guinea pigs with unchanged food intake and body weight 60 to 70% depression of 7α-hydroxylase activity.[12] A direct correlation between 7α-hydroxylase and ascorbate concentration in the liver showed a good agreement with the correlation between the rate of bile acid synthesis and hepatic ascorbic acid level.[20] The sharp depression of 7α-hydroxylase activity was not accompanied by distinct changes in the concentration of microsomal esterified or unesterified cholesterol, or by significantly changed levels of cytochromes P-420, P-450, or b_5 in hepatic microsomes of vitamin C-deficient animals.[12]

These data demonstrate the exceptional susceptibility of 7α-hydroxylase to vitamin C deficiency in guinea pigs. As the activity of intestinal and hepatic 3-hydroxy-3-methyl glutaryl coenzyme A reductase does not change markedly in vitamin C-deficient guinea pigs,[14,24] the decrease of 7α-hydroxylase activity results in cholesterol accumulation in blood and tissues of vitamin C-deficient animals. The combination of reduced bile acid synthesis and increased cholesterol accumulation results in the increase of gallbladder bile lithogenity, formation of gallstones, accumulation of cholesterol in the thoracic aorta, and in atherosclerotic changes in coronary arteries of guinea pigs with chronic marginal vitamin C deficiency.[4,6,11,20,25] Long-term ascorbic acid deficiency evokes hypercholesterolemia even in phylogenetically distant animals such as the marmoset monkey and rainbow trout.[26,27] In primates, however, this effect is less distinct than in guinea pigs.[28,29] Very limited information is available concerning the effects of chronic vitamin C deficiency upon cholesterol and bile acid metabolism in man. In a study of experimental human scurvy serum, cholesterol decreased during vitamin C depletion and rose again after ascorbic acid administration.[30] Nine week long vitamin C depletion had little effect on cholesterol and bile acid metabolism in five healthy men.[31] On the other hand, several research groups reported that long-term ascorbic acid administration to hypercholesterolemic individuals with a low initial vitamin C-status significantly reduced serum cholesterol levels.[10,20,32-34] No report has been published concerning the effect of vitamin C on 7α-hydroxylase in man.

The mechanism of 7α-hydroxylase regulation by ascorbic acid is unknown. It is known that vitamin C deficiency decreases the activity of the microsomal monooxygenase detoxication system. Vitamin C-deficient guinea pigs have prolonged pentobarbital sleeping time, increased plasma half-life of acetanilide, aniline, and antipyrine, decreased hydroxylation of many, but not all xenobiotics, and decreased levels of cytochrome P-450, b_5, and NADPH-cytochrome P-450 reductase in liver microsomes.[35-37] In spite of extensive research, the mechanism of this phenomenon has not been elucidated: in vitamin C-deficient animals *de novo* synthesis of microsomal proteins is operable, activities of key enzymes involved in heme synthesis and degradation are not changed, the turnover of cytochrome P-450 heme is unchanged, there is no significant decrease in the quantity of microsomal phosphatidyl choline, no change in the spin state of cytochrome P-450, and no significant change in cytochrome P-450 affinity for drug substrates.[36,37] In some experiments with marginally deficient guinea pigs the level of total cytochrome P-450 did not decrease, but simultaneously the activities of some monooxygenases including 7α-hydroxylase were significantly inhibited.[12,38] Cytochrome P-450 isolated from liver microsomes of vitamin C-deficient guinea pigs showed a considerably lower capacity to 7α-hydroxylate cholesterol, while other cytochrome P-450-dependent activities were unaffected.[9] The *p*-bromophenol pathway of bromobenzene and the 4-hydroxy pathway of biphenyl detoxication are substantially decreased in vitamin C-deficient microsomes, while the *o*-bromophenol and 2-hydroxy pathways of biphenyl detoxication are not significantly affected.[37] There are quantitative differences in polypeptides isolated from normal and vitamin C-deficient liver microsomes.[39] Therefore,

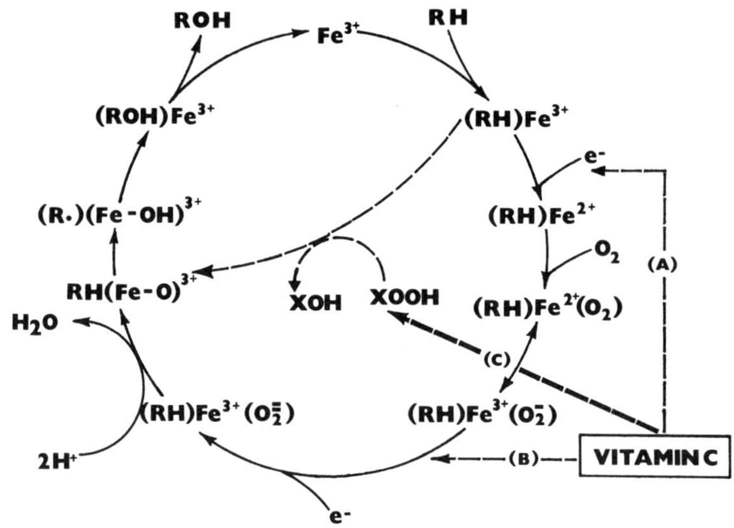

FIGURE 1. Cytochrome P-450 cycle in liver microsomes. RH: substrate (e.g., cholesterol), ROH: hydroxylated product (e.g., 7α-hydroxycholesterol), XOOH: oxidizing agent (e.g., hydroperoxide), and XOH: the corresponding alcohol. Arrows A, B, C: suggested possible interventions of vitamin C into the cycle. For details, see text.

vitamin C deficiency may selectively affect only a small portion of cytochrome P-450 isoenzymes with the subsequent decrease in the activity of some monooxygenases without any pronounced concomitant changes in the total cytochrome P-450 level. However, other explanations are possible.

During the hydroxylation of cholesterol or xenobiotics, the cytochrome P-450 system is thought to bind substrate and oxygen in a sequential manner — Figure 1.[40] Hydroxylation of substrate in this cycle is accomplished by an intermediate free radical mechanism. Nordblom et al.[41] have shown that the requirement for O_2, NADPH and NADPH-cytochrome P-450 reductase may be eliminated in the cytochrome P-450 cycle by substitution of H_2O_2, alkyl hydroperoxides, or peracids for these components. In the presence of these oxidizing agents (XOOH in Figure 1) the cycle may run along the shunt (broken line in the middle of the scheme, Figure 1) bypassing the need for O_2 and electron donors. Several mechanisms could exist through which vitamin C deficiency might affect the function of cytochrome P-450 cycle without simultaneously influencing the level of cytochrome P-450:

1. Vitamin C deficiency may selectively affect the transfer of electrons from NADPH-cytochrome P-450 reductase to the complex of oxidized cytochrome + substrate (arrow A in Figure 1).[42] Vitamin C-deficient microsomes contained altered forms of reductase which may be unable to reduce cytochrome P-450.[43]
2. Vitamin C deficiency may affect the transfer of electrons from NADPH-cytochrome P-450 reductase, or from cytochrome b_5 to the ferric-superoxide complex, by which a species at the oxidation level of hydrogen peroxide is produced (arrow B in Figure 1). Ascorbic acid may be an electron donor for this reaction, as the transfer of electrons between ferricytochrome b_5 and ascorbate has been proven.[44]
3. Vitamin C may stimulate the activity of the cycle by producing activated oxygen (free radicals or peroxides) due to which the cycle may proceed by shunt (arrow C in Figure 1). In the presence of Fe or Cu ions, ascorbic acid is able to produce hydroxyl radicals, superoxide anions, and hydrogen peroxide by nonenzymatic reduction of the molecular

> oxygen.[45-47] The ability of ascorbate to activate oxygen and to hydroxylate various substrates was demonstrated in various nonbiological model systems.[48,49] Therefore, it is possible that on the boundary of endoplasmic reticulum membranes and the cytosol, ascorbic acid activates molecular oxygen controlling thus the activity of monooxygenases.

This notion would explain the parallel biphase action of increasing vitamin C doses on the activity of 7α-hydroxylase (and other cytochrome P-450-dependent monooxygenases) and peroxidation reactions. In case of vitamin C deficiency, when the levels of ascorbic acid in the liver are extremely low, the activities of 7α-hydroxylase and other monooxygenases are low, and at the same time, lipoperoxidation in vitamin C-deficient microsomes is significantly slowed down.[12,18,37] With increasing vitamin C intake, the activity of 7α-hydroxylase and other monooxygenases increases and peroxidation in the liver microsomes is stimulated simultaneously. On the contrary, the extremely high intake of vitamin C inhibits the activity of 7α-hydroxylase and many other monooxygenases,[12,18,50,51] reduces lipoperoxidation,[52] and increases scavenging of oxygen-free radicals.[53] High concentrations of ascorbic acid show the same effects also in vitro: they inhibit 7α-hydroxylase activity,[54] have antiperoxidative effects,[55,56] and increase scavenging of oxygen free radicals. A number of other biologically important processes, e.g., steroidogenesis, ligand binding to neurotransmitter receptors, and lipid peroxidation in fatty acid micelles, microsomes, and various brain preparations show a similar parabolic response to increasing ascorbic acid concentrations.[57-59] In all such cases, the biphasic effect of increasing ascorbate doses may be explained by the fact that, as in vitamin C deficiency, at very high ascorbate inputs, oxygen activation becomes disrupted (depressed production or enhanced scavenging of oxygen free radicals). In good agreement with this notion is the fact that an extremely high intake of two other antioxidants, α-tocopherol and butylated hydroxytoluene, remarkably reduces 7α-hydroxylase activity in the liver of rats.[60] On the other hand, however, no stimulation effect of ascorbate on 7α-hydroxylation has been found in vitro as yet.[9,61] During the preparation of microsomes destruction of the hepatocyte compartmentalization takes place. Compartmentalization of the cell plays a key role in the control of production and scavenging of oxygen radicals. The use of hepatocytes in primary culture may be a valuable approach for further research on the mechanism of actin of vitamin C on 7α-hydroxylase.

III. DIETARY FIBER AND PROTEINS

Low levels of plasma cholesterol in native African populations as well as in vegetarians have indicated that plant food contains substances that are able to affect cholesterol metabolism. Intensive research in the past ten years has shown that there are at least three groups of substances with hypocholesterolemic effects: dietary fiber, saponins, and some vegetable proteins.[63-66] Dietary fiber is a generic term that includes a great number of substances, mostly carbohydrates in nature, that are resistant to digestive enzymes of human gastrointestinal tract (cellulose, lignin, hemicellulose, pectins, mucilages, gums, chitin, etc.). Saponins are steroid or triterpene glycosides that occur in many plant species. The mechanism of the hypocholesterolemic effect of plant materials is complex and consists, beside others, in the common ability of dietary fiber, saponins, and some vegetable proteins to disturb the enterohepatic cycle of bile acids. Some components of fiber (e.g., pectins, lignin, hemicellulose, and gums) and undigested soybean and other plant proteins may act in the intestine as a physiological chromatographic column, whose ion-exchange and gel filtration properties influence bile acid absorption. Saponins decrease bile acid absorption in such a way that they form large nonabsorbable mixed micelles with bile salts. The resulting metabolic effect is similar, although quantitatively lower, to that in cholestyramine treatment: increased excretion of bile acids in the feces and reduced return of bile acids to the liver via enterohepatic circulation. Disruption of the integrity of the hepatobiliary axis increases the activity of hepatic 7α-hydroxylase and evokes profound changes in cholesterol and lipoprotein metabolism.[67]

Theoretically, diets with high content of bile acid sequestering fiber should result in increased activity of hepatic 7α-hydroxylase. However, the actual situation is more complex. In several studies with rats, cholesterol 7α-hydroxylase did not show a consistent increase after feeding of various fiber sources (alfalfa, bran, or cellulose,[68] pectin,[69] cellulose, pectin, hemicellulose, or lignin,[70] wheat bran, alfalfa, yeast wall glycan, or cellulose[71]). Other authors, however, observed a remarkable increase in 7α-hydroxylase activity after administration of diets with increased content of fiber able to bind bile acids.[72-74] Controversion of these data can be caused by the fact that only some fractions and sources of dietary fiber bound bile acids.[75-77] Moreover, components isolated from various plant sources are often ill-defined and have little in common with the same material as it occurs in nature (e.g., protopectin vs. industrially manufactured pectin).

Proia et al.[78] compared mixed-function oxidation in rats fed a diet containing 10.4% of cellulose or the same quantity of pectin. Pectin has a considerably higher ability to bind bile acids than cellulose. Mean values for ethylmorphine *N*-demethylase, aniline hydroxylase, and cytochrome P-450 in liver microsomes were significantly higher in pectin-fed rats. These results suggest that a change in dietary fiber content, e.g., substituting pectin for cellulose, has the potential to alter host response, not only to cholesterol, but also to environmental toxins. In this connection it may be of interest that some xenobiotics, e.g., polychlorinated biphenyls, increase the activity of hepatal 7α-hydroxylase in rats.[79]

Substitution for animal proteins in the diet with vegetable proteins, particularly those derived from soybeans, reduces plasma cholesterol level in laboratory animals. Sirtori et al.[80] found in rats fed cholesterol diet with casein a marked increase of total and VLDL-cholesterol in plasma, accumulation of esterified cholesterol in the liver, marked reduction of hepatic high affinity receptors for β-VLDL, dramatic decline of the HMG-CoA reductase activity, and activation of 7α-hydroxylase in the hepatic microsomes. Substitution of soybean proteins for casein counteracted all these changes. It is possible that soybean proteins are involved in the regulation of hepatic lipoprotein receptors, but the mechanism of this effect is not clear.

Persistent enhancement of 7α-hydroxylase activity following stimulation of cholesterol catabolism by cholestyramine feeding in neonatal life was observed in laboratory animals.[81,82] It would be of great importance to find whether a similar ''biochemical memory'' exists also in people given in their childhood a diet with a high quantity of dietary fiber and/or soybean protein. Ginter et al.[83] found synergism of hypocholesterolemic effect of bile acid sequestrants and ascorbic acid. This could be the initial point for the production of a hypocholesterolemic ''drug'' on the basis of natural substances (e.g., the combination of pectin with vitamin C).[20]

IV. STRUCTURE OF DIETARY TRIGLYCERIDES

Epidemiological studies have suggested that the type of dietary lipids have an important influence on the serum cholesterol level and that diets rich in polyunsaturated triglycerides considerably reduce the serum cholesterol concentration. These findings have been confirmed and extended by numerous carefully controlled studies using laboratory animals and human subjects. In spite of countless biochemical studies, no decision can be reached with regard to the mechanism of the polyunsaturated fatty acid effect.

Even the study of the effect of fats of different composition on the activity of 7α-hydroxylase did not bring about the progress in knowledge of the mechanism of their effect. Two research groups found a significantly higher activity of hepatic 7α-hydroxylase in rats fed a diet with high amount of polyunsaturated fatty acids than in the rats fed saturated fats.[84,85] The rats fed a diet with 10% highly unsaturated safflower oil showed markedly increased rates of bile flow and excreted through bile and feces markedly higher amount of cholesterol and bile acids. At the same time the incorporation of (1-^{14}C)acetate and (2-^{14}C)mevalonate into the liver cholesterol and the conversion of (4-^{14}C)cholesterol into ^{14}C-bile acids were also higher in the rats.[85] It would be possible to make a conclusion from these data that the hypocholesterolemic effect of polyunsaturated fatty acids consists in activation of 7α-hydroxylase, which far outweighs the activation of cholesterol synthesis.

Unfortunately, two other groups of authors found a completely opposite situation when administering polyunsaturated triglycerides to rats, i.e., the reduction of the activity of hepatic 7α-hydroxylase.[86,87] Björkhem et al.,[87] in numerous carefully performed experiments, found that 7α-hydroxylation of exogenous as well as of endogenous cholesterol was significantly lower in the liver of rats fed trilinolein and triolein than in those fed tripalmitin and trierucin. The level of microsomal cytochrome P-450 was higher in animals fed trilinolein and triolein. The inhibitory effect of trilinolein and triolein on 7α-hydroxylase and the stimulatory effect of these triglycerides on cytochrome P-450 level was not due to the small amount of peroxides present in the unsaturated fats. The addition of the antioxidant (butylated hydroxyanisol) did not change the general pattern. The same results were obtained in experiments with lower concentrations of triglycerides in the diet and in experiments with different feeding patterns and lighting conditions. The effects obtained seem to be specific for 7α-hydroxylase. Several other monooxygenases (11- and 12-hydroxylation of laurate, 12α- and 26-hydroxylation of 5β-cholestane-3α, 7α-diol, 12α-hydroxylation of 7α-hydroxy-4-cholesten-3-one) were affected by the different dietary triglycerides in the opposite direction and their activity followed the changes in cytochrome P-450 levels. The changes of 7α-hydroxylase were better related to the degree of fat absorption than to the degree of fat unsaturation. The degree of absorption of fatty acids from rat intestine increases as follows: palmitic, erucic, oleic, and linoleic acid. On the other hand, 7α-hydroxylase activity in rats fed different triglycerides decreased as follows: tripalmitin, trierucin, triolein, and trilinolein. Björkhem et al.[87] suppose that reduced absorption of some fatty acids (e.g., palmitic acid) is coupled with a reduced absorption of bile acids. The higher loss of bile salts in stool may then activate 7α-hydroxylase due to a reduced feedback inhibition. Extreme controversions among these results suggest that the affect of the degree of dietary fat unsaturation on 7α-hydroxylase activity probably depends on the amount of other nutrients present in the diets.

With the aim to reduce the consumption of saturated animal fats the marketing of partially hydrogenated vegetable oils has been increased. However, during the hydrogenation of oils trans-isomers of fatty acids are produced. There are disputes if trans-isomers have unfavorable effects on cholesterol metabolism or not. Sugano et al.[88] compared cholesterol metabolism in rats fed diets containing either camellia oil or partially hydrogenated corn oil as a source of *cis*- and *trans*-octadecenoate, respectively. As long as the supply of linoleic acid was adequate, the *trans*-octadecenoate fed at relatively high dose in no way exhibited any unfavorable effects on cholesterol metabolism. Serum cholesterol level and cholesterol distribution in serum lipoproteins were comparable in both groups. Some parameters were even more favorable in the group fed *trans*-fat: lower cholesterol concentration in the liver, lower cholesterol absorption, much faster cholesterol turnover; rapidly exchangeable cholesterol pool was reduced as a consequence of stimulation of the removal rate from this pool and 7α-hydroxylase activity in liver microsomes were doubled. The effect on 7α-hydroxylase was specific as activities of other monooxygenases were not changed, including aminopyrine *N*-demethylase, *p*-aniline hydroxylase, and biphenyl hydroxylase. No differences were found in the activity of NADPH-cytochrome P-450 reductase and in the content of microsomal cytochrome P-450. It is probable that *trans*-fats are not more atherogenic than *cis*-fats.[89]

V. ETHANOL AND OTHER FACTORS

It is known that chronic ethanol administration to different animal species induces hypercholesterolemia and accumulation of esterified cholesterol in the liver. These changes are associated with both enhanced cholesterologenesis and decreased bile acid excretion in rats. Both mechanisms may play a role in ethanol-induced hypercholesterolemia, but the latter is probably predominant.[90] Neither HMG-CoA reductase nor 7α-hydroxylase were significantly affected within 2 hr after a single intraperitoneal injection of ethanol to rats.[91] Chronic ethanol feeding for 3 weeks did not affect the diurnal rhythm of 7α-hydroxylase, but caused a pronounced decrease in hepatic 7α-hydroxylase activity at the diurnal high point.[91] Accumulation of cholesterol in blood and liver of animals intoxicated with high doses of ethanol for a long time can, therefore, be explained by inhibition of 7α-hydroxylase activity with subsequent slowed-down degradation of cholesterol into bile acids. On the other hand in people with moderate alcohol intake (90 g in five divided doses daily for 4 weeks) no changes were observed in cholesterol absorption and synthesis and in bile acid excretion.[92] It is evident that the effects of ethanol on cholesterol degradation depend on the dose and on the time interval of ethanol consumption.

β-Sitosterol, a nonabsorbable plant sterol, reduces plasma cholesterol levels in animals and man by competitive inhibition of cholesterol absorption. The addition of 1 to 2% of sitosterol to the diet significantly increased hepatic 7α-hydroxylase activity in rat and hamster.[93-95] In the majority of these experiments the activity of HMG-CoA reductase increased simultaneously. It is possible that increased *de novo* synthesized cholesterol in hepatic microsomes resulted in 7α-hydroxylase activation in animals fed on the sitosterol diet.

There are reports on hypocholesterolemic effects of garlic in experimental animals and man. Administration of garlic oil to rats reduced 7α-hydroxylase activity.[96] Other hypocholesterolemic agents are discussed in Chapter 9.

REFERENCES

1. **Myant, N. B. and Mitropoulos, K. A.,** Cholesterol 7α-hydroxylase, *J. Lipid Res.,* 18, 135, 1977.
2. **Takeuchi, N., Ito, M., and Yamamura, Y.,** Regulation of cholesterol 7α-hydroxylation by cholesterol synthesis in rat liver, *Atherosclerosis,* 20, 481, 1974.
3. **Guchhait, R., Guha, B. C., and Ganguli, N. C.,** Metabolic studies on scorbutic guinea pigs. III. Catabolism of (4-^{14}C) cholesterol in vivo and in vitro, *Biochem. J.,* 86, 193, 1963.
4. **Ginter, E.,** Ascorbic acid in cholesterol and bile acid metabolism, *Ann. N.Y. Acad. Sci.,* 258, 410, 1975.
5. **Ginter, E., Bobek, P., and Ovecka, M.,** Model of chronic hypovitaminosis C in guinea-pigs, *Int. J. Vit. Res.,* 38, 104, 1968.
6. **Fujinami, T., Okado, K., Senda, K., Sugimura, M., and Kishikawa, M.,** Experimental atherosclerosis with ascorbic acid deficiency, *Jpn. Circ. J.,* 35, 1559, 1971.
7. **Nambisan, B. and Kurup, P. A.,** Ascorbic acid and glycosaminoglycan and lipid metabolism in guinea pigs fed normal and atherogenic diets, *Atherosclerosis,* 22, 447, 1975.
8. **Hornig, D. and Weiser, H.,** Ascorbic acid and cholesterol: effect of graded oral intakes on cholesterol conversion to bile acids in guinea-pigs, *Experientia,* 32, 687, 1976.
9. **Björkhem, I. and Kallner, A.,** Hepatic 7α-hydroxylation of cholesterol in ascorbate-deficient and ascorbate-supplemented guinea pigs, *J. Lipid Res.,* 17, 360, 1976.
10. **Hanck, A. and Weiser, H.,** The influence of vitamin C on lipid metabolism in man and animals, *J. Vit. Nutr. Res.,* Suppl. 19, 83, 1979.
11. **Jenkins, S. A.,** Vitamin C status, serum cholesterol levels and bile composition in the pregnant guinea-pig, *Br. J. Nutr.,* 43, 95, 1980.
12. **Holloway, D. E. and Rivers, J. M.,** Influence of chronic ascorbic acid deficiency and excessive ascorbic acid intake on bile acid metabolism and bile composition in the guinea pig, *J. Nutr.,* 111, 412, 1981.
13. **Yokota, F., Igarashi, Y., and Suzue, R.,** Hyperlipidemia in guinea-pig induced by ascorbic acid deficiency, *Atherosclerosis,* 38, 249, 1981.

14. **Holloway, D. E., Peterson, F. J., Prigge, W. F., and Gebhard, R. L.**, Influence of dietary ascorbic acid upon enzymes of sterol biosynthesis in the guinea pig, *Biochem. Biophys. Res. Commun.*, 102, 1283, 1981.
15. **Zloch, Z.**, Changes in the cholesterol content of high-density plasma lipoproteins in guinea pigs in the presence of a raised exogenous cholesterol intake and different degrees of vitamin C saturation, *Physiol. Bohemoslov.*, 31, 557, 1982.
16. **Holloway, D. E., Peterson, F. J., and Rivers, J. M.**, Effects of dietary ascorbic acid on bile acid metabolism in guinea pigs fed Krehl or Reid-Briggs diets, *Nutr. Rep. Int.*, 25, 941, 1982.
17. **Weight, M. J., Compton-James, K., Van Schalkwyk, D. J., and Seier, J. V.**, Cholesterol-ascorbic acid interactions in guinea pig liver homogenates, and in the in situ perfused liver: relationship between ascorbic acid status and cholesterol and bile acid synthesis from mevalonate, *Int. J. Vit. Nutr. Res.*, 52, 298, 1982.
18. **Ginter, E., Kosinova, A., Hudecova, A., and Mlynarcikova, A.**, Parabolic response of hepatic microsomal hydroxylating system and lipids to graded doses of ascorbic acid in guinea pigs on low and high α-tocopherol intake, *J. Nutr.*, 114, 485, 1984.
19. **Holloway, D. E. and Rivers, J. M.**, Long-term effects of inadequate and excessive dietary ascorbate on bile acid metabolism in the guinea pig, *J. Nutr.*, 114, 1370, 1984.
20. **Ginter, E. and Bobek, P.**, The influence of vitamin C on lipid metabolism, in *Vitamin C (Ascorbic Acid)*, Counsell, J. N. and Horning, D. H., Eds., Elsevier, New York, 1981, 299.
21. **Ginter, E.**, Cholesterol: vitamin C controls its transformation to bile acids, *Science*, 179, 702, 1973.
22. **Harris, W. S., Kottke, B. A., and Subbiah, M. T. R.**, Bile acid metabolism in ascorbic acid-deficient guinea pigs, *Am. J. Clin. Nutr.*, 32, 1837, 1979.
23. **Ginter, E., Ozdin, L., and Nemec, R.**, Localization of the interference of ascorbic acid deficiency with bile acid biogenesis, *Physiol. Bohemoslov.*, 26, 353, 1977.
24. **Björkhem, I., Kallner, A., and Löf, A.**, The activity of HMG coenzyme A reductase in ascorbate-deficient and ascorbate supplemented guinea pigs, *IRCS Med. Sci.*, 6, 230, 1978.
25. **Ginter, E.**, Marginal vitamin C deficiency, lipid metabolism, and atherogenesis, *Adv. Lipid Res.*, 16, 167, 1978.
26. **McIntosh, G. H., Richmond, W., and Himsworth, R. L.**, Vitamin C deficiency and hypercholesterolaemia in marmoset monkeys, *Nutr. Rep. Int.*, 23, 237, 1981.
27. **John, T. M., George, J. C., Hilton, J. W., and Slinger, S. J.**, Influence of dietary ascorbic acid on plasma lipid levels in the rainbow trout, *Int. J. Vit. Nutr. Res.*, 49, 400, 1979.
28. **Kotzé, J. P., Weight, M. J., DeKlerk, W. A., Menne, I. V., and Weight, M. J. A.**, Effect of ascorbic acid on serum cholesterol levels and on die-away curves of ^{14}C-4-cholesterol in baboons, *S.A. Pr. Med. J.*, 48, 1182, 1974.
29. **McIntosh, G. H., Bulman, F. H., Illman, R. J., and Topping, D. L.**, The influence of age, dietary cholesterol and vitamin C deficiency on plasma cholesterol concentration in the marmoset, *Nutr. Rep. Int.*, 29, 673, 1984.
30. **Hodges, R. E., Hood, J., Canham, J. E., Sauberlich, H. E., and Baker, E. M.**, Clinical manifestations of ascorbic acid deficiency in man, *Am. J. Clin. Nutr.*, 24, 432, 1971.
31. **Duane, W. C. and Hutton, S. W.**, Lack of effect of experimental ascorbic acid deficiency on bile acid metabolism, sterol balance, and biliary lipid composition in man, *J. Lipid Res.*, 24, 1186, 1983.
32. **Heine, H. and Norden, C.**, Vitamin C therapy in hyperlipoproteinemia, *Int. J. Vit. Nutr. Res.*, Suppl. 19, 45, 1979.
33. **Horsey, J., Livesley, B., and Dickerson, J. W. T.**, Ischaemic heart disease and aged patients: effects of ascorbic acid on lipoproteins, *J. Human Nutr.*, 35, 53, 1981.
34. **Fidanza, A., Audisio, M., and Mastroiacovo, P.**, Vitamin C and cholesterol, *Int. J. Vit. Nutr. Res.*, Suppl. 23, 153, 1982.
35. **Degkwitz, E., Walsch, S., Dubberstein, M., and Winter, J.**, Ascorbic acid and cytochromes, *Ann. N.Y. Acad. Sci.*, 258, 201, 1975.
36. **Omaye, S. T. and Turnbull, J. D.**, Effects of ascorbic acid on heme metabolism in hepatic microsomes, *Life Sci.*, 27, 441, 1980.
37. **Zannoni, V. G., Holsztynska, E. J., and Lau, S. S.**, Biochemical functions of ascorbic acid in drug metabolism, *Adv. Chem. Ser.*, 200, 349, 1982.
38. **Kuenzig, W., Tkaczevski, V., Kamm, J. J., Coney, A. A., and Burns, J. J.**, The effect of ascorbic acid deficiency on extrahepatic microsomal metabolism of drugs and carcinogens in the guinea pig, *J. Pharmacol. Exp. Ther.*, 201, 527, 1977.
39. **Rikans, L. E., Smith, C. R., and Zannoni, V. G.**, Ascorbic acid and cytochrome P-450, *J. Pharmacol. Exp. Ther.*, 204, 702, 1978.
40. **Coon, M. J.**, Oxygen activation in the metabolism of lipids, drugs and carcinogens, *Nutr. Rev.*, 36, 319, 1978.

41. **Nordblom, G. D., White, R. E., and Coon, M. J.,** Studies on hydroperoxide-dependent substrate hydroxylation by purified liver, microsomal cytochrome P-450, *Arch. Biochem. Biophys.*, 175, 524, 1976.
42. **Rikans, L. E.,** NADPH-dependent reduction of cytochrome P-450 in liver microsomes from vitamin C-deficient guinea pigs: effect of benzphetamine, *J. Nutr.*, 112, 1796, 1982.
43. **Rikans, L. E. and Kobayashi, S.,** personal communication, 1983.
44. **Weis, W.,** Ascorbic acid and electron transport, *Ann. N.Y. Acad. Sci.*, 258, 190, 1975.
45. **Winterbourn, C. C.,** Hydroxyl radical production in body fluids, *Biochem. J.*, 198, 125, 1981.
46. **Rowley, D. A. and Halliwell, B.,** Formation of hydroxyl radicals from hydrogen peroxide and iron salts by superoxide-and ascorbate-dependent mechanism: relevance to the pathology of rheumatoid disease, *Clin. Sci.*, 64, 649, 1983.
47. **Scarpa, M., Stevanato, R., Viglino, P., and Rigo, A.,** Superoxide ion as active intermediate in the autoxidation of ascorbate by molecular oxygen. Effect of superoxide dismutase, *J. Biol. Chem.*, 258, 6695, 1983.
48. **Udenfriend, S., Clarck, C. T., Axelrod, J., and Brodie, B. B.,** Ascorbic acid in aromatic hydroxylation. I. A model system for aromatic hydroxylation, *J. Biol. Chem.*, 208, 731, 1954.
49. **Ishimitsu, S., Fujimoto, S., and Ohara, A.,** Studies on the hydroxylation of phenylalanine by ascorbic acid — hydrogen peroxide system, *Chem. Pharm. Bull.*, 29, 2089, 1981.
50. **Peterson, F. J., Holloway, D. E., Duquette, P. H., and Rivers, J. M.,** Dietary ascorbic acid and hepatic mixed function oxidase activity in the guinea pig, *Biochem. Pharmacol.*, 32, 91, 1983.
51. **Sutton, J. L., Basu, T. K., and Dickerson, J. W. T.,** Effect of pharmacological doses of ascorbic acid on the hepatic microsomal haemoproteins in the guinea-pig, *Br. J. Nutr.*, 49, 27, 1983.
52. **Kunert, K. J. and Tappel, A. L.,** The effect of vitamin C on in vivo lipid peroxidation in guinea pigs as measured by pentane and ethane production, *Lipids*, 18, 271, 1983.
53. **Perkins, R. C., Beth, A. H., Wilkerson, L. S., Serafin, W., Dalton, L. R., Park, C. R., and Park, J. H.,** Enhancement of free radical reduction by elevated concentrations of ascorbic acid in avian dystrophic muscle, *Proc. Natl. Acad. Sci. U.S.A.*, 77, 790, 1980.
54. **Schwartz, M. A. and Margolis, S.,** Effects of drugs and sterols on cholesterol 7α-hydroxylase activity in rat liver microsomes, *J. Lipid Res.*, 24, 28, 1983.
55. **Yau, T. M. and Mencl, J.,** A study of the peroxidation of fatty acid micelles promoted by ionizing radiation, hydrogen peroxide and ascorbate, *Int. J. Rad. Biol.*, 40, 47, 1981.
56. **Brogan, W. C., Miles, P. R., and Colby, H. D.,** Lipid peroxidation in adrenal and testicular microsomes, *Environ. Health Persp.*, 38, 105, 1981.
57. **Shimizu, K.,** Effects of ascorbic acid on the side-chain cleavage of cholesterol, *Biochim. Biophys. Acta*, 210, 333, 1970.
58. **Pintauro, S. J. and Bergan, J. G.,** Effects of ascorbic acid on in vitro steroidogenesis in guinea pigs, *J. Nutr.*, 112, 584, 1982.
59. **Heikkila, R. E., Cabbat, F. S., and Manzino, L.,** Inhibitory effects of ascorbic acid on the binding of (^{3}H) dopamine antagonists to neostriatal membrane preparations: relationship to lipid peroxidation, *J. Neurochem.*, 38, 1000, 1982.
60. **Kritchevsky, D., Nitzsche, C., Czarnecki, S. K., and Story, J. A.,** Influence of vitamin E supplementation on cholesterol metabolism in rats, *Nutr. Rep. Int.*, 22, 339, 1980.
61. **Kritchevsky, D., Tepper, S. A., and Story, J. A.,** Influence of vitamin C on hydroxylation and side chain oxidation of cholesterol in vitro, *Lipids*, 8, 482, 1973.
62. **Bissell, D. M. and Guzelian, P. S.,** Ascorbic acid deficiency and cytochrome P-450 in adult rat hepatocytes in primary monolayer culture, *Arch. Biochem. Biophys.*, 192, 569, 1979.
63. **Story, J. A.,** Dietary fiber and lipid metabolism, in *Medical Aspects of Dietary Fiber*, Spiller, G. A. and McPherson Kay, R., Eds., Plenum Press, New York, 1980, 137.
64. **Terpstra, A. H. M., Hermus, R. J. J., and West, C. E.,** Dietary protein and cholesterol metabolism in rabbits and rats, in *Animal and Vegetable Proteins in Lipid Metabolism and Atherosclerosis*, Alan R. Liss, New York, 1983, 19.
65. **Oakenfull, D. G. and Sidhu, G. S.,** A physico-chemical explanation for the effects of dietary saponins on cholesterol and bile salt metabolism, *Nutr. Rep. Int.*, 27, 1253, 1983.
66. **Story, J. A., LePage, S. L., Petro, M. S., West, L. G., Cassidy, M. M., Lightfoot, F. G., and Vahouny, G. V.,** Interactions of alfalfa plant and sprout saponins with cholesterol in vitro and in cholesterol-fed rats, *Am. J. Clin. Nutr.*, 39, 917, 1984.
67. **Packard, C. J. and Shepherd, J.,** The hepatobiliary axis and lipoprotein metabolism: effects of bile acid sequestrants and ileal bypass surgery, *J. Lipid Res.*, 23, 1081, 1982.
68. **Story, J. A., Tepper, S. A., and Kritchevsky, D.,** Influence of dietary alfalfa, bran, or cellulose on cholesterol metabolism in rats, *Artery*, 3, 154, 1977.
69. **O'Brien, B. C. and Reiser, R.,** Comparative effects of purified and human-type diets on cholesterol metabolism in the rat, *J. Nutr.*, 109, 98, 1979.

70. **Meuller, M. A., Cleary, M. P., and Kritchevsky, D.,** Influence of dietary fiber on lipid metabolism in meal-fed rats, *J. Nutr.*, 113, 2229, 1983.
71. **Vahouny, G. V., Roy, T., Gallo, L. L., Story, J. A., Kritchevsky, D., and Cassidy, M.,** Dietary fibers III. Effects of chronic intake on cholesterol absorption and metabolism in the rat, *Am. J. Clin. Nutr.*, 33, 2182, 1980.
72. **Morgan, B., Heald, M., Atkin, S. D., Green, J., and Chain, E. B.,** Dietary fibre and sterol metabolism in the rat, *Br. J. Nutr.*, 32, 447, 1974.
73. **Brydon, W. G., Tadesse, K., Eastwood, M. A., and Lawson, M. E.,** The effect of dietary fibre on bile acid metabolism in rats, *Br. J. Nutr.*, 43, 101, 1980.
74. **Sablé-Amplis, R. and Dupouy, D.,** Increase in cholesterol 7α-hydroxylase activity in hamsters fed with a fruit-supplemented diet, *IRCS Med. Sci.*, 11, 69, 1983.
75. **Selvendran, R. R.,** The binding of bile salts by vegetable fibre, *Qual. Plant.-Plant Foods. Hum. Nutr.*, 29, 109, 1979.
76. **Vahouny, G. V., Tombes, R., Cassidy, M. M., Kritchevsky, D., and Gallo, L. L.,** Dietary fibers V. Binding of bile salts, phospholipids and cholesterol from mixed micelles by bile acid sequestrants and dietary fibers, *Lipids*, 15, 1012, 1980.
77. **Chen, M. L., Chang, S. C., and Guoo, J. Y.,** Fiber contents of some chinese vegetables and their in vitro binding capacity of bile acids, *Nutr. Rep. Int.*, 26, 1053, 1982.
78. **Proia, A. D., McNamara, D. J., Edwards, K. D. G., and Anderson, K. E.,** Effects of dietary pectin and cellulose on hepatic and intestinal mixed-function oxidations and hepatic 3-hydroxy-3-methylglutaryl-coenzyme A reductase in the rat, *Biochem. Pharmacol.*, 30, 2553, 1981.
79. **Quazi, S., Takahata, M., Horio, F., and Yoshida, A.,** Hepatic 3-hydroxy-3-methylglutaryl coenzyme A reductase and cholesterol 7α-hydroxylase activities in rats fed PCB, *Nutr. Rep. Int.*, 30, 617, 1984.
80. **Sirtori, C. R., Galli, G., Lovati, M. R., Carrara, P., Bosisio, E., and Kienle, M. G.,** Effect of dietary proteins on the regulation of liver lipoprotein receptors in rats, *J. Nutr.*, 114, 1493, 1984.
81. **Hassan, A. S., Gallon, L. S., Zimmer, L. A., Balistreri, W. F., and Subbiah, M. T. R.,** Persistent enhancement of bile acid synthesis in guinea pigs following stimulation of cholesterol catabolism in neonatal life, *Steroids*, 38, 477, 1981.
82. **Subbiah, M. T. R., Deitmeyer, D., and Yunker, R. L.,** Decreased atherogenic response to dietary cholesterol in pigeons after stimulation of cholesterol catabolism in early life, *J. Clin. Invest.*, 71, 1509, 1983.
83. **Ginter, E., Bobek, P., Ozdin, L., Kubec, F., and Vozar, J.,** Synergism between ascorbic acid and bile acid binding agents, *Adv. Physiol. Sci.*, 35, 67, 1981.
84. **Mayer, G. and Mayer, D.,** Secretion of bile acids and cholesterol-7α-hydroxylase activity in dependence of the fatty acid composition in rats on a fat-rich diet, in *Advances in Bile Acid Research*, Matern, S., Hackenschmidt, J., Back, P., and Gerok, W., Eds., F. K. Schattauer Verlag, New York, 1975, 443.
85. **Ramesha, C. S., Paul, B., and Ganguly, J.,** Effect of dietary unsaturated oils on the biosynthesis of cholesterol, and on biliary and fecal excretion of cholesterol and bile acids in rats, *J. Nutr.*, 110, 2149, 1980.
86. **Kritchevsky, D., Tepper, S. A., and Story, J. A.,** Influence of dietary fats on lipid metabolism in rats fed a semipurified diet, *Nutr. Rep. Int.*, 15, 65, 1977.
87. **Björkhem, I., Blomstrand, R., and Svensson, L.,** Effect of different dietary triglycerides on 7α-hydroxylation of cholesterol and other mixed-function oxidations, *J. Lipid Res.*, 19, 359, 1978.
88. **Sugano, M., Ryu, K., and Ide, T.,** Cholesterol dynamics in rats fed cis- and trans-octadecenoate in the form of triglyceride, *J. Lipid Res.*, 25, 474, 1984.
89. **Ruttenberg, H., Davidson, L. M., Little, N. A., Klurfeld, D. M., and Kritchevsky, D.,** Influence of trans unsaturated fats on experimental atherosclerosis in rabbits, *J. Nutr.*, 113, 835, 1983.
90. **Lefevre, A. F., DeCarli, L. M., and Lieber, C. S.,** Effect of ethanol on cholesterol and bile acid metabolism, *J. Lipid Res.*, 13, 48, 1972.
91. **Lakshmanan, M. R. and Veech, R. L.,** Short- and long-term effects of ethanol administration in vivo on rat liver HMG-CoA reductase and cholesterol 7α-hydroxylase activities, *J. Lipid Res.*, 18, 325, 1977.
92. **Crouse, J. R. and Grundy, S. M.,** Effects of alcohol on plasma lipoproteins and cholesterol and triglyceride metabolism in man, *J. Lipid Res.*, 25, 486, 1984.
93. **Mosbach, E. H.,** Regulation of bile acid synthesis, in *Bile Acids in Human Diseases*, Back, P. and Gerok, W., Eds., F. K. Schattauer Verlag, New York, 1972, 89.
94. **Shefer, S., Hauser, S., Lapar, V., and Mosbach, E. H.,** Regulatory effects of sterols and bile acids on hepatic 3-hydroxy-3-methylglutaryl CoA reductase and cholesterol 7α-hydroxylase in the rat, *J. Lipid Res.*, 14, 573, 1973.
95. **Singhal, A. K., Cohen, B. I., Finver-Sadowsky, J., McSherry, C. K., and Mosbach, E. H.,** Role of hydrophilic bile acids and of sterols on cholelithiasis in the hamster, *J. Lipid Res.*, 25, 564, 1984.
96. **Kritchevsky, D., Tepper, S. A., Morrisey, R., and Klurfeld, D.,** Influence of garlic oil on cholesterol metabolism in rats, *Nutr. Rep. Int.*, 22, 641, 1980.

Chapter 9

PHARMACOLOGICAL CONTROL OF CHOLESTEROL 7α-HYDROXYLASE ACTIVITY

R. Fears

TABLE OF CONTENTS

I. INTRODUCTION

The study of the regulation of enzyme activity by drugs may be worthwhile for several reasons. First, pharmacological intervention, whether inhibition or enhancement, can provide useful information about the enzyme, e.g., regarding the role of co-factors and the nature of the cellular environment, thus contributing to the basic understanding of the biochemistry and physiology of the catalytic mechanism. Secondly, variation in the reaction rate of an individual enzyme in response to a synthetic compound or isolated natural product could have important clinical consequences. The net effect may be therapeutic by design, or, if contingent on attempting some other action the outcome could be either beneficial or deleterious. Each of these aspects can be illustrated in a review of the pharmacological control of cholesterol 7α-hydroxylase (7α-monooxygenase, EC 1.14.13.17), the enzyme usually believed to be rate-limiting for bile acid synthesis.

The pharmacological control of cholesterol 7α-hydroxylase is now recognized as clinically valuable for it has become clear that an elevation in activity can lead both to a lowering of the circulating cholesterol concentration and to a depletion of certain tissue sterol pools. This action is favorable for there is considerable epidemiological and experimental evidence supporting a contribution by hyperlipidemia and particularly by low-density lipoprotein (LDL) cholesterol to the pathogenesis of atheroma and coronary heart disease (CHD).[1] While the use of drugs and modified diets to influence lipoprotein levels has had a long and partly disappointing history, there is now unequivocal evidence to show that correction of dyslipoproteinemia does retard and reverse atherosclerosis with a concomitant decline in morbidity and mortality.[1] The activation of cholesterol 7α-hydroxylase has been cardinal to some of the most successful of these attempts.

Conversely, pharmacological inhibition of the enzyme contributes to the iatrogenic hypercholesterolemia observed during therapy with chenodeoxycholic acid for the dissolution of cholesterol gallstones. While 7α- and 12α-hydroxylase activities may already be low in gallstone patients,[2] administration of exogenous bile acid further inhibits cholesterol 7α-hydroxylation.[2,3] The accumulating hepatic sterol represses cholesterol biosynthesis,[3] but LDL high-affinity receptor synthesis may also be suppressed. The resultant rise in circulating LDL cholesterol is one of the most significant side-effects of chenodeoxycholic acid therapy and has contributed to the increased use of an alternative bile acid, ursodeoxycholic acid, that may act by a different mechanism.

II. LIPID-LOWERING AGENTS THAT ACT ON CHOLESTEROL 7α-HYDROXYLASE

A. General Mechanism of Action

Unless the cholesterol is merely to be transferred to another tissue pool the deposition of sterol in an atheromatous plaque can only be prevented or reversed by altering the net cholesterol balance of the body. Theoretically, this can be achieved by one of several mechanisms (Table 1). Catabolism to bile acids accounts for up to 60% of the cholesterol output in man[4] and although this may be a smaller proportion than in some laboratory animal species the enterohepatic recirculation is susceptible to intervention and such an approach may not be associated with some of the potential problems arising from the other traditional approaches to lowering cholesterol levels (Table 1). Although recent studies in man suggest that a small proportion of cholic acid can be obtained by a pathway that does not involve an initial 7α-hydroxylation of cholesterol,[5] it remains likely that the enzyme cholesterol 7α-hydroxylase is the rate-limiting step for the major part of the bile acid pool, as discussed elsewhere in this volume (see Chapters 2 and 3).

In general, the economy of the enterohepatic circulation seems to be dependent on the

Table 1
PRINCIPAL METHODS TO LOWER THE CIRCULATING CONCENTRATION OF CHOLESTEROL

General approach	Specific target	Drug	Possible disadvantages of this mechanism
Inhibition of absorption	Micellar stability	Neomycin	Endogenous biosynthesis may rise to compensate
	Sterol receptor in small intestine	β-Sitosterol	
Inhibition of synthesis	Provision of 2C units	Nicotinic acid	
	HMG-CoA reductase	Compactin, mevinolin	Depletion of other isoprenoids
Enhancement of sterol excretion	Various	Clofibrate	Supersaturation of bile with cholesterol → gallstones
Enhancement of bile acid excretion	Cholesterol 7α-hydroxylase		See text for details

Table 2
PHARMACOLOGICAL MEANS TO ACTIVATE CHOLESTEROL 7α-HYDROXYLASE

1. Derepression secondary to increased fecal bile acid excretion.
2. Alteration of phosphorylation/dephosphorylation status.
3. Modification of endoplasmic reticulum microenvironment.
4. Induction of cytochrome P-450.
5. Variation in origin or rate of supply of substrate.

interrelationship between bile acid input, cycling frequency, and intestinal loss from the pool,[6] so stimulation of bile acid synthesis will only be clinically useful if accompanied by a net increase in fecal excretion, e.g., as achieved by bile acid sequestration using the anion-exchange resins. While the enterohepatic recirculation of bile acids is normally efficient, ileal reabsorption is controlled by saturable high-affinity receptor-mediated binding and this rate may already be at Vmax at the usual ileal concentration of bile acid[7] so that it is reasonable to expect augmented synthesis to result in increased disposal. It should, however, be appreciated that if induction of 7α-hydroxylase activity elevates biliary bile acid concentration, then the efficiency of cholesterol absorption may be improved for this is proportional to cholic acid flux. Furthermore, any mechanism to increase enzyme activity that did not depend on derepression arising from promotion of fecal loss of bile acids must also be able to persist in the presence of an amplified feedback inhibition. Because of the tendency to homeostasis and because of the various methodological problems that have been identified in the assay of the enzyme (Chapter 4) changes in activity should not be accepted as physiologically significant without evidence for associated changes in formation of product or turnover of substrate pool. Taking account of these admonitions, a pharmacological variation of cholesterol 7α-hydroxylase activity might be achieved by one of several distinct mechanisms (Table 2) although, ultimately, each mechanism is dependent on changes either in enzyme synthesis or catabolism or a post-translational modification, e.g., activation of inactive precursor form. Recent research suggests that the enzyme may be stimulated by several incompletely characterized factors including a cytosolic protein[8] (perhaps facilitating reduction of protein disulfide bonds) and, in diabetic rats,[9] a circulating metabolite. It is not known whether these natural controls could be subject to pharmacological regulation so they will not be discussed further.

Derepression of the enzyme following sequestration of bile acids in the gut lumen has been the most successful therapeutic approach and the consequences will be described in detail for cholestyramine and colestipol (Section II.B). Net enzyme activity can also be raised by increasing that proportion phosphorylated (see Chapter 7), but hormones rather than pharmacological agents are the only regulators of the phosphorylation/dephosphorylation status known at present. Cholesterol 7α-hydroxylase is a microsomal enzyme and thus might be responsive to changes in the membrane microenvironment, e.g., fluidity, but the evidence available from Arrhenius plots[10] suggests there may be little such interaction by contrast with the well-characterized control for hepatic β-hydroxy-β-methylglutaryl-CoA (HMG-CoA) reductase. Further investigation of the possible influence of membrane fluidity or structural changes might be valuable. Modern analytical techniques such as electron spin resonance and fluorescence polarization can compare changes in mobility within the different layers of a membrane and standard agents are available that promote (benzyl alcohol) or restrict (dimethyl nitrosamine) fluidity. Whether or not cholesterol 7α-hydroxylase activity does respond to pharmacological perturbations in membranes, membrane phospholipid changes may explain the decrement in activity of this enzyme — in common with other P-450-requiring enzymes — in obesity.[11] The role of cytochrome P-450 in the control of the enzyme (a monooxygenase) is discussed extensively in Chapters 5 and 6, but there are therapeutic implications and these will be illustrated by the standard inducer SKF 525A (Section III). Finally (Table 2), the response of cholesterol 7α-hydroxylase to changes in substrate supply has been of use in defining the physiological origin of the substrate pool; in the future this aspect may be particularly relevant to an increased understanding of the function of high-density lipoprotein (HDL) a putative negative risk factor for atherosclerosis and CHD,[1] and a target for current attempts to correct dyslipoproteinemias.

B. Cholestyramine, Colestipol, and Other Bile Acid-Binding Agents

The mechanism of action of the various anion-exchange resins is similar and depends on the binding of bile acids in the lumen of the small intestine so that fecal disposal is increased with a consequent derepression of bile acid synthesis, mediated by activation of cholesterol 7α-hydroxylase.[12] In response to the accelerated catabolism, part of the homeostatic mechanism for the repletion of hepatic sterol pools is the induction of LDL (apoprotein B) high-affinity receptor synthesis so that the LDL fractional catabolic rate is increased. The biochemistry of this response has received detailed attention[13] and is reviewed in Chapter 11; it is the biological consequences that will be discussed here, together with the implications for the next generation of hypolipidemic agents. The recent results from the Lipid Research Clinics multicenter, randomized, double-blind trial with cholestyramine for asymptomatic type II hyperlipidemics[14] have demonstrated — perhaps unambiguously for the first time — that LDL-lowering can lead to amelioration of CHD (Table 3). The changes can be dramatic: for those subjects taking the recommended dose (24 g daily) the reduction of LDL cholesterol, 35%, was associated with 49% decrease in CHD.[15] The cholesterol-lowering response to colestipol has also been related to clinical endpoints in a combined primary and secondary prevention trial[16] although the trial design has been criticized.[1] A long-term, uncontrolled study[17] related LDL-lowering to biological endpoints, in particular regression of xanthoma and improvements in arterial function (Table 3). The progression of coronary atherosclerosis can certainly be retarded,[17,18] regression may be possible.[1]

A compensatory rise in hepatic lipogenesis sometimes also occurs during therapy with the resins. Any increase in triglyceride synthesis can usually be controlled by diet, but the rise in cholesterogenesis may be sufficient to diminish the clinical response so that large doses of agent are usually thought to be necessary for efficacy. There have been various attempts to limit the increase in cholesterol biosynthesis by co-administering a second lipid-lowering drug (Table 4). Patients with heterozygous familial hyperlipidemia (FH, type II)

Table 3
PRINCIPAL CLINICAL STUDIES USING ANION EXCHANGE RESINS

Cholestyramine: copolymer of styrene and divinylbenzene
Colestipol: copolymer of diethylene triamine and 1-chloro-2,3-epoxypropane

Study design	Lipid response	Effect on clinical and biological endpoints
Cholestyramine: 1,905 hyperlipidemic men received up to 24 g/day for 7 years (placebo-controlled)[14,15]	Plasma total cholesterol ↓ 9%; LDL cholesterol ↓ 13%; response proportional to dose	CHD death ↓ 24% Nonfatal MI ↓ 19%
Colestipol: 600 men, 600 women (combined primary and secondary prevention trial) received 15 g/day for up to 3 years (placebo-controlled)[16]	Total cholesterol ↓ 15%	Total and CHD mortality ↓ in men <50 years old
Colestipol: 25 familial hypercholesterolemics received 30 g/day for 7 years (uncontrolled)[17]	Total cholesterol ↓ 35%; LDL cholesterol ↓ 45%	Resolution of xanthoma; improved angina; stabilized atheroma
Colestipol: 25 high-risk subjects received 20—30 g/day for 2 years (placebo-controlled using nonresponders)[18]	Total cholesterol ↓ 21%	Retardation in progression of coronary atherosclerosis

have been treated with high doses of colestipol plus nicotinic acid for up to 9 years and it is agreed[19-21] that the net decrease in total and LDL cholesterol and triglyceride, and the rise in HDL (assumed beneficial) are clearly superior to the effect of colestipol alone. This normalisation of LDL cholesterol concentration leads to regression of xanthoma[21] and cessation of coronary atherogenesis.[20] A current secondary prevention study with men following coronary bypass surgery,[22] is yielding similar results (e.g., LDL cholesterol ↓ 33%, apoprotein B ↓ 31%, HDL cholesterol ↑ 33% at one year) even though the starting LDL levels were low.

The combination of resin with the anti-lipolytic agent nicotinic acid is logical for the latter should act to decrease fatty acid flux to the liver and hence curtail provision of substrate for lipogenesis (Table 1). Direct proof for this interaction, which may be synergistic,[1] has been obtained recently.[23] When cholesterogenesis was monitored in mononuclear leukocytes (considered representative of other tissues) from patients with heterozygous FH, the rate was increased 40% by colestipol alone (20 to 30 g daily) but returned to the basal level on co-administering nicotinic acid (3 to 8 g daily).

While this combination of drugs represents the best existing treatment for heterozygous FH, the high doses used are unpalatable and the success of the therapy is limited by patient compliance. One solution may be to use compounds that are much more effective than nicotinic acid as inhibitors of cholesterogenesis. Compactin and mevinolin are specific inhibitors of HMG-CoA reductase[24] (Table 1) and compactin has been used with cholestyramine in a well-tolerated, low-dose combination[25] to achieve excellent clinical results (Table 4). It may also be possible to achieve a more efficient binding of bile acid, and hence derepression of cholesterol 7α-hydroxylase, for the particulate resins act at only approximately 2% of their theoretical binding capacity. Other, lower molecular weight, soluble resins have been examined in animal models.[26] For example, poly-[(dimethylimino)-tri-

Table 4
RESIN COMBINATION THERAPY IN HETEROZYGOUS FAMILIAL HYPERCHOLESTEROLEMICS

Study design	Clinical response
30 g Colestipol + 3—7 g nicotinic acid for up to 8 years (n = 32)[19]	Compared to diet alone: total cholesterol ↓ 34%; triglycerides ↓ 34%; LDL cholesterol ↓ 45%; HDL cholesterol ↑ 22%. Progression of coronary atherosclerosis arrested
20 g Colestipol + 3—8 g nicotinic acid for up to 4 years (n = 13)[20]	Compared to diet alone: total cholesterol ↓ 40%; triglycerides ↓ 35%; LDL cholesterol ↓ 47%; HDL cholesterol ↑ 32%.
30 g Colestipol + up to 7.5 g nicotinic acid for 8 months (n = 25)[21]	Compared to diet alone: total cholesterol ↓ 45%; triglycerides ↓ 35%; LDL cholesterol ↓ 55%; VLDL cholesterol ↓ 50%; HDL cholesterol ↑ 15%. Xanthoma ↓ 6%.
12 g Cholestyramine + 90 mg compactin for 12 weeks (n = 10)[25]	Compared to diet alone: total cholesterol ↓ 39%; LDL cholesterol ↓ 53%.

methylene chloride] (MW 37,000) and poly-diallyldimethyl-ammonium chloride (MW 200,000) have superior in vitro bile acid-binding properties to cholestyramine and decreased cholesterol levels in dogs at a lower dose. DEAE-Sephadex also has a higher binding capacity for cholic acid in vitro, but the clinical results have been relatively modest.[1] The soluble polycationic derivative DEAE-Dextran (MW 500,000) may be more useful for it has been used with success in much smaller amounts (3 g daily) and seems to be well-tolerated even on long-term administration.[27] Certainly, there is considerable potential for an agent that binds bile acids efficiently, particularly if triglyceride levels are also lowered,[27] for then less attention need be paid to diet and therapy can be extended to type IV hyperlipidemics.

Some types of dietary fiber can lower the serum concentration of VLDL and LDL cholesterol. This action is also mediated by bile acid-binding and interruption of the enterohepatic recirculation with derepression of cholesterol 7α-hydroxylase. A comprehensive discussion of the properties of dietary fiber is beyond the scope of this review, but it is worth noting that it is the saponin content that may be responsible for the sequestration of bile acids. One of the most potent saponin preparations in this respect[28] is obtained from Quillaia, a permitted food additive for stabilizing ice cream and ginger beer foam. In studies on hyperlipidemic rats, fed a semisynthetic diet (Table 5), a large dietary supplement of Quillaia decreased the cholesterol concentration in the liver and in serum, the principal effect being exerted on LDL.

Table 5
EFFECT OF QUILLAIA SAPONINS ON CHOLESTEROL LEVELS IN RATS

	Control[a]	*Quillaia saponins*[a] (1000 mg/kg body weight)
Gain in body weight (g)	55 ± 2.0	49 ± 1.8
Serum cholesterol (mg/100 mℓ):		
Total	137 ± 7.5	112 ± 4.3[b]
VLDL	46 ± 8.2	32 ± 2.8
LDL	34 ± 3.4	17 ± 0.9[b]
HDL	57 ± 2.1	62 ± 1.8
Liver cholesterol (mg/g)[c]	9.1	6.7

[a] Mean values ± standard error for 8 rats maintained on semisynthetic diet for 7 days.
[b] $p < 0.05$.
[c] Pooled samples.

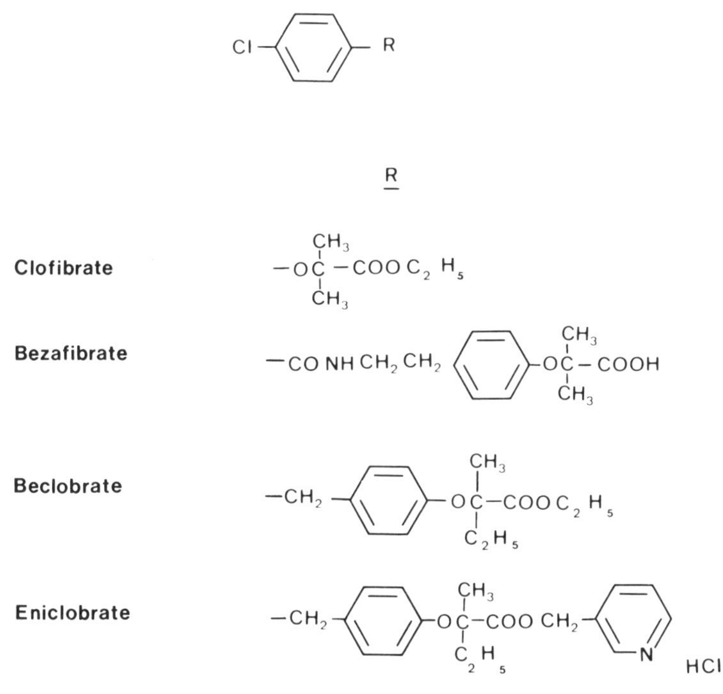

FIGURE 1. Clofibrate and its analogues.

C. Other Commercial Hypocholesterolemic Agents

There is no conclusive evidence that an effect on cholesterol 7α-hydroxylase contributes to the mode of action of the other principal therapeutic agents. While clofibrate (Figure 1) may increase hepatic cytochrome P-450 levels and has been reported to elevate 12α-hydroxylation and lithocholate 6β-hydroxylation in rats,[29] the earlier findings of augmented 7α-hydroxylation have not been confirmed. The newer analogue bezafibrate (Figure 1) increases high affinity receptor-mediated LDL catabolism in patients,[30] but it has not yet been established whether this action is translated into an amplification of bile acid synthesis. Results for two other new analogues, beclobrate and eniclobrate (Figure 1), have recently

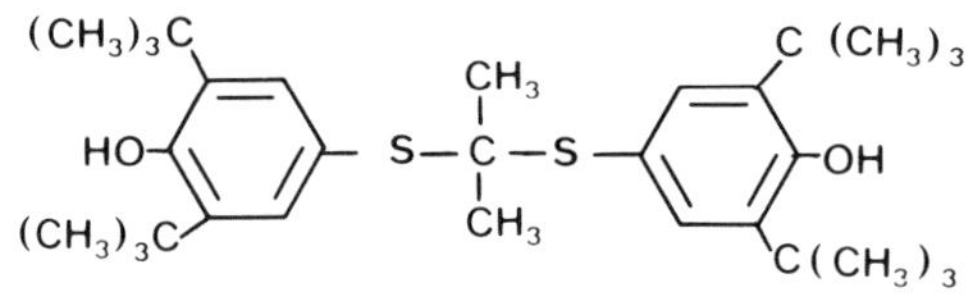

FIGURE 2. Probucol.

been published.[31] When given to rats by daily oral dosing, eniclobrate (12 mg/kg) significantly decreased cholesterol 7α-hydroxylase activity (by 40%) with a smaller decrement (15%) in fecal bile acid excretion. In a direct comparison, beclobrate neither decreased enzyme activity nor steroid excretion; for both compounds the mechanism of cholesterol-lowering remains to be established.

Several studies have indicated that probucol (Figure 2) increases the fecal elimination of bile acids[32] in conjunction with promotion of LDL turnover and possibly as a result of impaired bile acid absorption. A specific effect on cholesterol 7α-hydroxylase has not been substantiated and other actions, e.g., on lipoprotein synthesis, have been discussed.

III. ENZYME INDUCTION

The various classes of drugs that act to induce cytochrome P-450 levels do not exert a consistent effect on cholesterol 7α-hydroxylase activity. The discrimination between the response to the different types of monooxygenase inducer has been useful in aiding our understanding of the mechanism of 7α-hydroxylation (Chapter 6). Of particular relevance to the present discussion is the observation that administration of SKF 525A to rats consistently enhanced cholesterol 7α-hydroxylase activity (Table 6) although other effects on cholesterol metabolism, e.g., hepatic synthesis, were influenced by the nature of the dietary regimen. LDL levels were lowered selectively and, as for the resins, this action might be taken as evidence for an induction of hepatic apoprotein B receptor synthesis.[33] Because of a multiplicity of effects on intermediary metabolism, however, and the possibility of toxicity on chronic administration, SKF 525A is no longer of direct interest as a lipid-lowering agent.

Any enhancement of cholesterol 7α-hydroxylase activity by other inducers of cytochrome P-450 has not always been so clearly related to lipid-lowering, although this correlation has also been made for 16α-isothiocyanopregnenolone-3-acetate (Figure 3), which stimulated 7α-hydroxylation and normalized total and LDL cholesterol in rats given an atherogenic diet.[34]

The identification of the specific fraction of cytochrome P-450 responsible for 7α-hydroxylation (Chapter 5) suggests that it may be feasible to seek more selective inducers. It is known that 7α-hydroxylation can be stimulated when other P-450-dependent enzymes are inhibited, e.g., after subcutaneous estradiol administration to rats.[35] It is interesting to note that not only does this action of estradiol on cholesterol catabolism appear synergistic with cholestyramine,[35] but also that estrogens can increase hepatic LDL receptor activity in rats.[36] The association of receptor synthesis with activation of cholesterol 7α-hydroxylase that is found for the resins, SKF-525A, and the estrogens is convincing evidence that a primary action on cholesterol catabolism is an effective stimulus for increased LDL turnover. Estrogens also lower LDL (and raise HDL) in women, but they are no longer used therapeutically to modify lipoprotein levels because of the side-effects disclosed in the Coronary Drug Project trial.

Table 6
EFFECT OF SKF 525A ON CHOLESTEROL METABOLISM IN RATS

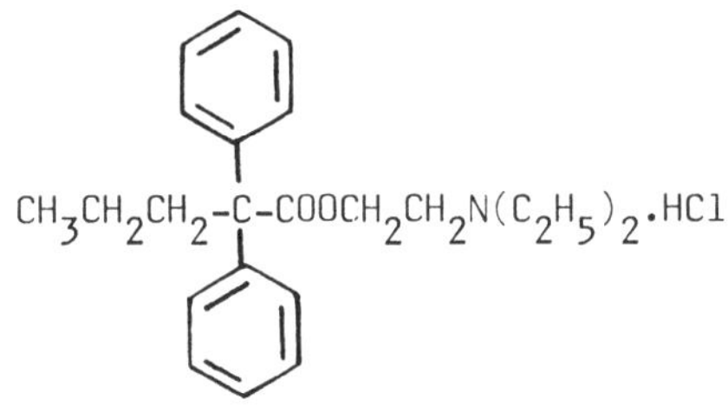

	Normolipidemic[a]	Hyperlipidemic[a]
Serum cholesterol (mg/100 mℓ)		
Total	−4	−22[b]
LDL	−45[b]	−40[b]
HDL	−4	−4
Hepatic cholesterol metabolism (nmol/min/g)		
7α-Hydroxylase	+98[b]	+160[b]
Cholesterogenesis	+80[b]	−2

[a] Results are expressed as % change from a comparable control group (8 rats per group). SKF 525A (50 mg/kg body weight) was administered by i.p. injection (2 mℓ/kg) once daily for 2 days and rats were killed 2 hr after the second dose.

[b] $p < 0.05$.

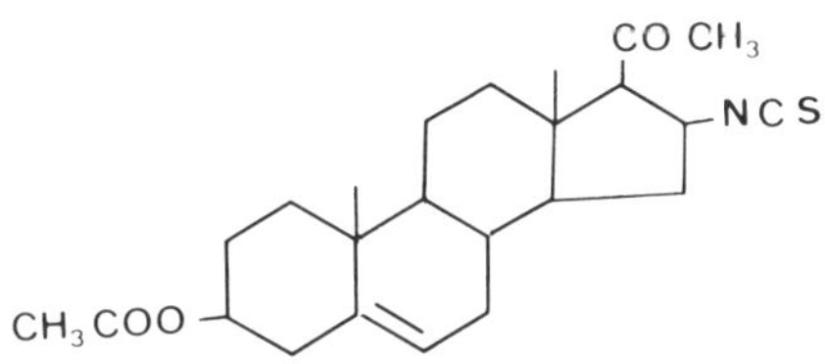

FIGURE 3. 16α-Isothiocyanopregnenolone 3-acetate.

IV. CHOLERETICS

The choleretic agent Bay a 4798 (Figure 4) increased cholesterol 7α-hydroxylase activity in rats when given by intraperitoneal injection (100 mg/kg body weight) for 5 days.[37] The biliary secretion of bile acids was increased and cholesterol decreased when Bay a 4798 was administered by intraduodenal infusion: these changes are likely to affect the physiochemical properties of the bile sufficiently to account for the observed inhibition of experimental gallstone formation. Other choleretics may not act in this manner. For example, dihydroxy-dibutyl ether exerted minimal effect on cholesterol in rats.[38] The proprietory choleretic Rowachol, a mixture of six monoterpenes, appeared to prevent gallstone formation by inhibiting HMG-CoA reductase,[39] that is, diminishing cholesterol production rather than amplifying catabolism.

V. MISCELLANEOUS EXPERIMENTAL AGENTS

4-*O*-Methylascochlorin, a modified fungal prenylphenol, decreased serum cholesterol levels when given as an oral dose (up to 100 mg/kg body weight) to rats.[40] The biliary

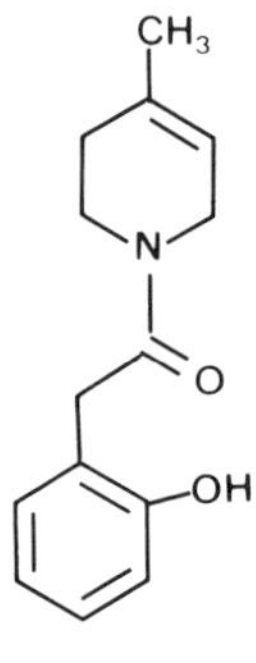

FIGURE 4. Bay a 4798.

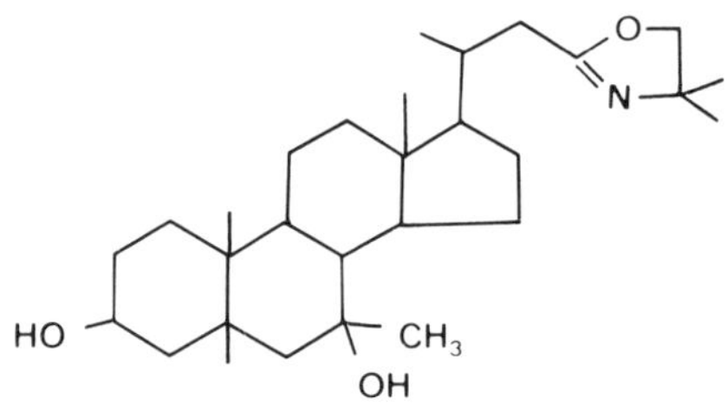

FIGURE 5. Oxazoline.

excretion of cholic and chenodeoxycholic acids in bile duct-cannulated rats was enhanced and cholesterol 7α-hydroxylase activity was raised significantly, if modestly (by up to 30%). It was claimed[40] that 4-*O*-methylascochlorin affected the specific activity of the enzyme without influencing the rate of new protein synthesis but the mechanism has not been elucidated. Cholesterogenesis was not induced in these experiments, presumably because the extent of the change in cholesterol catabolism was insufficient to cause a homeostatic reaction.

A rise in HMG-CoA reductase activity did accompany the marked increment in 7α-hydroxylation observed for the synthetic bile acid "oxazoline" (2-[3α-7ξ-dihydroxy-7ξ-methyl-24-nor-5β-cholanyl]-4,4-dimethyl-2-oxazoline) (Figure 5) given as a supplement in the diet (0.1% for 2 weeks) to rats.[41] There was no corresponding change in bile acid output although the composition of the bile acid pool was altered. Surprisingly, the concentration of serum cholesterol was raised but the study was too small to be definitive.

S-Methylmethionine sulfonium chloride increased 7α-hydroxylation by up to 80% in mice after oral administration (2 g/kg body weight for 7 days)[42] and this response was related to increased fecal excretion of bile acids and to hypocholesterolemic activity in parallel studies.

VI. USE OF PHARMACOLOGICAL AGENTS AS TOOLS TO STUDY ENZYME CHARACTERISTICS

Mention has already been made of how the drug inducers of cytochrome P-450 can be used to study the role of the coenzyme in the maintenance of enzyme activity and how the anion-exchange resins have provided detail on the interrelationship between the activities of cholesterol 7α-hydroxylase, HMG-CoA reductase, and the LDL receptor in the homeostatic regulation of intermediary metabolism and cell function. It is now becoming clear that pharmacological agents can also contribute to the identification of the origin of the cholesterol substrate for 7α-hydroxylation (other aspects concerning the nature of the physiological substrate are discussed in Chapter 4).

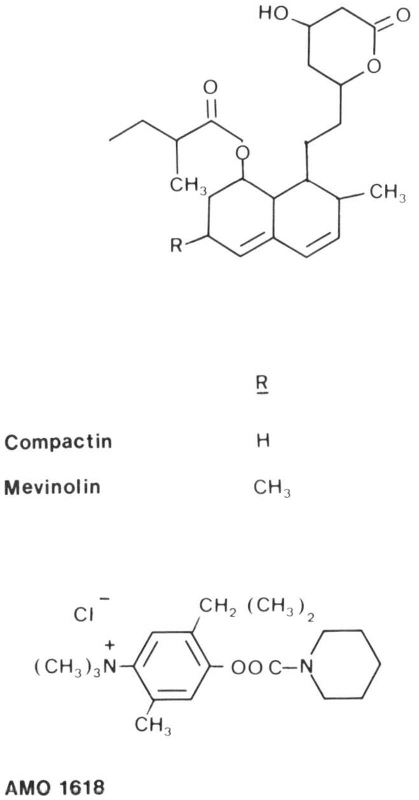

FIGURE 6. Inhibitors of HMG-CoA reductase.

A. Role of Newly Synthesized Cholesterol as Substrate

The evidence from turnover studies indicates that newly synthesized hepatic cholesterol might be the preferred substrate for cholic acid production in rats,[43-45] but that chenodeoxycholic acid can be formed from another discrete pool[43] that is possibly derived from plasma cholesterol.[45]

The role of newly synthesized cholesterol as a substrate for the 7α-hydroxylase can now be evaluated using the specific inhibitors of HMG-CoA reductase, compactin, and mevinolin (Figure 6), providing, of course, that bile acid synthesis is studied in experimental systems where cholesterogenesis occurs. Compactin has no direct action on cholesterol 7α-hydroxylase in either rat or dog microsomes.[24]

Inhibition of cholesterol biosynthesis in cultured rat hepatocytes (Table 7) impaired the synthesis of both cholic and β-muricholic (formed in vitro by 6-hydroxylation of chenodeoxycholic) acids.[46,47] Using compactin, there was relatively greater inhibition of cholic acid production[46] and this observation is consistent with the concept of partly separate substrate pools.[43,45] The relatively greater inhibition of β-muricholic acid output in another experiment[47] is more difficult to explain, but may be related to the sustained presence of very high levels of the more potent inhibitor, mevinolin. In studies ex vivo in rats (Table 7), a moderate daily dose of compactin had little effect on the fecal excretion of total bile acids,[48] but a much larger dose significantly decreased excretion in association with significant inhibition of cholesterol 7α-hydroxylase.[49] When administered to hamsters receiving a lithogenic diet, compactin suppressed 7α-hydroxylase activity to barely detectable levels[50] and HMG-CoA reductase activity and bile acid flux were moderately inhibited. By contrast, compactin promoted bile acid excretion in dogs[51] and cynomolgus monkeys[52] and this re-

Table 7
EFFECT OF INHIBITORS OF HMG-CoA REDUCTASE ON CHOLESTEROL 7α-HYDROXYLASE ACTIVITY AND BILE ACID TURNOVER

Type of study	Cholesterol 7α-hydroxylase activity	Bile acid concentration
1. Isolated rat hepatocytes	—	1 Hr incubation:
1.1 12 μM compactin (cholestyramine pretreatment)[46]		Cholic acid ↓ 25%
		β-Muricholic acid 0%
		2 Hr incubation:
		Cholic acid ↓ 71%
		β-Muricholic acid ↓ 52%
1.2 120 μM mevinolin[47]	—	Cholic acid ↓ 20%
		β-Muricholic acid ↓ 40%
2. Oral dosing		
2.1 Normolipidemic rat (no change in serum cholesterol):		
50 mg/kg body wt for 7—28 days[48]	—	Fecal bile acids ↓ 20%
500 mg/kg body wt for 11 days[49]	↓ 31%	Fecal bile acids ↓ 53%
2.2 Hypercholesteroemic hamster: 100 mg/kg body wt for 3 weeks (plasma cholesterol ↓ 15%)[50]	↓ 100%	Biliary bile acids ↓ 35%
2.3 Normolipidemic dog: 100 mg/kg body wt for 5 weeks (serum cholesterol ↓ 45%)[51]	↑ 15%	Fecal bile acids ↑ 80%
2.4 Normolipidemic cynomolgus monkey: 50 mg/kg body wt for 11 days (serum cholesterol ↓ 26%)[52]	—	Fecal bile acids ↑ 40%

sponse was greater than can be accounted for by the minor change in 7α-hydroxylase activity (Table 7).

It seems reasonable to infer from the dosing studies on rats and hamsters and from the work using cultured cells that a marked inhibition of HMG-CoA reductase will lead to inhibition of cholesterol 7α-hydroxylase so confirming that newly synthesized sterol is an important physiological substrate for bile acid, particularly cholic acid, synthesis. A coordinated pharmacological control of the two rate-limiting enzymes has also been observed for the hypolipidemic agent AMO 1618 (2-iso-propyl-4-dimethylamino-5-methylphenyl-1-piperidine carboxylate methyl chloride) (Figure 6) in rats and chickens.[53] It should be appreciated, however, that the interpretation of the response to a compound like compactin can be complicated for the inhibitor acts also to induce synthesis of HMG-CoA reductase, and other enzymes, and this additional activity is expressed when the cellular concentration of inhibitor is diminished by metabolism.[24] There may be an alternative explanation though for the unexpected response by dogs and monkeys (Table 7). It is conceivable that severe inhibition of HMG-CoA reductase in the small intestine could impair absorptive function so that the enterohepatic recirculation is interrupted and fecal loss of bile acids contributes to the observed hypocholesterolemia.[24]

B. Role of Lipoprotein Cholesterol as Substrate

Although hepatic synthesis is unlikely to be the only source of substrate for cholesterol 7α-hydroxylation, it has not been established whether there is a preferential circulating lipoprotein donor. The correlation between the activities of cholesterol 7α-hydroxylase and the hepatic LDL receptor for both man and laboratory animals suggests that circulating LDL could provide substrate for bile acid synthesis but there is little evidence for a common,

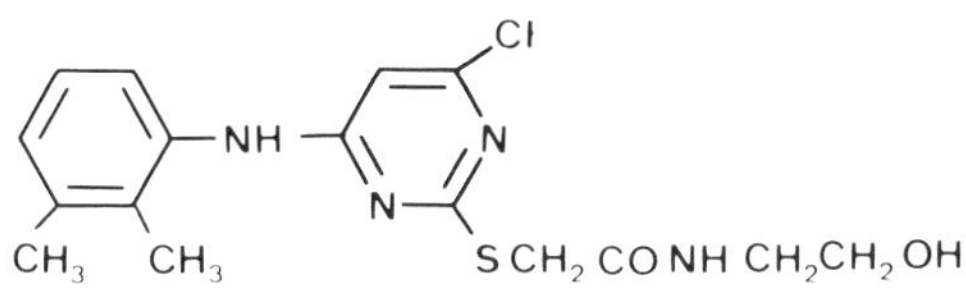

FIGURE 7. BR 931.

mutual sterol pool. VLDL and intermediate-density lipoprotein (IDL), isolated from cholesterol-fed rats, stimulated bile acid synthesis in rat hepatocytes[47] whereas LDL and HDL did not. This provision of cholesterol possibly involves apoprotein E receptor-mediated recognition of cholesteryl ester-rich chylomicron remnants but it remains to be ascertained to what extent this contribution is relevant to normal metabolism when circulating remnant levels are low. By contrast, a recent publication describing[54] a new method for the measurement of cholesterol 7α-hydroxylase activity in the rat — measurement of product by gas chromatography/mass spectrometry after incubation of a microsomal acetone powder — suggests that HDL is a better substrate than VLDL (in terms of lower Km, higher Vmax) with LDL intermediate in efficacy. This conclusion is of doubtful validity, for no circulating lipoprotein can be a direct substrate for the enzyme, but other evidence supports a role for HDL and it has recently been suggested[55] that the apoprotein E-rich subfraction, HDL_1, may donate cholesterol to the hepatocyte in man. Further corroboration has been claimed from a study in patients[56] where the HDL content of free cholesterol was suggested to be the primary precursor of chenodeoxycholic acid synthesis, although this conclusion was based on results from a very small group and the methodological problem of rapid isotope exchange between lipoprotein pools had not entirely been circumvented. There is no doubt that a contribution by HDL to the supply of sterol precursor for catabolism is consistent with the postulated role as a negative risk factor for atherosclerosis, mediating the centripetal transport from peripheral tissues to the liver, although the store in transit might be expected to be esterified rather than free cholesterol.

The recent identification of agents that influence HDL levels relatively specifically may allow a clearer delineation of the function of the lipoprotein as a donor of cholesterol for 7α-hydroxylation. BR 931 (Figure 7) raised HDL in rats, rabbits with experimental atherosclerosis,[57] and patients,[58] but significantly decreased cholesterol 7α-hydroxylase activity in rats.[59] It was claimed[59] that in rats the action of agents on cholesterol 7α-hydroxylase is generally not representative of that found in other species, including man, but that assertion is not consistent with other findings as described elsewhere in this review. In fact, because the action of BR 931 is not confined to HDL, but also encompasses depletion of hepatic cholesterol[59] and LDL-lowering,[57,58] it is not really feasible to comment on the significance of the interference with 7α-hydroxylation.

BRL 26314 (Table 8) was found recently to elevate the concentration of HDL in rats in association with an enhanced centripetal transport, depletion of aortic sterol and promotion of cholic acid secretion.[60] A slight rise in cholesterol 7α-hydroxylase activity ex vivo was probably negligible (Table 8) although we studied the enzyme at the nadir of the circadian rhythm and a diurnal variation in response to BRL 26314 is conceivable. A slight activation of the enzyme, in vitro, on incubation with two soluble salts of BRL 26314 (at a concentration of 0.1 m*M*) was also probably insignificant (54.2, 53.2 pmol · min^{-1} · mg^{-1} microsomal protein by comparison with the control rate of 49.0; n = 3). The mechanism of action of BRL 26314 is not established, but it may include promotion of the cholesterol 12α-hydroxylation step and this possibility is of particular interest in view of the observation of attenuated 12α-hydroxylation in patients with very low plasma HDL.[61]

Not all agents that elevate the concentration of HDL in rats accelerate cholesterol turnover. For example, the cholestatic agent α naphthyl isothiocyanate (ANIT) increased HDL cho-

Table 8
COMPARISON OF EFFECTS OF BRL 26314 AND 1-NAPHTHYL ISOTHIOCYANATE (ANIT) ON HDL LEVELS AND CHOLESTEROL METABOLISM

BRL 26314

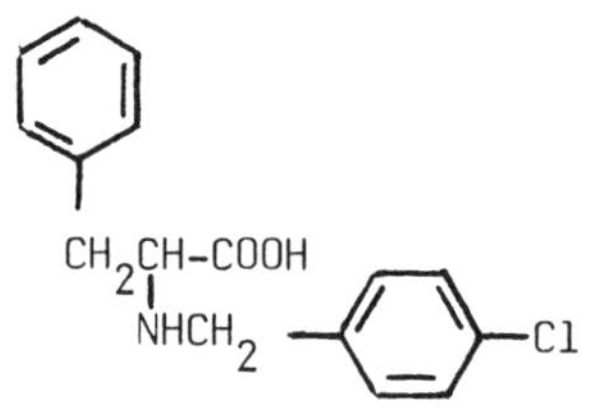

ANIT

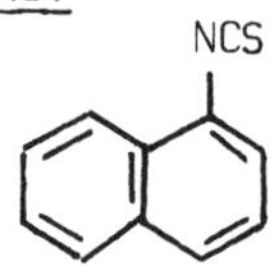

	BRL 26314[a] (100 mg/kg body wt)	ANIT[a] (25 mg/kg body wt)
Normolipidemic rats:		
HDL cholesterol (mg/100 mℓ)	+35[b]	+60[b]
Bile acid excretion (mg/day)[c]	+30	−30
Aortic cholesterol (mg/g)	−45[b]	+20
Adipose tissue cholesterol (mg/g)	−40[b]	+250[b]
Hyperlipidemic rats:		
HDL cholesterol (mg/100 mℓ)	+40[b]	+35[b]
Cholesterol 7α-hydroxylase (nmole/min per liver)	+7	−30[b]

[a] Results are expressed as % change from a comparable control group (8 rats/group). Compounds were administered as a supplement to the experimental diets for 7 days.
[b] $p < 0.05$.
[c] Samples were pooled, being collected daily for the last three days of the experiment from each cage of 4 rats.

lesterol to a similar extent as BRL 26314 (Table 8), but bile acid excretion declined and cholesterol accumulated in the peripheral tissues.[60] ANIT significantly inhibited cholesterol 7α-hydroxylase and this response can be attributed to feedback inhibition by the accumulating serum and tissue trihydroxy bile acids. In related studies,[38] another hepatotoxin, 2,4-dimethylxylene, also increased HDL cholesterol concomitantly with a rise in serum bile acids. It may also be relevant to note that in chronic biliary obstruction, abnormal HDL metabolism (acquired deficiency of lecithin:cholesterol acyltransferase) is associated with the appearance of xanthoma and with premature atherosclerosis.

It cannot be concluded, therefore, that a pharmacological increase in the concentration of HDL cholesterol is consistently correlated with an alteration in bile acid metabolism, for this will depend, in part, on whether the rise in HDL concentration represents an amplification of the shuttling of sterol to the liver (as is found with BRL 26314) or its curtailment (as in cholestasis). When HDL turnover is promoted then bile acid flux may also increase but the role of cholesterol 7α-hydroxylase in the process has yet to be defined.

VII. CONCLUSIONS

The use of lipid-lowering agents to attempt to decrease atherosclerosis and CHD has had a relatively long history — the first clinical reports on cholestyramine appeared in 1959 — yet it has been difficult to achieve a consensus of opinion on the merits of this approach by comparison, for example, with the general assumptions relating hypertension to stroke. It now seems possible to demonstrate the therapeutic benefits to an extent not feasible previously; among the reasons for this recent improvement is the clearer perception concerning which lipoprotein fractions should be modulated and which regulatory steps are responsive to external control. Traditionally, lipid-lowering drugs have been employed on an empirical basis and the mechanism of action has often been poorly defined. For the future, it seems certain that increasing emphasis on selective interference with key enzymes or receptors will replace the general screening tests for the identification of new agents. Some of the most effective of the drugs used to treat metabolic disease are believed to act essentially by inhibition of a single enzyme, for example, the nonsteroidal anti-inflammatory agents (cyclooxygenase), captopril (angiotensin-converting enzyme), methyldopa (Dopa decarboxylase), the methyl xanthines (phosphodiesterase), methotrexate (dihydrofolate reductase), physostigmine (cholinesterase), and allopurinol (xanthine oxidase). These clinical successes give encouragement to the search for even more specific agents and to the identification of other potential sites for inhibition, for example, HMG-CoA reductase.[24] Therapeutic agents that act by increasing the activity of a key enzyme are much rarer and apart from the vitamins probably less selective. In fact the ability of the bile acid-sequestering resins to remove the normal allosteric feedback control of cholesterol 7α-hydroxylase is one of the best examples of clinical efficacy associated with specific activation of an enzyme.

The considerable progress achieved in the last decade — reviewed elsewhere in this volume — has led to recognition of the pivotal role of the 7α-hydroxylation step in the regulation of cholesterol metabolism and substantiates the diverse pharmacological attempts at intervention. Augmentation of the activity may be particularly apposite for those groups where the basal level is low, for example, gallstone patients,[2] patients with cirrhosis[62] and perhaps other types of liver disease, patients with heterozygous familial hypercholesterolemia where there is a marked and consistent decrease in the LDL fractional catabolic rate[63] and those hyperlipidemics with an attenuated turnover of HDL cholesterol. While some of these metabolic defects are clearly genetic in origin it is becoming evident that the level of LDL receptor production is susceptible to the influence of subtle polygenic abnormalities aggravated by environmental factors, so that the common multifactorial hypercholesterolemias will also be treatable by stratagems designed to induce LDL receptor synthesis.

Thus, the pharmacological induction of cholesterol 7α-hydroxylase activity is mechanistically appropriate to the reversal of the underlying defect in some of the most common hyperlipidemias, particularly if combined with an inhibitor of HMG-CoA reductase to prevent the homeostatic reaction and particularly if achieved by a local effect in the small intestine so that the types of toxicity associated with other classes of hypolipidemic agent are circumvented.

REFERENCES

1. **Fears, R.,** Drug treatment of hyperlipidaemia, *Drugs of Today,* 20, 257, 1984.
2. **Coyne, M. J., Bonorris, G. G., Goldstein, L. I., and Schoenfield, L. J.,** Effect of chenodeoxycholic acid and phenobarbital on the rate-limiting enzymes of hepatic cholesterol and bile acid synthesis in patients with gallstones, *J. Lab. Clin. Med.,* 87, 281, 1976.
3. **Key, P. H., Bonorris, G. G., Marks, J. W., Chung, A., and Schoenfield, L. J.,** Biliary lipid synthesis and secretion in gallstone patients before and during treatment with chenodeoxycholic acid, *J. Lab. Clin. Med.,* 95, 816, 1980.
4. **Lewis, B.,** *The Hyperlipidaemias,* Blackwell, Oxford, 1976.
5. **Schwartz, C. C., Vlahcevic, Z. R., and Swell, L.,** Pathways of cholesterol removal via bile acid synthesis and biliary excretion in man, in *Bile Acids and Lipids,* Paumgartner, G., Stiehl, A., and Gerok, W., Eds., MTP Press, Lancaster, 1981, 79.

6. **Redinger, R. N., Hawkins, J. W., and Grace, D. M.,** The economy of the enterohepatic circulation of bile acids in the baboon. I. Studies of controlled enterohepatic circulation of bile acids, *J. Lipid Res.*, 25, 428, 1984.
7. **Dietschy, J. M.,** Mechanisms for the intestinal absorption of bile acids, *J. Lipid Res.*, 9, 297, 1968.
8. **Danielsson, H., Kallas, I., and Wikvall, K.,** Regulation of hydroxylations in biosynthesis of bile acids, *J. Biol. Chem.*, 259, 4258, 1984.
9. **Subbiah, M. T. R. and Yunker, R. L.,** Evidence for the presence of non-lipoprotein factors in diabetic serum capable of stimulating rat hepatic cholesterol-7α-hydroxylase in vitro, *Biochem. Biophys. Res. Commun.*, 121, 743, 1984.
10. **Sipat, A. B. and Sabine, J. R.,** Membrane-mediated control of hepatic β-hydroxy-β-methylglutaryl-coenzyme A reductase, *Biochem. J.*, 194, 889, 1981.
11. **Rush, W. R., Fears, R., Green, J., and Parke, D. V.,** Liver microsomal mixed-function oxidase activity (MFO) and cholesterol metabolism in the obese, arthritic STR/IN strain of mouse, *Comp. Biochem. Physiol.*, 69B, 493, 1981.
12. **Myant, N. B. and Mitropoulos, K. A.,** Cholesterol 7α-hydroxylase, *J. Lipid Res.*, 18, 135, 1977.
13. **Packard, C. J. and Shepherd, J.,** The hepatobiliary axis and lipoprotein metabolism: effects of bile acid sequestrants and ileal bypass surgery, *J. Lipid Res.*, 23, 1081, 1982.
14. Lipid Research Clinics, The Lipid Research Clinics Coronary Primary Prevention Trial results. I. Reduction in incidence of coronary heart disease, *JAMA*, 251, 351, 1984.
15. Lipid Research Clinics, The Lipid Research Clinics Coronary Primary Prevention Trial results. II. The relationship of reduction in incidence of coronary heart disease to cholesterol lowering, *JAMA*, 251, 365, 1984.
16. **Dorr, A. E., Gundersen, K., Schneider, J. C., Jr., Spencer, T. W., and Martin, W. B.,** Colestipol hydrochloride in hypercholesterolemic patients — effect on serum cholesterol and mortality, *J. Chron. Dis.*, 31, 5, 1978.
17. **Kuo, P. T., Hayase, K., Kostis, J. B., and Moreyra, A. E.,** Use of combined diet and colestipol in long-term (7-7 1/2 years) treatment of patients with Type II hyperlipoproteinaemia, *Circulation*, 59, 199, 1979.
18. **Nash, D. T., Gensini, G., and Esente, P.,** Effect of lipid-lowering therapy on the progression of coronary atherosclerosis assessed by scheduled repetitive coronary arteriography, *Int. J. Cardiol.*, 2, 43, 1982.
19. **Kuo, P. T., Kostis, J. B., Moreyra, A. E., and Hayes, J. A.,** Familial type II hyperlipoproteinemia with coronary heart disease. Effect of diet-colestipol-nicotinic acid treatment, *Chest*, 79, 286, 1981.
20. **Illingworth, D. R., Phillipson, B. E., Rapp, J. H., and Connor, W. E.,** Colestipol plus nicotinic acid in treatment of heterozygous familial hypercholesterolaemia, *Lancet*, 1, 296, 1981.
21. **Kane, J. P., Malloy, M. J., Tun, P., Phillips, N. R., Freedman, D. D., Williams, M. L., Rowe, J. S., and Havel, R. J.,** Normalization of low-density-lipoprotein levels in heterozygous familial hypercholesterolemia with a combined drug regimen, *N. Engl. J. Med.*, 304, 251, 1981.
22. **Nessim, S. A., Chin, H. P., Alaupovic, P., and Blankenhorn, D. H.,** Combined therapy of niacin, colestipol and fat-controlled diet in men with coronary bypass, *Arteriosclerosis*, 3, 568, 1983.
23. **Sundberg, E. E. and Illingworth, D. R.,** Effects of hypolipidemic therapy on cholesterol homeostasis is freshly isolated mononuclear cells from patients with heterozygous familial hypercholesterolemia, *Proc. Natl. Acad. Sci. U.S.A.*, 80, 7631, 1983.
24. **Fears, R.,** Pharmacological control of 3-hydroxy-3-methylglutaryl-coenzyme A reductase activity, in *3-Hydroxy-3-Methylglutaryl Coenzyme A Reductase*, Sabine, J. R., Ed., CRC Press, Boca Raton, Fla., 1983, 189.
25. **Mabuchi, H., Sakai, T., Sakai, Y., Yoshimura, A., Watanabe, A., Wakasugi, T., Koizumi, J., and Takeda, R.,** Reduction of serum cholesterol in heterozygous patients with familial hypercholesterolemia, *N. Engl. J. Med.*, 308, 609, 1983.
26. **Kuron, G. W., Grier, N., and Huff, J. W.,** The bile acid binding and hypocholesterolemic action of two water-soluble polymers, *Atherosclerosis*, 37, 353, 1980.
27. **Pupita, F. and Barone, A.,** Clinical pharmacology of DEAE-Dextran for long-term administration (one year), *Int. J. Clin. Pharmacol. Res.*, 3, 287, 1983.
28. **Topping, D. L., Illman, R. J., Fenwick, D. E., and Oakenfull, D. G.,** Effects of quillaja and soya saponins on plasma cholesterol and fecal steroid excretion in the rat, *Proc. Nutr. Soc. Aust.*, 5, 195, 1980.
29. **Angelin, B., Björkhem, I., and Einarsson, K.,** Effects of clofibrate on some microsomal hydroxylations involved in the formation and metabolism of bile acids in rat liver, *Biochem. J.*, 156, 445, 1976.
30. **Stewart, J. M., Packard, C. J., Lorimer, A. R., Boag, D. E., and Shepherd, J.,** Effects of bezafibrate on receptor-mediated and receptor-independent low density lipoprotein catabolism, *Atherosclerosis*, 44, 355, 1982.
31. **Kritchevsky, D., Tepper, S. A., Mueller, M. A., and Klurfeld, D. M.,** Influence of beclobrate and eniclobrate on cholesterol metabolism in rats, *Arzneim-Forsch.*, 33, 1469, 1983.

32. **Glueck, C. J.,** Colestipol and probucol: treatment of primary and familial hypercholesterolemia and amelioration of atherosclerosis, *Ann. Intern. Med.*, 96, 475, 1982.
33. **Rush, W. R. and Fears, R.,** The role of cholesterol 7α-hydroxylase in the hypobetalipoproteinaemic activity of SKF-525A in rats, *Biochem. Pharmacol.*, 31, 2423, 1982.
34. **Salganik, R. I., Manonkova, N. M., and Semenova, L. A.,** Hypocholesterolemic effect of the induction of microsomal enzymes in rat liver tissue, *Vopr. Med. Khim.*, 23, 468, 1977.
35. **Deliconstantinos, G. H., Livanou, T., and Margioris, A. N.,** Cholesterol 7α-hydroxylase and drug oxidation in castrated rats after oestradiol treatment, *IRCS Med. Sci.*, 8, 136, 1980.
36. **Kovanen, P. T., Brown, M. S., and Goldstein, J. L.,** Increased binding of low density lipoproteins to liver membranes from rats treated with 17α-ethinyl estradiol, *J. Biol. Chem.*, 254, 11367, 1979.
37. **Brandau, K. and Keup, U.,** Studies on the determination and induction of cholesterol 7α-hydroxylase, *Arzneim.-Forsch.*, 26, 1837, 1976.
38. **Fears, R.,** unpublished results.
39. **Bell, G. D., Clegg, R. J., Cohn, M. R., Duggleby, J. E., Ellis, W. R., Macdonald, I. A., Middleton, B., and White, D. A.,** Terpene therapy for gallstones — effects of individual monoterpenes on bile flow, bile composition and hepatic cholesterogenesis in the rat, *Br. J. Pharmacol.*, 72, 104P, 1980.
40. **Hosokawa, T., Sawada, M., Ando, K., and Tamura, G.,** Alteration of cholesterol metabolism by 4-O-methylascochlorin in rats, *Lipids,* 16, 433, 1981.
41. **Mosbach, E. H., Singhal, A. K., Ayengar, N. K. N., May, P. S., and McSherry, C. K.,** Effect of a bile alcohol and a bile acid analogue on sterol metabolism in the rat, in *Bile Acids and Lipids*, Paumgartmer, G., Stiehl, A., and Gerok, W., Eds., MTP Press, Lancaster, 1981, 73.
42. **Matsuo, T., Seri, K., and Kato, T.,** A possible activation of cholesterol 7α-hydroxylation by S-methylmethionine (vitamin U), *Arzneim.-Forsch.*, 30, 258, 1980.
43. **Mitropoulos, K. A., Myant, N. B., Gibbons, G. F., Balasubramaniam, S., and Reeves, B. E. A.,** Cholesterol precursor pools for the synthesis of cholic and chenodeoxycholic acids in rats, *J. Biol. Chem.*, 249, 6052, 1974.
44. **Normann, P. T. and Norum, K. R.,** Newly synthesized hepatic cholesterol as precursor for cholesterol and bile acids in rat bile, *Scand. J. Gastroenterol.*, 11, 427, 1976.
45. **Ayaki, Y., Ogura, Y., Kitayama, S., Endo, S., and Ogura, M.,** Difference between cholic and chenodeoxycholic acid in dependence upon cholesterol of hepatic and plasmatic sources as the precursor in rats, *Steroids,* 41, 509, 1983.
46. **Kempen, H. J., Holstein, M. V-v., and deLange, J.,** Bile acids and lipids in isolated rat hepatocytes. II. Source of cholesterol used for bile acid formation, estimated by incorporation of tritium from tritiated water, and by the effect of ML-236B, *J. Lipid Res.*, 24, 316, 1983.
47. **Davis, R. A., Hyde, P. M., Kuan, J.-C. W., Malone-McNeal, M., and Archambault-Schexnayder, J.,** Bile acid secretion by cultured rat hepatocytes, *J. Biol. Chem.*, 258, 3661, 1983.
48. **Fears, R., Richards, D. H., and Ferres, H.,** The effect of compactin, a potent inhibitor of 3-hydroxy-3-methylglutaryl coenzyme-A reductase activity, on cholesterogenesis and serum cholesterol levels in rats and chicks, *Atherosclerosis,* 35, 439, 1980.
49. **Endo, A., Tsujita, Y., Kuroda, M., and Tanzawa, K.,** Effects of ML-236B on cholesterol metabolism in mice and rats: lack of hypocholesterolemic activity in normal animals, *Biochim. Biophys. Acta,* 575, 266, 1979.
50. **Bosisio, E., Cighetti, G., Di Padova, C., Ravagnati, P., Galli Kienle, M., Galli, G., and Paoletti, R.,** Effects of compactin (ML-236B) on biliary lipid composition and cholesterol catabolism in the hamster, *Pharmacol. Res. Commun.*, 14, 577, 1982.
51. **Tsujita, Y., Kuroda, M., Tanzawa, K., Kitano, N., and Endo, A.,** Hypolipidaemic effects in dogs of ML-236B, a competitive inhibitor of 3-hydroxy-3-methylglutaryl coenzyme A reductase, *Atherosclerosis,* 32, 307, 1979.
52. **Kuroda, M., Tsujita, Y., Tanzawa, K., and Endo, A.,** Hypolipidemic effects in monkeys of ML-236B, a competitive inhibitor of 3-hydroxy-3-methylglutaryl coenzyme A reductase, *Lipids,* 14, 585, 1979.
53. **Qureshi, A. A., Abuirmeileh, N., Burger, W. C., Din, Z. Z., and Elson, C. E.,** Effect of AMO 1618 on cholesterol and fatty acid metabolism in chickens and rats, *Atherosclerosis,* 46, 203, 1983.
54. **Sanghvi, A., Diven, W. F., Seltman, H. J., Paul, R., and Rizk, M.,** Properties of cholesterol 7α-hydroxylase in rat liver microsomal acetone powder, *Metabolism,* 33, 443, 1984.
55. **Assmann, A. and Jabs, H.-U.,** The enzymology of lecithin:cholesterol acyl-transferase, *Biochem. Soc. Transact.*, 13, 19, 1985.
56. **Halloran, L. G., Schwartz, C. C., Vlahcevic, Z. R., Nisman, R. M., and Swell, L.,** Evidence for high-density lipoprotein-free cholesterol as the primary precursor for bile acid synthesis in man, *Surgery,* 84, 1, 1978.
57. **Sirtori, C. R., Gomarasca, P., D'Atri, G., Cerutti, S., Tronconi, G., and Scolastico, C.,** Pharmacological profile of BR-931, a new hypolipidaemic agent that increases high-density lipoproteins, *Atherosclerosis,* 30, 45, 1978.

58. **Nunes, G.,** Preliminary evaluation of the lipid lowering effect of BR 931 (pirinixil) in different types of hyperlipoproteinemia, *Curr. Ther. Res.,* 31, 193, 1982.
59. **Kritchevsky, D., Klurfeld, D. M., Tepper, S. A., Mueller, M. A., Puglisi, L., and Sirtori, C. R.,** Increased plasma cholesterol and decreased body lipid levels in Wistar rats following pirinixil (BR 931) treatment, *Pharmacol. Res. Commun.,* 11, 475, 1979.
60. **Fears, R., Rush, W. R., Walker, P., and Ferres, H.,** Hyperalphalipoproteinaemic activity of BRL 26314. I. Enhanced cholesterol turnover in rats, *Biochem. Pharmacol.,* 33, 209, 1984.
61. **Beher, W. T., Gabbard, A., Norum, R. A., and Stradnieks, S.,** Effect of blood high density lipoprotein cholesterol on fecal steroid excretion in humans, *Life Sci.,* 32, 2933, 1983.
62. **Goldman, M., Vlahcevic, Z. R., Schwartz, C. C., Gustafsson, J., and Swell, L.,** Bile acid metabolism in cirrhosis. VIII. Quantitative evaluation of bile acid synthesis from [7β-^{3}H]7α-hydroxycholesterol and [G-^{3}H]26-hydroxycholesterol, *Hepatology,* 2, 59, 1982.
63. **Lewis, B.,** Turnover of apoprotein B in VLDL and LDL in primary hyperlipidaemia, in *Treatment of Hyperlipoproteinaemia,* Carlson, L. A. and Olsson, A. G., Eds., Raven Press, New York, 1984, 35.

Chapter 10

DEVELOPMENTAL ASPECTS OF CHOLESTEROL 7α-HYDROXYLASE

M. T. R. Subbiah

TABLE OF CONTENTS

I. INTRODUCTION

High serum cholesterol levels during adult life is believed to be a major risk factor for the development of atherosclerosis.[1] Consequently, there is interest in developing potential means through which cholesterol catabolism and elimination from the body can be enhanced. Therefore, studies (1) of developmental aspects of cholesterol catabolism, and (2) to determine whether pediatric manipulations can enhance cholesterol catabolism to bile acids and successfully improve the ability of the organism to handle dietary cholesterol are of utmost importance. Since cholesterol 7α-hydroxylase is the rate-limiting enzyme of cholesterol catabolism,[2] details of its developmental pattern, influence of hormones, and pharmacological agents would be of interest. The current data available on these aspects of cholesterol 7α-hydroxylase during development is at best minimal and fragmentary. In this review an attempt is made to summarize current information available on the developmental changes of hepatic cholesterol 7α-hydroxylase and some evidence is presented to suggest that pharmacological and dietary induction of the enzyme is possible during development.

II. RELATIONSHIP OF CHANGES IN BILE ACID POOL AND CHOLESTEROL 7α-HYDROXYLASE ACTIVITY DURING DEVELOPMENT

A number of studies have shown that bile acid synthesis occurs during late gestation and increases rapidly after birth, reaching adult levels by weaning.[3] Lambiotte[4] showed that rat fetuses up to 21 days of gestation contain extremely small amounts of cholic acid (<10 μg), with a 20-fold increase on the twenty-second day. Following birth, the bile acid pool increases further and reaches adult levels by weaning. Croizat and Lambiotte[5] identified the earliest period during gestation when bile acids were first noticed in the fetuses of several animal species. They arrived at day 19 for rats, day 22 for rabbits, days 55 to 65 for pigs, days 45 to 50 for guinea pigs, and day 11 for chickens. Using radioimmunoassay, Little et al.[6] showed that the bile acid pool in 19-day-old rat fetuses is composed of cholate, chenodeoxycholate, hyodeoxycholate, and β-muricholate in the proportions 34:26:7:16:17, respectively. Following birth, however, there is a drop in the concentration of chenodeoxycholate and β-muricholate to negligible amounts. In the rat, at day 19 of gestation, $>95\%$ of the taurocholate pool is found in the liver. Soon after birth, approximately 83.5% of the taurocholate pool is found in the gastrointestinal tract. In the rabbit[7] during the 26th day of gestation, total bile acid pool (mg/g) is around 169.6 ± 40.1, with cholate, chenodeoxycholate, and deoxycholate in the ratios of 45:32:23. The concentration of chenodeoxycholic acid decreases markedly after birth and the total bile acid pool in newborn rabbits nearly doubles when compared to the amount noted in fetal rabbits. Li et al.[8] carried out a systematic study of the postnatal changes in total bile acid pool with age in guinea pigs. They noted a linear increase from day 1 to day 17. In fetal and neonatal guinea pigs chenodeoxycholic acid is the major bile acid, with significant amounts of 3β-hydroxy Δ^5-cholenoic acid.[8] The total bile acid pool in dog fetuses is approximately one-half of the adult pool.[9,10] In monkey fetuses, the presence of taurocholate, taurodeoxycholate, and glycocholate have been demonstrated.[11] Studies by Hardy et al.[12] have shown that the bile acid pool in fetal sheep is less than that in neonates. These authors further showed that bile acid synthesis in response to bile acid pool depletion is less marked in the fetal sheep. In the bile of human fetuses, the presence of taurine and glycine conjugates of cholate and chenodeoxycholate, with a predominance of the latter bile acid have been noted.[13] Watkins et al.[14] measured the bile acid pool size in newborn infants using deuterium-labeled bile salts and found that the bile acid pool in infants is lower than in adults. Whether the age-related increase in bile acid pool is due to changes in the hepatic activities of cholesterol-catabolizing enzymes or due to some other factors is not known. Studies by Danielsson and Rutter[15] on enzymes

Table 1
HEPATIC CHOLESTEROL 7α-HYDROXYLASE ACTIVITY IN NEONATAL GUINEA PIGS (ISOTOPIC INCORPORATION METHOD)

Age (days)	Cholesterol 7α-hydroxylase activity ($pmol \cdot min^{-1} \cdot mg^{-1}$)
1	2.64 ± 0.80
3	1.93 ± 0.39
5	3.38 ± 1.64
8	0.85 ± 0.54
10	1.03 ± 0.08
17	2.23 ± 0.77
22	1.19 ± 0.03
Adult	8.46 ± 1.82

Note: Data shown as mean ± SEM; 3 to 4 guinea pigs were used in each group.

After Li, J. R., Subbiah, M. T. R., and Kottke, B. A., *Steroids,* 34, 47, 1979.

concerned with bile acid conjugation and 12α-hydroxylation in rats showed that there was a progressive increase in enzyme activity from late fetal stage to 5 days after birth. However, change in the activity of cholesterol 7α-hydroxylase (rate-limiting enzyme of cholesterol catabolism, EC 1.14.13.17) was not examined in that study. Studies from our laboratory[16] examined cholesterol 7α-hydroxylase activity in heptic microsomal fractions of guinea pigs from days 1 to 45 of life (Table 1). It was noted that while the activity of cholesterol 7α-hydroxylase in newborn guinea pigs was significantly lower than noted in adults, no clear cut increase with age could be demonstrated in early neonatal life (days 1 to 45). It is possible that true differences in enzyme activity with age were masked due to changing concentration of substrate in microsomes, which is a known factor influencing activity of this enzyme.[17] Furthermore, whether differences in enzyme activity noted during development is due to true changes in enzyme activity itself or due to lack of postmicrosomal supernatant factors known to stimulate the enzyme activity[18] could not be differentiated in previous studies.[16] Therefore, studies of developmental changes in hepatic cholesterol 7α-hydroxylase activity in guinea pigs was determined using the microsomal acetone powder method of Shefer et al.[17] to eliminate differences due to changes in microsomal substrate pool, and enzyme activities were related to bile acid pool[19] (Table 2). Bile acid pool showed a 50-fold increase from day 1 to day 15 (from 86.5 ± 7.1 in day 1 to 3031 ± 151 in day 15). However, the activity ($pmol^{-1} \cdot mg^{-1} \cdot min^{-1}$) of cholesterol 7α-hydroxylase showed an inverse relationship with the bile acid pool, with high values (10.70 ± 0.69) in day 1 decreasing to 3.31 ± 0.41 by day 15. This was a totally unexpected finding.

The response of the cholesterol 7α-hydroxylase of the neonatal liver to 100,000 × g supernatant factors (SF) is shown in Table 3. While the enzyme showed about 27% increase in response to SF from neonatal liver, its response to adult SF fraction was negligible.

Table 4 demonstrates the response of injecting puromycin to newborn guinea pigs on cholesterol 7α-hydroxylase activity in day 2. While the enzyme activity normally declined from 11.81 ± 68 in day 1 to 8.0 ± 1.1 in day 2, puromycin injection in day 1 virtually abolished enzyme activity (0.46 ± 0.3) in day 2. This indicates enzyme activity levels are maintained by protein synthesis.

Table 2
DEVELOPMENTAL CHANGES IN THE ACTIVITY OF HEPATIC CHOLESTEROL 7α-HYDROXYLASE (ACETONE POWDER METHOD) AND TOTAL BILE ACID POOL IN GUINEA PIGS[a]

Age	Cholesterol-7α-hydroxylase activity (pmol · mg^{-1} · min^{-1})	Total bile acid pool (μg · 100 g^{-1})
1	10.70 ± 0.69 (6)	86.5 ± 7.1 (5)
2	7.08 ± 0.67 (6)	401.0 ± 37.0 (4)
5	3.03 ± 0.53 (4)	838.0 ± 91.0 (5)
10	3.22 ± 0.24 (6)	1436.0 ± 107.0 (4)
15	3.02 ± 0.24 (6)	2351.0 ± 151.0 (4)
60	2.42 ± 0.20 (6)	3008.0 ± 210.0 (4)

[a] Bile acid pool at various ages was measured by GLC. Cholesterol 7α-hydroxylase activity was determined in acetone extracted microsomes as described in Reference 19.

After Yunker, R. L. and Subbiah, M. T. R., *Biochem. Biophys. Res. Commun.*, 132, 702, 1985.

Table 3
EFFECT OF POSTMICROSOMAL SUPERNATANT FRACTIONS (SF) OF LIVER FROM NEWBORN AND ADULT GUINEA PIGS ON CHOLESTEROL 7α-HYDROXYLASE ACTIVITY IN NEWBORN GUINEA PIG LIVER

Experimental conditions (n)	Cholesterol 7α-hydroxylase activity[a] (pmol · mg^{-1} · min^{-1})
1. Control newborn livers (6)	10.70 ± 0.69
2. Control newborn liver + SF fraction of newborn liver (6)	14.51 ± 0.48[a]
3. Control newborn liver + SF from adult liver (6)	10.36 ± 0.46
4. Control adult liver (6)	2.42 ± 0.20
5. Control adult liver + SF from adult liver (6)	9.46 ± 0.49[b]
6. Control adult liver + SF from newborn livers (6)	6.38 ± 0.29[c]

[a] Enzyme activity was determined under Methods in the presence of absence of supernatant factor (SP).
[b] Significantly different (p <0.05) from control newborn liver.
[c] Significantly different (p <0.05) from control adult liver values.

After Yunker, R. L. and Subbiah, M. T. R., *Biochem. Biophys. Res. Commun.*, 132, 702, 1985.

The results of this study using acetone powder method to measure enzyme activity clearly shows that in the guinea pig there is a progressive decrease in the activity of cholesterol 7α-hydroxylase from birth to adult life. In previous studies with the guinea pig using isotope incorporation method,[16] no clear cut increase (in fact, a slight decrease) was noted at certain

Table 4
EFFECT OF PUROMYCIN INJECTION TO NEWBORN GUINEA PIGS ON SUBSEQUENT HEPATIC CHOLESTEROL 7α-HYDROXYLASE ACTIVITY

Group	Cholesterol 7α-hydroxylase activity[a]
Day 1	10.70 ± 0.69
Day 2	
Control (6)	7.08 ± 0.69[b]
Puromycin injected (6)	0.46 ± 0.06

[a] Enzyme activity measured 18 hr later after the intraperitoneal injection of puromycin (5 mg · 100 g^{-1}) as described under Methods.
[b] Significantly different ($p < 0.05$) from day 1 and puromycin injected (day 2) values.

After Yunker, R. L. and Subbiah, M. T. R., *Biochem. Biophys. Res. Commun.*, 132, 702, 1985.

ages, although the results were inconsistent. The results of guinea pigs do not parallel results obtained in rats where an increase with age has been noted,[19] although isotope incorporation technique was used in that study. To date, no study of developmental changes in rats using acetone extraction method has been done. It is also possible that the guinea pig, because of its advanced development at birth compared to rats and humans, might have a different developmental pattern. Further studies of developmental changes in hepatic cholesterol 7α-hydroxylase activity in various animal species is badly needed.

III. MANIPULATION OF CHOLESTEROL 7α-HYDROXYLASE DURING DEVELOPMENT

Since the activity of cholesterol 7α-hydroxylase is low at birth and subsequently increases, it provides an ideal set-up to determine whether the enzyme activity can be permanently modulated by stimulation in early life or by weaning or by alterations in hormonal status.

A. Effect of Preweaning

Preweaning itself has been shown[20-22] to cause delayed changes in cholesterol metabolic parameters, some of which persist into adult life. Hahn et al.[21] showed that preweaning the rats caused an increase in serum cholesterol that was evident in adult life. Mott et al.[22] noted that breast fed and formula fed baboons showed differences in serum HDL and Apo I levels when fed saturated or unsaturated fat diets at 4 years of age. In the rat, Kris-Etherton et al.[23] noted that preweaned rats had a relative inability to handle a cholesterol challenge diet in adult life. The mechanisms responsible for preweaning induced changes in cholesterol metabolism are not known. Whether preweaning can modify the developmental pattern of cholesterol 7α-hydroxylase activity has not been investigated until recently. Studies from our laboratory[24] showed that preweaning the guinea pigs caused a decrease in hepatic cholesterol 7α-hydroxylase activity which persists even after one month (Table 5). Addition of cholesterol to the diet did not negate preweaning induced changes in cholesterol 7α-hydroxylase activity (Table 5).

B. Postnatal Manipulation

Studies from our laboratory[25] showed that neonatal guinea pigs, when pretreated with cholestyramine (which binds bile acids and increases the activity of cholesterol 7α-hydrox-

Table 5
EFFECT OF PREWEANING PLUS CHOLESTEROL FEEDING ON HEPATIC CHOLESTEROL 7α-HYDROXYLASE ACTIVITY

	Cholesterol 7α-hydroxylase activity ($pmol \cdot mg^{-1} \cdot min^{-1}$)	
Group	**At 10 days of age**	**At 45 days of age[a]**
Normally weaned (6)	3.91 ± 0.07	4.50 ± 0.14
Preweaned (6)	3.37 ± 0.12[a]	3.71 ± 0.23[b]
Preweaned + cholesterol (6)	2.98 ± 0.05[a]	3.95 ± 0.09[b]

[a] Measured at 4 weeks after all the groups were switched to stock guinea pig chow diet to determine persistent effects.
[b] Significantly different (p <0.05) when compared to normally weaned group.

After Subbiah, M. T. R., Yunker, R. L., Menkhaus, A., and Poe, B., *Am. J. Physiol.*, 249, E251, 1985.

Table 6
FECAL CHOLESTEROL AND BILE ACID EXCRETION IN CONTROL AND CHOLESTYRAMINE-PRETREATED GUINEA PIGS[a]

		Steroid excretion ($mg \cdot 100\ g^{-1}$ body weight 3 days)		
Dietary period[b]	**Group**	**Neutral sterols**	**Bile acids**	**Total fecal sterols**
1				
	Control (5)[c]	5.5 ± 0.4	6.3 ± 0.7	11.8 ± 0.9
	Cholestyramine (5)	9.5 ± 0.9[d]	16.6 ± 2.4	25.8 ± 3.0[d]
2				
	Control (3)	4.7 ± 0.4	7.9 ± 0.5	12.5 ± 0.8
	Cholestyramine (5)	5.6 ± 0.3	14.0 ± 1.9[d]	19.6 ± 2.0[d]

[a] Data reported as mean ± SEM.
[b] In dietary period 1, control group was fed stock diet while cholestyramine group was fed stock diet plus 2.5% cholestyramine. In dietary period 2, both groups were fed stock diet. Fecal steroid excretion was measured after 4 weeks on stock diet.
[c] Numbers in parenthesis indicate the number of animals studied.
[d] Significantly different from corresponding control, p <0.05.

After Hassan, A. S., Gallon, L. S., Zimmer, L. A., Balistreri, W. F., and Subbiah, M. T. R., *Steroids*, 38, 477, 1981.

ylase), show an increased excretion of bile acids even when the animals are switched to laboratory chow diet (Table 6). Furthermore, when challenged with cholesterol, these animals display a lower serum cholesterol level when compared to normally weaned controls (Figure 1). We have shown[26] that stimulation of cholesterol catabolism even after weaning is still as effective in influencing the subsequent response to dietary cholesterol in magnitude as noted following stimulation at the neonatal stage. However, stimulation of cholesterol catabolism following maturity has no significant effect on subsequent response to dietary

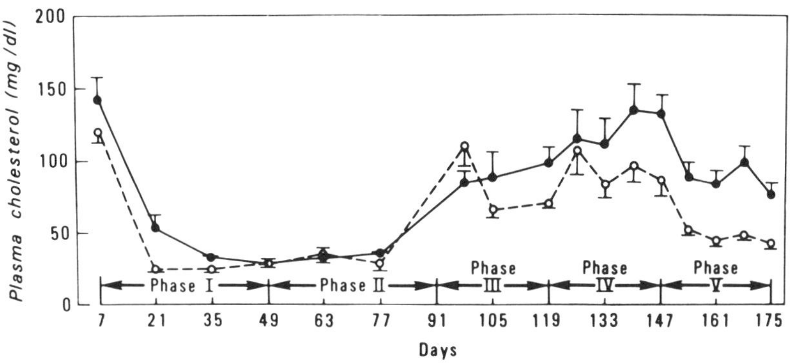

FIGURE 1. Plasma cholesterol levels in newborn guinea pigs treated with cholestyramine (0) and control group (0). Phase I: Guinea pigs on 2.5% cholestyramine or regular chow diet; Phase II: both groups of guinea pigs on regular chow diet; Phase III: both groups of guinea pigs on 0.25% cholesterol diet; Phase IV: both groups of guinea pigs on 0.5% cholesterol diet; Phase V: both groups of guinea pigs on regular chow diet. Each point shows mean ± SEM. (After Li, J. R., Bale, L. K., and Subbiah, M. T. R., *Atherosclerosis,* 32, 93, 1979.)

Table 7
CHOLESTEROL 7α-HYDROXYLASE ACTIVITY IN CONTROL AND CHOLESTYRAMINE-PRETREATED GUINEA PIGS[a]

Group (n)	Cholesterol 7α-hydroxylase activity[b] ($pmol \cdot mg^{-1} \cdot 20\ min^{-1}$)
Control	(4) 9.34 ± 0.57
Cholestyramine-pretreated	(4) 21.27 ± 1.70[c]

[a] Data reported as mean ± SEM.
[b] Enzyme activity assayed after 4 weeks on stock diet.
[c] Significantly different from control, $p < 0.01$.

After Hassan, A. S., Gallon, L. S., Zimmer, L. A., Balistreri, W. F., and Subbiah, M. T. R., *Steroids,* 38, 477, 1981.

cholesterol indicating the importance of the neonatal period (until maturity) for exploring potential maneuvers to improve handling of cholesterol in adult life. Hassan et al.[27] exploring possible mechanisms of the phenomenon have also shown that early stimulation with cholestyramine causes an increase in cholesterol 7α-hydroxylase activity that persists even after removal of the stimulus (Table 7). Li et al.[28] noted that guinea pigs pretreated with cholestyramine in neonatal life have a higher cholesterol 7α-hydroxylase activity during cholesterol challenge period as well. These studies show that cholesterol 7α-hydroxylase activity stimulated in early life can persist into adult life. The enhanced enzyme activity might be responsible for the noted resistance to hypercholesterolemia and atherosclerosis when challenged with a high cholesterol diet in adulthood.[29] How long this enhanced cholesterol 7α-hydroxylase activity will persist is not known. Studies[30] have also been done to determine whether the activity of cholesterol 7α-hydroxylase can be suppressed permanently by neonatal manipulation. Hassan et al.[30] showed that chronic feeding of chenodeoxycholic acid to

Table 8
EFFECT OF SHORT-TERM AND LONG-TERM CHENODEOXYCHOLIC ACID FEEDING AND WITHDRAWAL IN NEONATAL GUINEA PIG ON HEPATIC CHOLESTEROL 7α-HYDROXYLASE ACTIVITY

Group	Cholesterol 7α-hydroxylase activity ($pmol \cdot mg^{-1} \cdot min^{-1}$) Study period A[a]	Study period B[b]
Control	1.83 ± 0.43 (4)[c]	0.47 ± 0.03 (4)
CDCA	0.02 ± 0.02 (5)[d]	0.67 ± 0.05 (4)[d]

[a] Enzyme activity assayed immediately following 10-day feeding of 0.25% chenodeoxycholic acid diet. Data reported as mean ± SEM.
[b] Enzyme activity assayed 4 weeks after cessation of a 6-week feeding of 0.25% chenodeoxycholic acid diet. Data reported as mean ± SEM.
[c] Numbers in parenthesis indicate the number of animals studied.
[d] Significantly different from control, $p < 0.05$.

After Hassan, A. S., Yunker, R. L., Sprinkle, D. J., and Subbiah, M. T. R., *J. Nutr.*, 113, 986, 1983.

neonatal guinea pigs suppresses cholesterol 7α-hydroxylase activity (Table 8). Following withdrawal of chenodeoxycholic acid, this enzyme activity does not remain suppressed. Indeed, the chenodeoxycholic acid-treated animals have a higher cholesterol 7α-hydroxylase activity when compared to controls, suggesting a "rebound" increase in the enzyme activity following chronic suppression.

C. Prenatal Manipulation

Prenatal manipulation of cholesterol metabolism through maternal treatment with diet and drugs is another potential maneuver to be considered for inducing permanent changes in cholesterol metabolism in the neonate. Studies by Naseem et al.[31,32] showed that feeding of cholesterol during pregnancy to rats causes a marked increase in the cholesterol 7α-hydroxylase activity in the neonate. However, plasma cholesterol levels remain high in these neonates. Our studies using guinea pigs showed that feeding cholesterol during pregnancy has detrimental effects in the neonate and causes a decrease in the neonatal bile acid pool and an increase in plasma cholesterol.[33] Innis [34,35] investigated the effect of cholestyramine feeding during pregnancy on HMG-CoA reductase and cholesterol 7α-hydroxylase in the rat. While maternal HMG-CoA reductase activity increased, cholesterol 7α-hydroxylase activity showed no change. Hassan et al.[36] repeated these studies and found that cholestyramine did increase the activity of cholesterol 7α-hydroxylase as expected in the pregnant rat. Differences in the methods used to assay the enzyme seems to be the primary reason for this discrepancy in results. Feeding chenodeoxycholic acid during gestation in the rat did not influence neonatal cholesterol 7α-hydroxylase activity.[37]

IV. EFFECT OF ENDOCRINE FACTORS ON CHOLESTEROL 7α-HYDROXYLASE DURING DEVELOPMENT

Although glucocorticoids[38,39] and diabetes[40] during pregnancy have been shown to influence bile acid pool size in the neonate, no studies have been done on the effect of endocrine factors on fetal or neonatal cholesterol 7α-hydroxylase activity. Such studies are badly needed.

V. CONCLUSIONS

It is obvious from this chapter that very little is known regarding the developmental aspects of hepatic cholesterol 7α-hydroxylase. Special attention should be paid to the factors which can modulate the activity during development and perhaps set it at a higher level throughout life. Since this enzyme is key to the cholesterol degradation process, knowledge of its control should be applicable to treating diseases like atherosclerosis and gallstones (see also Chapters 9 and 11).

ACKNOWLEDGMENT

The authors are indebted to the following colleagues for their help and collaboration in carrying out the investigations reported in this review: Drs. A. S. Hassan, J. R. Li, L. Marai, W. F. Balistreri, Mr. R. L. Yunker, and Ms. Linda Zimmer. Sincere thanks are due to Wilma Roll for typing the manuscript. The work reported in this review was supported by grants HL-24263 and HL-24071 from the National Heart, Lung and Blood Institute.

REFERENCES

1. **Enos, W. F., Holmes, R. H., and Beyer, S. J.,** Coronary artery disease among United States soldiers killed in action in Korea, *JAMA,* 152, 1090, 1953.
2. **Danielsson, H.,** in *Bile Acids,* Vol. II, *Physiology and Metabolism,* Nair, P. P. and Kritchevsky, D., Eds., Plenum Press, New York, 1971, 1.
3. **Subbiah, M. T. R. and Hassan, A. S.,** Development of bile acid biogenesis and its significance in cholesterol homeostasis, *Adv. Lipid Res.,* 19, 137, 1982.
4. **Lambiotte, M.,** Donnees preliminary sur les accides biliares au cours de la vie foetale et neonatale du ratbbac, *C. R. Acad. Sci.,* 259, 444, 1964.
5. **Croizat, B. and Lambiotte, M.,** Les Acides biliares au cours de la vie embroyonnaire, *Arch. Sci. Physiol.,* 25, 303, 1971.
6. **Little, J. M., Richey, J. E., Van Thiel, D. H., and Lester, R.,** Taurocholate pool size and distribution in the fetal rat, *J. Clin. Invest.,* 63, 1042, 1979.
7. **Subbiah, M. T. R., Marai, L., Dinh, D. H., and Penner, J. W.,** Sterol and bile acid metabolism during development. 1. Studies on the gall bladder and intestinal bile acids of newborn and fetal rabbit, *Steroids,* 29, 83, 1977.
8. **Li, J. R., Dinh, D. M., Ellefson, R. D., and Subbiah, M. T. R.,** Sterol and bile acid metabolism during development. 3. Occurrence of neonatal hypercholesterolemia in the guinea pig and its relation to bile acid pool, *Metabolism,* 28, 151, 1979.

9. **Jackson, B. T., Smallwood, R. A., Piasecki, G. T., Brown, A. S., and Lester, R.,** Fetal bile salt metabolism. I. The metabolism of sodium cholate-C^{14} in the fetal dog, *J. Clin. Invest.*, 50, 1286, 1971.
10. **Smallwood, R. D., Lester, R., Piasecki, G. J., Klein, P. D., and Jackson, B. T.,** Fetal bile salt metabolism. II. Hepatic excretion of endogenous bile salt and of a taurocholate load, *J. Clin. Invest.*, 51, 1388, 1972.
11. **Little, J. M., Smallwood, R. A., Lester, R., Piasecki, G. J., and Jackson, B. T.,** Bile salt metabolism in the primate fetus, *Gastroenterology*, 69, 1315, 1975.
12. **Hardy, K. J., Hoffman, N. E., Mitally, G., Sewell, R. B., and Smallwood, R. A.,** Bile acid metabolism in fetal sheep: perinatal changes in the bile acid pool, *J. Physiol.*, 309, 1, 1980.
13. **Sharp, H. L., Peller, J., Carey, J. B., and Kriwit, W.,** Primary and secondary bile acids in meconium, *Pediatr. Res.*, 5, 274, 1971.
14. **Watkins, J. B., Ingall, D., Szczepanik, P., Klein, P. D., and Lester, R.,** Bile salt metabolism in the newborn. Measurement of pool size and synthesis by stable isotope technique, *N. Engl. J. Med.*, 288, 431, 1973.
15. **Danielsson, H. and Rutter, W. J.,** The metabolism of bile acids in the developing rat liver, *Biochemistry*, 7, 346, 1968.
16. **Li, J. R., Subbiah, M. T. R., and Kottke, B. A.,** Hepatic HMG-CoA reductase and cholesterol 7α-hydroxylase activity in neonatal guinea pigs, *Steroids*, 34, 47, 1979.
17. **Shefer, S., Cheng, F. W., Hauser, S., Bhatta, A., and Salen, G.,** Regulation of bile acid synthesis: measurement of cholesterol 7α-hydroxylase activity in rat liver microsomal preparations in the absence of endogenous cholesterol, *J. Lipid Res.*, 22, 532, 1981.
18. **Kwok, C. T., Burnett, W., and Hardie, I. R.,** Regulation of rat liver microsomal cholesterol 7α-hydroxylase: presence of a cytosolic activator, *J. Lipid Res.*, 22, 570, 1981.
19. **Yunker, R. L. and Subbiah, M. T. R.** Ontogeny of guinea pig liver cholesterol 7α-hydroxylase, *Biochem. Biophys. Res. Commun.*, 132, 702, 1985.
20. **Hahn, P. and Innis, S.,** Cholesterol oxidation and 7α-hydroxylation during postnatal development of the rat, *Biol. Neonat.*, 46, 48, 1984.
21. **Hahn, P., Girard, J., Assan, J., Kervran, A., and Koldovsky, O.,** Late effects of premature weaning to different diets in the rat, *J. Nutr.*, 108, 1783, 1978.
22. **Mott, G. E., McMahan, C. A., Kelly, J. L., Farley, C. M., and McGill, H. C., Jr.,** Influence of infant diet and juvenile diet on serum cholesterol, lipoprotein cholesterol and apoprotin concentration in juvenile baboons, *Atherosclerosis*, 49, 191, 1982.
23. **Kris-Etherton, P. M., Layman, D. K., York, P. V., and Frantz, I. D., Jr.,** The influence of early nutrition on the serum cholesterol of the adult rat, *J. Nutr.*, 109, 1244, 1979.
24. **Subbiah, M. T. R., Yunker, R. L., Menkhaus, A., and Poe, B.,** Preweaning induced changes of cholesterol metabollism in the guinea pig, *Am. J. Physiol.*, 249, E251, 1985.
25. **Li, J. R., Bale, L. K., and Subbiah, M. T. R.,** Effect of enhancement of cholesterol degradation during neonatal life of guinea pig on its subsequent response to dietary cholesterol, *Atherosclerosis*, 32, 93, 1979.
26. **Hassan, A. S., Gallon, L. S., Yunker, R. L., and Subbiah, M. T. R.,** Effect of enhancement of cholesterol catabolism in guinea pigs after weaning on subsequent response to dietary cholesterol, *Am. J. Clin. Nutr.*, 35, 546, 1982.
27. **Hassan, A. S., Gallon, L. S., Zimmer, L. A., Balistreri, W. F., and Subbiah, M. T. R.,** Persistent enhancement of bile acid synthesis in guinea pigs following stimulation of cholesterol catabolism in neonatal life, *Steroids*, 38, 477, 1981.
28. **Li, J. R., Bale, L. K., and Kottke, B. A.,** Effect of neonatal modulation of cholesterol homeostasis on subsequent response to cholesterol challenge in adult guinea pig, *J. Clin. Invest.*, 65, 1060, 1980.
29. **Subbiah, M. T. R., Dietemeyer, D., and Yunker, R. L.,** Decreased atherogenic response to dietary cholesterol following stimulation of cholesterol catabolism in early life, *J. Clin. Invest.*, 71, 1509, 1983.
30. **Hassan, A. S., Yunker, R. L., Sprinkle, D. J., and Subbiah, M. T. R.,** Chronic suppression of cholesterol 7α-hydroxylase by dietary chenodeoxycholic acid in neonatal guinea pigs: its effect on subsequent bile acid metabolism in the adult, *J. Nutr.*, 113, 986, 1983.
31. **Naseem, S. Y., Khan, M. A., Jacobson, M. S., Nair, P. P., and Heald, F. P.,** The influence of dietary cholesterol and fat on the homeostasis of cholesterol metabolism in early life in the rat, *Pediatr. Res.*, 14, 1061, 1980.
32. **Naseem, S. Y., Khan, M. A., Heald, F. P., and Nair, P. P.,** The influence of cholesterol and fat in maternal diet of rats in the development of hepatic cholesterol metabolism in the offspring, *Atherosclerosis*, 36, 1, 1980.
33. **Hassan, A. S., Yunker, R. L., and Subbiah, M. T. R.,** Decreased bile acid pool in neonates of guinea pigs fed cholesterol during pregnancy, *J. Nutr.*, 111, 2030, 1981.
34. **Innis, S. M.,** Effect of cholestyramine administration during pregnancy in the rat, *Am. J. Obstet. Gynecol.*, 146, 13, 1983.

35. **Innis, S. M.,** Influence of maternal cholestyramine treatment on cholesterol and bile acid metabolism in adult offspring, *J. Nutr.*, 113, 2404, 1983.
36. **Hassan, A. S., Hackley, J. J., and Johnson, L. L.,** Cholestyramine treatment during pregnancy in the rat results in hypercholesterolemia, *Atherosclerosis*, 57, 139, 1985.
37. **Sprinkle, D. J., Hassan, A. S., and Subbiah, M. T. R.,** Effect of chenodeoxycholic acid feeding during gestation in the rat on bile acid metabolism and liver morphology, *Proc. Soc. Exp. Biol. Med.*, 175, 386, 1984.
38. **Graham, T. O., Van Thiel, D. H., Little, J. R., and Lester, R.,** Synthesis of taurocholate by rat liver in organ culture: Effects of cortisol in vivo, *Am. J. Physiol.*, 237, E177, 1979.
39. **Haber, L. R., Vauphas, V., Vitullo, B. B., Seemayer, T. A., and De Bella, R. C.,** Bile acid conjugation in organ culture of human fetal liver, *Gastroenterology*, 74, 1214, 1978.
40. **Subbiah, M. T. R., Yunker, R. L., Hassan, A. S., and Thibert, P.,** Abnormal bile acid pool and composition in neonates of spontaneously diabetic Wistar BB rats and its change during development, *Biochim. Biophys. Acta*, 794, 355, 1984.

Contribution to the Regulation of Lipoprotein and Bile Acid Metabolism in Health and Disease

Chapter 11

INVOLVEMENT IN HEPATOBILIARY AXIS AND REGULATION OF PLASMA LIPOPROTEIN LEVELS

C. J. Packard and J. Shepherd

TABLE OF CONTENTS

I. INTRODUCTION

The liver and intestine act together to control cholesterol and bile acid metabolism.[1,2] They possess the pathways of synthesis, secretion, and absorption of both steroids and mediate their continuous recycling through the blood plasma and enterohepatic compartments. As Figure 1 shows, hepatic cholesterol may be directed either into the systemic circulation in association with lipoprotein or into the intestinal lumen via the bile. The same sterol may return to the liver in the course of plasma lipoprotein catabolism or as a result of absorption in the gut and packaging into chylomicrons which are ultimately taken up by hepatic parenchymal cells.[3] Bile acids, synthesized from cholesterol, accompany the sterol into the gut and facilitate its reassimilation by their detergent-like properties before being themselves absorbed and recycled in a process of conservation designed to spare the synthetic machinery of the liver the burden of producing the large quantities of these molecules which are needed each day.[4] The rates of synthesis of cholesterol and bile acids are governed by two pacemaker enzymes, 3-hydroxy-3-methylglutaryl coenzyme A reductase (HMG CoA reductase) and cholesterol 7α-hydroxylase. These are subject to loose coordinate control in that the product of one pathway is substrate for the other, but in the main they operate independently in the hepatocyte.[4-6] In the discussion that follows, we elaborate upon key events in the body's response to interruption of the enterohepatic circulation and attempt to describe the role that cholesterol 7α-hydroxylase plays in this situation. The chapter is divided into two sections which deal with, first, the delivery of substrate sterol to the 7α-hydroxylase reaction and secondly, the transport and metabolism of the bile acids whose synthesis is initiated by it.

II. CHOLESTEROL METABOLISM IN PLASMA AND LIVER

The lipoproteins that circulate in the blood may be conveniently divided into two families on the basis of their protein content. The first of these, containing apolipoproteins A (AI and AII) encompass the high density lipoprotein fraction (d = 1.063 to 1.21 kg/ℓ) and may be separated by density gradient centrigufation into two principal subclasses, HDL_2 (d = 1.063 to 1.125 kg/ℓ) an HDL_3 (d = 1.125 to 1.21 kg/ℓ). The second family, containing apolipoprotein B as its major protein component, comes from two sources, liver and intestine. These organs have recently been shown to elaborate different isoforms of the protein which are classified according to their relative rates of migration in SDS polyacrylamide gels. Liver cells secrete the larger B_{100} isoform in association with very low density, intermediate density, and (according to some reports) low density lipoproteins (VLDL, d <1.006 kg/ℓ; IDL, d = 1.006 to 1.019 kg/ℓ; LDL d = 1.019 to 1.063 kg/ℓ)[3]. The smaller, B_{48} protein is elaborated in enterocytes and appears in the plasma as a component of chylomicrons (d <0.95 kg/ℓ) and intestinal VLDL.[8,9] These gut-derived triglyceride-rich particles also possess A apolipoproteins on their surfaces, but lose them into the HDL density interval soon after emerging from the thoracic duct into the systemic circulation (see References 3 and 11 for a detailed discussion of lipoprotein structure and function).

The apo A- and apo B-containing lipoproteins differ in their basic functions. The latter are responsible for the transport of triglyceride and cholesterol from their sites of synthesis in liver and intestine to their points of utilization in peripheral tissues.[3] Under normal circumstances these tissues are capable of degrading triglyceride fully to glycerol and acetate, even in times of surfeit. However, the cholesterol nucleus, once formed, cannot be degraded in significant quantities by man and therefore must be excreted intact.[2] Recent reports[12,13] suggest that HDL may play a role here by binding to the surfaces of cells replete with the sterol and assimilating it from their membranes. The dynamic of this process is preserved by the action of LCAT which maintains a low cholesterol: phospholipid ratio on the surface

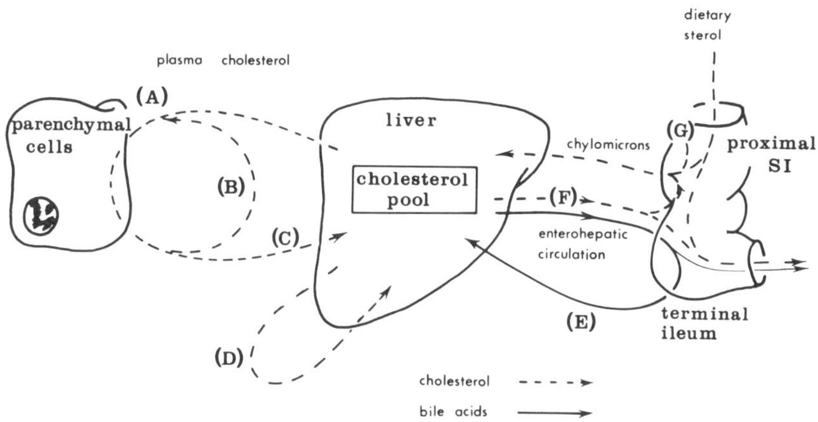

(A) Sterol exported from the liver is taken up by peripheral cells.

(B) Cholesterol released from these sites may recycle or be

(C) returned to the liver.

(D) Much of the exported hepatic sterol is reassimilated by absorptive endocytosis.

(E) Bile acids return to the liver in the enterohepatic circulation.

(F) Biliary cholesterol or

(G) intestinally synthesized sterol can join the chylomicron flow back to the liver.

FIGURE 1. Steroid cycling through the enterohepatic and systemic compartments.

of HDL.[2,3] Cholesteryl ester formed from the action of the enzyme on the particle rapidly migrates to the hydrophobic HDL core, creating a vacancy on the particle surface that allows more sterol to be scavenged from cells. In addition, the esters that accumulate in HDL are rapidly transferred to other less dense lipoproteins by the action of plasma neutral lipid transfer proteins,[14,15] thereby permitting further cycles of sterol capture and esterification. So, by this postulated mechanism, excess sterol deposited in peripheral cells by the uptake of LDL (see below) can be redirected into circulating VLDL and LDL. Such a process would be futile were LDL catabolism restricted to peripheral tissues. However, it is now realized that the liver is a major site of LDL degradation,[16] removing the particle from the circulation and excreting its cholesterol into the bile. In fact it is the only organ capable of eliminating significant quantities of the lipid from the body and, therefore, the mechanisms (Figure 2) whereby sterol is returned to the liver for export are of prime importance.

A. Mechanisms of LDL Catabolism

The metabolism of LDL in vivo may be followed using a variety of isotopic techniques to trace the plasma clearance of its apo B moiety and if the radiolabel happens to be associated with a nondegradable moiety such as sucrose[16] then the sites of LDL breakdown can also be determined by examining the tissue distribution of the accumulated marker. The concentration of LDL in human plasma is a function of its relative rates of synthesis and catabolism and measurement of these parameters in normal and hyperlipidemic subjects has revealed that both processes are of equal importance in determining the circulating level of the lipoprotein.[18] Also, both are located primarily in the liver making this organ the predominant site for the control of plasma LDL levels.[3,16] Early studies of LDL turnover in normal and hypercholesterolemic patients revealed that the clearance rate of the lipoprotein (as measured by the fraction of the plasma pool catabolized daily) was severely restricted

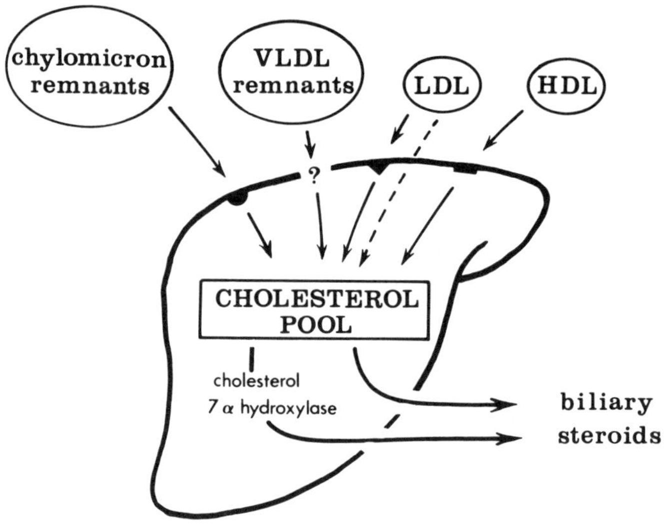

FIGURE 2. Delivery of plasma lipoprotein cholesterol to the liver.

in patients with heterozygous or homozygous familial hypercholesterolemia (FH).[17-19] In a search for the genetic defect in this disease, Goldstein and Brown[20] compared the metabolism of LDL in fibroblasts cultured from skin biopsies from normal and homozygous FH individuals. The outcome of these pioneering studies was the description of an autoregulated pathway for the facilitated absorptive endocytosis of LDL which was absent or severely defective in FH. The key component is a cell membrane receptor of approximately mol. wt. 100,000 daltons,[21] which is able to bind LDL and other apoprotein B-containing ligands and concentrate them at specialized areas of the cell surface membrane (coated pits) which undergo rapid internalization to form intracellular vesicles or ''endosomes''. The interaction of the acidic glycoprotein receptor[22] with its ligands is dependent on a number of functionally significant arginine and lysine residues on the apolipoprotein surface that interact with complementary negatively charged domains on the receptor.[23] Following internalization the ligand and receptor part in the ''endosome'' possibly as a result of an ATP-dependent proton pump that drives the intravesicular pH down to levels no longer consistent with the electrostatic attraction that operated at the cell surface. Receptor proteins appear at this stage to segregate from the mass of the vesicle and by some as yet unknown mechanism recycle back to the cell membrane. It has been calculated that each receptor can mediate the internalization of 150 LDL particles before succumbing itself to degradation.[24] The vesicle with its contents proceeds to fuse with lysosomes and digestion of LDL occurs with the release of free cholesterol into the cell.[20] This lipid is used for membrane synthesis in rapidly dividing cells (e.g., crypt cells of the intestinal microvilli), as substrate for steroid hormone production (in adrenals, ovary, or testis) and, as we shall see later, also contributes to the sterol pool required for bile acid synthesis in the liver. When the cell no longer requires cholesterol, synthesis of the receptor is downregulated, so that overaccumulation of this potentially toxic compound is prevented.[20]

The contribution of the receptor pathway to LDL catabolism can be measured in the whole organism by following simultaneously the rates of clearance of native LDL and of a chemically modified tracer which is no longer recognized by the receptor binding sites.[25] Mahley and co-workers[23] demonstrated that it is possible to produce chemical derivatives of lipoproteins whose gross physical and chemical properties are not altered but which fail to be recognized by the LDL receptor. For example, conversion of 30 to 50% of the lysine residues on apo LDL to dimethylysine by reaction with formaldehyde and $NaBH_4$ renders the lipoprotein incapable of competing with a labeled LDL tracer for the receptor binding sites on cultured fibroblasts. Similarly, reaction of arginine residues with the agent 1,2 cyclohexandione blocks ligand-receptor interaction. We[26] have used this latter modification in a series of studies designed to investigate the control of LDL catabolism in vivo. When ^{125}I-native LDL and ^{131}I-cyclohexandione-treated LDL are injected into the plasma of a normal volunteer the ^{131}I-tracer is cleared more slowly than its native counterpart. If it is accepted that the ^{125}I-LDL traces all catabolic routes and the only activity retarded by cyclohexandione treatment is receptor binding, then the difference in clearance rates of the two tracers is a measure of whole body receptor activity. In normal subjects about 30 to 50% of total LDL catabolism is receptor-mediated, the remainder being carried out by processes acting independently of the high affinity receptor.[25] FH heterozygotes catabolize LDL by the receptor route at half the normal rate while the two homozygous patients examined so far cleared native and chemically modified LDL at the same rate.[25,27] These findings are consistent with the cell culture studies that demonstrated half the normal receptor complement in heterozygotes and a near complete lack of receptors in homozygotes.[20] The sites of receptor-mediated catabolism in man are difficult to locate, but two approaches have been used in animal studies to determine where receptors are most abundant. In the first, microsomal membranes were prepared from a number of tissues and LDL equilibrium binding measured as an index of receptor activity.[28] In the second, comparison was made of the uptakes in vivo of native and chemically modified LDL labeled either with iodine[30] or with a nondegradable marker.[16] The amount of native LDL assimilated by a tissue in excess of the amount of chemically modified LDL appearing there was deemed to indicate the extent of involvement of receptors in the uptake process. Qualitatively, the results of both techniques were in agreement. Adrenal glands were the most active organs at assimilating LDL; and the fact that they could be stimulated to bind even more by the administration of ACTH to the animal[20,28] indicated that the LDL receptor route is subject to regulation in vivo. Other tissues including ovary, testis, liver, and intestine were also active at binding and assimilating LDL.[16,28,29] However, when organ weights were taken into account, it was the liver that was found to contain the main component of LDL receptor-mediated catabolism. Using sucrose labeled tracers, Carew et al.,[16] and later, Dietschy and co-workers[30,31] found that 60 to 70% of total LDL catabolism occurred in the liver; and, of that, a high percentage was due to receptors. Dietschy[31] has compared LDL receptor-mediated uptake and HMG CoA reductase activity in a number of tissues from different animals and has postulated that body tissues can be divided into two broad categories. First there are those that depend heavily on lipoprotein cholesterol and so in response to suppression of plasma cholesterol levels (by administration of 4-aminopyrazolopyrimidine) enhance their rates of endogenous cholesterol synthesis to compensate for the loss of the plasma sterol source. These include *liver*, adrenal gland, *small intestine*, lung, and kidney. Other tissues like skin, the central nervous system, and striated muscle, that normally take up little LDL, hardly change their rates of cholesterologenesis when faced with a plasma deficit of LDL. This kind of information further emphasizes the quantitative importance of the gut-liver axis not only as a source of lipoproteins, but also as the site where LDL catabolism and endogenous cholesterol production are closely balanced.

It is clear from in vivo turnover studies that the receptor pathway is not responsible for all LDL catabolism. In fact the tissues of receptor-deficient humans and animals[32] (Watanabe Heritable Hyperlipemic (WHHL) rabbits) are able to take up and degrade LDL by receptor-independent mechanisms at high rates, but this can only occur at the expense of a greatly increased plasma concentration of the lipoprotein. Expansion of the circulating LDL pool drives these inefficient mechanisms so that in net terms, supranormal amounts of cholesterol are delivered to organs like the liver and spleen despite their lack of receptors.[32] Thus this nonreceptor pathway can meet some of the cell's need for cholesterol in FH. However, endogenous cholesterol synthesis, particularly in younger patients with the disease,

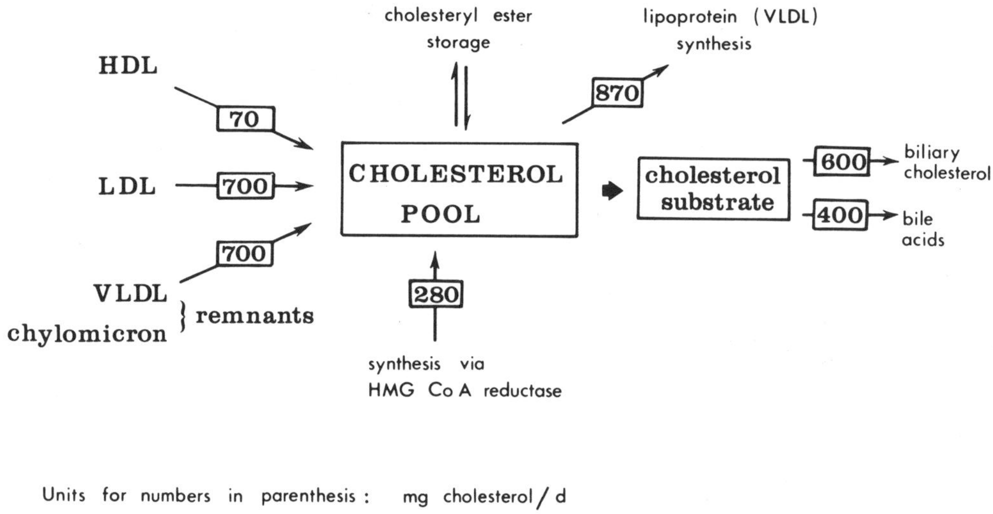

FIGURE 3. Sterol flux within the hepatocyte.

proceeds at a greater than normal rate, perhaps indicating some shortfall in cholesterol supply by receptor-independent routes.[33] In normal subjects too, receptor-independent uptake of LDL accounts for a substantial proportion of overall catabolism.[18] Since, in addition to its paramount role in receptor-mediated LDL catabolism the liver also is a major site of receptor-independent catabolic processes, it is likely that the sterol delivered by the latter route will also contribute to the pool that feeds bile acid synthesis. Some studies[34] suggest that the hepatic parenchyma is primarily responsible for this activity but other workers have speculated that nonparenchymal cells also take up LDL by receptor-independent pathways. Certainly, LDL cholesterol ingested by these cells can be transferred into hepatocytes and thence into the biliary canaliculi.[36] However, it appears that before LDL is endocytosed by such cells in the liver it has to undergo modification in the circulation to render it susceptible to degradation.[36] If the animal data can be extrapolated to man then it is possible to calculate that 60% of the average total LDL apoprotein clearance of 12 $mg \cdot kg^{-1} \cdot day^{-1}$ occurs in the liver. This corresponds to a delivery of 700 $mg \cdot day^{-1}$ of sterol, given that the cholesterol/protein ratio in the lipoprotein is 1.4 (Figures 2 and 3).

B. Metabolism of Chylomicron Remnants

Dietary fat appearing in the upper portion of the small intestine is hydrolyzed and its constituent mono- and diglycerides emulsified along with any cholesterol present, by the action of bile acids. It is thought that these hydrolytic products are incorporated into biliary micelles containing bile acids, lecithin, and cholesterol, which facilitate their passage across the diffusion barrier of the ''unstirred water layer'' into villus cells.[2,3] The cholesterol that appears in the absorptive enterocyte can arise from three sources — from bile, from dietary sterol, or from mucosal cells shed during renewal of the intestinal lining. Bile acids, especially those containing three hydroxyl groups (cholic acid in man), are a prerequisite for efficient sterol absorption which takes place mainly in the upper small intestine along with the absorption of polar lipids.[3,37] The bile acids themselves do not appear to be taken up in this process and their major site of reabsorption is the distal one-third of the ileum.[4] Normally only 30 to 60% of cholesterol present in the gut contents is assimilated by mucosal cells.[3,38] This may be due, in part, to the relatively rapid uptake of the more polar mono- and diglycerides in the proximal part of the duodenum-jejunum, which prevents their participation with bile acids in the absorption of cholesterol from more distal sites. Diets containing large amounts of fat increase the supply of these polar lipids and so favor the absorption of sterol by the intestine.

The daily intake of cholesterol in most "Western" diets is of the order of 300 to 750 mg. To this is added 700 to 1250 mg of biliary sterol so that 1.0 to 2.0 g of the lipid is presented each day to the intestinal tract.[2,31] As mentioned above, absorption is incomplete and only about 45% is taken up into the mucosal cells, the rest appearing in feces as neutral steroid at the rate of approximately 0.5 to 1.0 g · day^{-1}.[2,31,38] The large amount of cholesterol which is absorbed cannot remain in the intestinal cell and so uptake is balanced by:

1. Utilization for membrane synthesis (e.g., in rapidly turning over crypt villus cells)
2. Release, possibly to HDL, by the mechanism discussed in Section II
3. Incorporation as free or esterified cholesterol into chylomicrons

Large chylomicron particles with a diameter of 750 to 10,000 Å, containing on average 1% by weight each of free cholesterol and cholesteryl ester and up to 95% triglyceride,[3,39,40] are secreted by the intestine during dietary fat absorption. Soon after they appear in the systemic circulation, their triglyceride content rapidly falls due to the hydrolytic action of lipoprotein lipase and a smaller, denser particle is produced which is no longer susceptible to hydrolytic attack. This "remnant" is enriched in apolipoproteins B and E and, according to current concepts, contains at least all of the cholesteryl ester present in the original nascent particle.[3] Some sterol ester may also have been incorporated into the core of the lipoprotein by exchange from HDL via cholesteryl ester transfer protein,[14,15] while a portion of the surface phospholipid and free cholesterol will have been lost pari passu into the HDL range. If for the purposes of the present discussion these last two processes are disregarded then the cholesteryl ester content of the chylomicron remnant would be the same as that of its parent particle initially elaborated by the gut.[3,39,40] The remnant with its encapsuled sterol is cleared rapidly from the circulation by a sinusoidal transport mechanism in the liver,[43] which appears to recognize apoprotein E on the surface of the particle.[42,43] Uptake is mediated by a receptor on hepatocyte membranes,[42,43] which is genetically and physiologically distinct from the LDL receptor.[3,44] From the above argument it can be postulated that this mechanism has the potential of depositing 250 to 500 mg of sterol each day in the liver parenchymal cells, an appreciable fraction of the total cholesterol (Figure 3) acquired by the organ.

C. Contribution of VLDL Remnants

The transport of endogenously produced triglyceride is mediated largely through the elaboration of VLDL by the liver. These triglyceride-rich particles undergo peripheral catabolism with removal of their triglyceride by the action of lipoprotein lipase situated on capillary endothelium.[3] After a series of lipolytic steps, a remnant physiologically similar to that derived from chylomicrons is produced. This remnant (with a nominal flotation rate (Sf) of 12 to 100), when compared with its parent nascent particle, is relatively poorer in triglyceride and C apolipoproteins and richer in cholesteryl ester, apolipoprotein B and apolipoprotein E.[45] Early metabolic studies indicated that in normal subjects all VLDL particles were ultimately converted into LDL.[46] However, recent work[47] from our laboratory has shown that in normal and hypertriglyceridemic volunteers, most large triglyceride-rich VLDL remnants are cleared directly from the circulation, presumably into the liver. Thus, their metabolism follows a pattern similar to that of the gut-derived chylomicron. The cholesteryl ester found in the core of the VLDL remnant can arise from two sources. Some undoubtedly will have been synthesized as cholesteryl oleate in the liver via acyl CoA:cholesterol acyltransferase (ACAT)[48] and will have been present within the nascent particle. However the majority of the remnant sterol is condensed with linoleate, good evidence that it originated from the action of LCAT on HDL and was subsequently transferred in the plasma to VLDL by cholesteryl ester transfer protein.[2] The contribution made by VLDL remnant catabolism to the hepatic sterol pool can be estimated from the observations that (1) VLDL with the flotation properties of remnants has a cholesterol: apo B ratio of about 2.5:1[49] and (2) about half of the total circulating VLDL B protein in normal subjects (approximately 200 mg) is cleared directly from the plasma without being converted to LDL.[47] If we assume that the liver is responsible for this clearance, which occurs at a rate of approximately 1.3 pools per day, then over 300 mg of cholesterol may be delivered every day to the organ by this route. There is evidence to suggest that this uptake like that of the chylomicron remnant is receptor mediated. For example, VLDL remnant clearance is delayed in familial hypercholesterolemic patients,[50] and in type III hyperlipoproteinemic subjects,[47] consistent with the view that interaction of apo E on the particle with the LDL receptor is important for its efficient catabolism by the liver. Additional support for this concept comes from our observation that cyclohexanedione treatment of VLDL delays remnant clearance without interfering with initial hydrolysis of the triglyceride-rich particle.[51]

D. Delivery of Cholesterol to the Liver by HDL

HDL secreted by the liver and small intestine has a discoidal structure that is different from the spherical form isolated from plasma.[3,52,53] The transition occurs during its participation in cholesterol transport. Nascent HDL from the liver contains primarily apoprotein E, phospholipid, and some free cholesterol,[52] whereas that derived from the gut has apoprotein AI as its major protein component.[53] LCAT plays a central role in the maturation of these nascent particles. It depletes the lipoprotein of lecithin and cholesterol and generates cholesteryl esters which migrate from the surface of the bilayer to its hydrophobic core producing in the process a thermodynamically stable spherical particle. From the known turnover rates of human apoproteins AI and AII (15 $mg \cdot kg^{-1} \cdot day^{-1}$ and 3 $mg \cdot kg^{-1} \cdot day^{-1}$, respectively[54]), and the cholesterol/protein (0.32) ratio in HDL, it is possible to calculate that the total delivery of sterol by this vehicle to all catabolic sites is about 400 $mg \cdot day^{-1}$. While little is known of the particular tissues involved in the process of HDL clearance, rat liver perfusion experiments[55] indicate that less than 10% occurs in that organ, although a slightly higher value is obtained when the metabolic fate of ^{125}I-cellobiose-tyramine-labeled AI is measured in these animals.[56] Their livers accumulate 26% of the nondegradable radioactive marker, mainly in hepatocytes. If we extrapolate these values to man we must limit the amount of sterol delivered to the liver by HDL degradation to approximately 40 to 100 $mg \cdot day^{-1}$. However, this is not inconsistent with current concepts of centripetal cholesterol transport which, although implicating HDL in the initiation of the process, suggest that it subsequently transfers its sterol load to apo B-containing lipoproteins whose responsibility is to deposit it in the liver.

From the above discussion it is clear that several lipoproteins are responsible for the delivery of cholesterol to the sterol substrate pool in the liver. Receptor-mediated mechanisms seem to be primarily involved although nonreceptor routes also play a role. It is noteworthy that all three of the apo B-containing, cholesterol-enriched lipoproteins (chylomicron remnants, VLDL remnants, and LDL) mentioned above have been invoked as possible agents responsible for the deposition of sterol in the foam cells of growing atherosclerotic plaques. LDL has long been the favorite potential donor of cholesterol in this situation, and intimal concentrations of this lipoprotein correlate with its level in the circulation.[57] The particle will not, however, promote cholesteryl ester deposition in either smooth muscle cells or macrophages grown in vitro.[58] Chylomicron and VLDL remnants, on the other hand, when added to the medium of cultured macrophages will produce intracellular lipid accumulation[58] and these lipoproteins have been cited as good candidates for lipid delivery in the initial stages of formation of atheromatous lesions when macrophages are believed to play a critical role.[59] It is of benefit to the whole organism if this flow of sterol can be directed to the liver rather than to the vessel wall.

E. Sterol Flux in the Hepatocyte

The hepatocyte occupies a central role in the maintenance of corporeal cholesterol homeostasis and appreciation of this has led many investigators to question how the flow of sterol is regulated between the many sites of import and export within the liver cell. Figure 3 presents some estimated rates based on the calculations described above and on the known rates of biliary cholesterol and bile acid output in man. To complete the picture, we need to be able to assess the flux of sterol in two other metabolic pathways — *de novo* synthesis from acetate and secretion in the form of lipoprotein cholesterol. Total body production of the sterol is reported to be in the range 700 to 900 mg · day^{-1},[2,31,38] and recent data from the studies of Dietschy and co-workers[31] indicate that in species with a cholesterol metabolism similar to that operating in man the hepatic contribution to this total is of the order of 30 to 40%. The amount of sterol which leaves the human liver in association with nascent lipoproteins (mainly VLDL) has not been measured. However, it is known that VLDL apo B is synthesized at a rate of 10 to 15 mg · kg · day^{-1};[60] and the free cholesterol/apo B ratio in the largest (i.e., newly produced) VLDL is about 1/1.[3,49] If we assume that the surface of VLDL requires to be stabilized by free sterol to the extent seen in nascent particles then approximately 870 mg per day of free cholesterol must be released into the circulation in association with the lipoprotein. So the VLDL secretory pathway is a major drain on the hepatic sterol pool and this fact may explain the observed close temporal relationship between the diurnal rhythms of VLDL release and HMG CoA reductase activity in rat liver.[61]

If the data summarized in Figure 3 are representative of the situation in man, then it may be concluded (1) that there are two major pathways for the export of hepatic sterol — VLDL synthesis and bile formation and (2) that the liver balances this cholesterol loss by the uptake of remnant lipoproteins and LDL. Any perturbation in hepatic sterol excretion such as occurs following medical or surgical interruption of the enterohepatic circulation has a profound effect on the other pathways of cholesterol metabolism. In these circumstances, the liver has to cope with a six- to sevenfold increase[62] in cholesterol 7α-hydroxylase activity, which increments bile acid production from about 400 mg · day^{-1} to 4 to 5 times that value.[4,63,64]

III. REGULATORY INFLUENCE OF CHOLESTEROL 7α-HYDROXYLASE AND BILE ACIDS ON LIPOPROTEIN METABOLISM

Bile acids have an important modulating effect on the metabolism of cholesterol and plasma lipoproteins at two levels. First, they control their own synthesis by suppressing the activity of cholesterol 7α-hydroxylase, and so under normal conditions minimize the loss of hepatocyte sterol by that route. Secondly, they are essential for the absorption of cholesterol from the lumen of the upper intestine and hence exert influence on cholesterol and lipoprotein synthesis in that organ.[2] Other regulatory effects have been ascribed to the bile acids, including direct inhibition of HMG CoA reductase in the liver and intestine[5,7] and inhibition of hepatic phosphatidic acid phosphatase, the rate-limiting enzyme of triglyceride synthesis.[65] The evidence for these two phenomena is not yet complete, but their existence does offer possible explanations for a number of observations.[66] For example, when cholic acid is fed to rats it suppresses hepatic HMG CoA reductase activity even when the return of cholesterol to the liver is diverted by a thoracic duct fistula or blocked by coadministration of β-sitosterol. This suggests that bile acids themselves have a regulatory influence on cholesterol synthesis.[67,68] However, other experiments performed in the rat seem to indicate that only cholesterol returning to the liver can depress sterologenesis, while bile acids themselves have little influence.[69] This controversy is at present unresolved (see also Chapter 3). The possibility that triglyceride and bile acid synthesis are metabolically linked[65] may account for the repeatedly demonstrated interdependence of these two processes.[70-73] This topic is discussed elsewhere in this monograph (Chapter 12).

Table 1
CHOLESTEROL METABOLISM FOLLOWING INTERRUPTION OF THE ENTEROHEPATIC CIRCULATION

A. Changes in cholesterol balance

All subjects	Plasma cholesterol (mg/dℓ)	Output (mg/day) Fecal bile acids	Neutral sterols
Before therapy	546 ± 58 (n = 13)	193 ± 31	648 ± 81
Cholestyramine	453 ± 66	1434 ± 176	668 ± 87
Ileal bypass	368 ± 46	2414 ± 264	602 ± 62
Homozygous FH			
Before	1196	6 26	263
Cholestyramine	1203	532	296
Ileal bypass	925	741	236

B. Effects on cholesterol turnover

Treatment	Production (g/day)	Ma (g)	Ka (day^{-1})
Control	1.4 ± 0.14	29.1 ± 6.3	0.049 ± 0.010
Resin	2.3 ± 0.40	25.4 ± 3.8	0.089 ± 0.015

A. Effects of Interruption of the Enterohepatic Circulation on Cholesterol and Triglyceride Turnover

Within days of administering a bile acid sequestrant resin to patients, plasma cholesterol falls to a new steady state value. Cholesterol balance studies reveal that the most probable cause of this is the greatly increased fecal loss of bile acids, which prompts the liver to replenish the bile acid pool by cholesterol oxidation and make up its own sterol needs by extracting the lipid from the plasma. Table 1 contains data from a set of hypercholesterolemic subjects[64] who were subjected to both cholestyramine therapy and subsequent ileal bypass. In the group as a whole, the resin was able to produce a 17% fall in plasma cholesterol, apparently by increasing fecal bile acid output by more than 1 $g \cdot day^{-1}$ (i.e. over sevenfold). The total output of neutral steroid of dietary, biliary, or intestinal mucosal origin remained unchanged by the treatment and in fact cholesterol malabsorption is not a feature of resin therapy. McNamara et al.[74] studied the influence of cholestyramine on sterol absorption in a large population of hyperlipidemic individuals and found that, if anything, the mean percentage cholesterol uptake increased during drug therapy. However, immediately after ingestion of the resin there was a transient interference with absorption. The failure to develop general lipid malabsorption in the face of interruption of the enterohepatic circulation of bile acids[66] is good evidence for the substantial reserve capacity of the liver to increase bile acid synthesis and maintain an adequate detergent pool in the upper small intestine. The small but consistent increase[74] in cholesterol absorption during cholestyramine therapy may be explained by either (1) a change in the saturation level of the bile, increasing its solubilization capacity for the sterol[75] or (2) an increase in the relative concentration of cholic acid in bile which could be predicted to promote its cholesterol-carrying capacity and enhance its ability to facilitate sterol absorption.[2]

The fall in plasma cholesterol which attends sequestrant resin therapy or (more strikingly) partial ileal bypass surgery correlates well with the effectiveness of these maneuvers in promoting fecal bile acid loss. However, there is an important exception to this rule in the case of patients with homozygous FH. Their plasma cholesterol levels are refractory to sequestrant resin therapy and in most cases they also do not respond to ileal bypass.[63,64,76] In fact, it is possible to remove gram quantities of steroid in the form of fecal bile acids without observing any fall in their plasma cholesterol. The reason for this aberrant behavior became clear when the mechanism of the hypocholesterolemic effect of these treatments was discovered as discussed in Section III.C.

In the new steady-state situation which follows interruption of the enterohepatic circulation, the increased loss of steroids in feces must be balanced by augmented cholesterol production or depletion of body cholesterol pools since the dietary sterol intake is not markedly altered. Isotopic studies of cholesterol turnover have revealed a near doubling of production rate in subjects given a sequestrant resin.[77,78] Table 1B presents typical results from such a study.[77] It can be seen that the resin caused a fall in the ''rapidly miscible pool'' of body cholesterol which was attributed[77] to a decrement in the plasma component of this pool. Other body sterol pools were not detectably altered during the course of the investigation. The elimination rate of isotopically labeled cholesterol from this rapidly turning-over pool was also increased, in accord with the idea that sterol in the liver and plasma compartments was being synthesized and excreted more quickly than normal. In the study of Miller and his colleagues[77] the stimulation of cholesterol clearance correlated significantly with the elevated rate at which bile acids were lost in the feces. However, these two phenomena are not inevitably linked since, as mentioned above, the administration of resins to FH homozygotes raises their fecal bile acid loss without evoking any change in the turnover of cholesterol in their plasma.[76] The increase in sterol synthesis that follows interruption of the enterohepatic circulation occurs both in the liver and the intestine. Animals given cholestyramine show a severalfold stimulation of hepatic HMG CoA reductase activity.[5,62,66] The mechanism of this stimulation is not clear, although it appears not to derive from changes in phosphorylation status[79] or immunological characteristics of the enzyme.[80] One likely explanation is that the drug-induced fall in a regulatory intracellular sterol pool (secondary to increased bile acid production and possibly also to raised VLDL secretion, see Section II.B) promotes transcription of the enzyme.[6] An alternative possibility, noted earlier, is that medical or surgical biliary diversion may reverse a direct suppressant action of the bile acids (particularly chenodeoxycholate) or hepatic HMG CoA reductase itself.[66] Most of the substrate requirement for enhanced 7α-hydroxylase activity during resin treatment is met by the increase in cholesterologenesis and this induction of synthesis actually blunts the potential hypocholesterolemic action of sequestrant resins. Addition of a second pharmacologic agent to inhibit the rise in HMG CoA reductase activity gives a marked improvement in cholesterol lowering over that obtained using the resin alone.[81]

Triglyceride turnover is also perturbed when the return of bile acids to the liver is blocked. In studies of hyperlipidemic patients given cholestyramine, Angelin and collaborators[82] were able to demonstrate an increase in both the synthetic and catabolic rates of this lipid. Chenodeoxycholic acid administration[82] on the other hand had the opposite effect. The end result in terms of plasma triglyceride levels seems to vary from patient to patient. Most initially normotriglyceridemic subjects have no long-term change in their plasma concentration, but those whose basal levels are above normal tend to show an increment in response to cholestyramine or colestipol.[66] If the stimulation of hepatic triglyceride synthesis is a general phenomenon that remains in force throughout the treatment period then the concentration of triglyceride in the plasma will depend on the patient's ability to handle this excess lipid, presumably, by increasing lipoprotein lipase activity. Subjects with a reserve capacity for lipolysis can accommodate the change in secretion rate while others with already overburdened triglyceride removal mechanisms accumulate the lipid in their circulation.

B. Effects of Interruption of the Enterohepatic Bile Acid Circulation on Lipoprotein Synthesis in the Intestine

Cholesterol synthesis in the intestine of many species appears to be downregulated by the presence of both cholesterol and bile acids in the mucosal cell.[2,7,85] However, in the rat and in man[86] the latter appear to be the predominant regulators of this pathway. Sequestrant resins and ileal bypass surgery lower the concentration of bile acids in the enterocyte and consequent increases in sterol production are observed. Panini et al.[87] have shown that the cholestyramine-induced rise in ileal HMG CoA reductase activity is particularly evident in apical villus cells which are normally associated with cholesterol and bile acid resorption. Indirect evidence for a similar phenomenon in resin-treated humans was obtained by Grundy et al.[63] who biopsied the intestine shortly after feeding patients cholestyramine. They found that the specific activity of mucosal cholesterol was decreased below that of the sterol in plasma or bile, suggesting that synthesis was promoted in that tissue.

During the absorption of a fatty meal the intestine packages triglyceride and cholesterol into chylomicrons which are released in a wave into the lymphatic channels.[3,53] Since fat absorption is not compromised by interruption of the enterohepatic circulation and lipid does not accumulate in the intestinal mucosa, it is possible to infer that chylomicron synthesis and secretion is not altered by sequestrant resin therapy or ileal bypass. The same may not be true of the low level of "small chylomicron" or "VLDL" production that continues even in the fasting state. Starved rats, when given cholestyramine or when subjected to biliary diversion, exhibit a reduced concentration of triglyceride and cholesterol in their intestinal lymph.[84] Thus the small contribution that the intestine makes to VLDL triglyceride in fasting individuals may be even less during resin administration. Conversely, there is evidence to suggest that the gut's production of HDL is stimulated when bile acid resorption is blocked. The plasma HDL_2 concentration is increased in subjects during cholestyramine therapy raising the HDL_2/HDL_3 ratio.[89] This increment is associated with a stimulation of the synthetic rate of apolipoprotein AI so that the concentration of this peptide relative to apo AII rises in the HDL_2 density range.[89] These findings have been interpreted as indicating that the drug promotes the secretion of HDL particles containing AI but no AII[90] and animal studies suggest that the source of these particles may be the intestine. Bile diverted rats secrete more apo AI into the HDL of mesenteric lymph than do those with an intact enterohepatic circulation; and the apo AI which is released appears to be associated with phospholipid in discoidal HDL particles.[91]

C. Metabolic Consequences of Bile Acid Diversion on Lipoprotein Synthesis and Catabolism in the Liver

It has already been noted that derepression of bile acid production unbalances cholesterol homeostasis within the hepatic parenchymal cell and leads to stimulation of HMG CoA reductase activity. However, according to our present understanding of cholesterol metabolism in cells[20] this is not the only change that should occur, not would it produce a hypocholesterolemic effect as evidenced by the response of resin-treated FH homozygotes. Other pathways of sterol flux are also disturbed when the enterohepatic circulation is blocked. Hepatic triglyceride synthesis increases, and this is associated with the appearance in the plasma of large VLDL that has a higher triglyceride/protein and triglyceride/cholesterol ratio than is observed in the basal state.[92] These changes occur in virtually all subjects given colestipol regardless of their initial plasma lipid levels,[92] suggesting that this is a direct response to the resin therapy. Adler et al.[93] have confirmed the generality of this phenomenon by showing that monkeys treated with Triton WR1339 accumulate VLDL in their plasma more rapidly if they have previously been subjected to biliary diversion. After a few weeks of resin administration to humans the concentration, composition, and size of VLDL usually return to their baseline values, presumably due to the adjustment of triglyceride-clearing mechanisms to the altered rate of synthesis so that a new "steady state" is achieved. This increase in lipoprotein synthesis means that more unesterified cholesterol leaves the hepatocyte in association with VLDL, possibly as an essential component of the "coat" of the particle (Figure 3). Clifton-Bligh et al.[94] were able to show that colestipol given to subjects during a cholesterol turnover study caused a decrease in the specific activity of free cholesterol in VLDL at a time when the mass of the free sterol in that fraction was increasing. This indicated that there was an influx of newly synthesized cholesterol into this density interval. Therefore, the liver, at a time when it needs to meet a substrate demand from cholesterol 7α-hydroxylase, exports cholesterol in association with plasma VLDL, and so produces an additional drain on the hepatic sterol pool.

When the cholesterol requirement of hepatocytes is increased by interruption of the enterohepatic circulation the organ responds not only by raising endogenous synthesis, but also by importing lipoproteins from the circulation. Indeed, it has been shown that rats receiving cholestyramine can have their cholesterol production returned to control levels by the intravenous injection of appropriate quantities of chylomicrons.[69] However, in the normal situation in man virtually all of the cholesterol absorbed from the diet that appears in chylomicrons is removed by the liver anyway, so there is little to be gained from stimulating hepatic chylomicron remnant clearance to fulfill sterol requirements. Measurements of the activity of the chylomicron remnant or ("apo E") receptor show that it is relatively unperturbed by manipulation of the hepatic cholesterol pool.[42,95]

Assimilation of VLDL remnants provides an alternative source of sterol for the liver, and their uptake may be stimulated in resin-treated subjects (see Section II.C). Indeed we have found[96] that the concentration of Sf 12-100 intermediate density lipoproteins falls in the plasma of FH heterozygotes when they are placed on resin. However, the metabolic reasons for this change are not yet known.

Most of the hypocholesterolemic action of bile acid sequestrant resins and ileal bypass derives from their influence on LDL cholesterol, which can be reduced by 25 to 35% in patients consuming the drugs and by up to 45% following surgery.[66,97] Apolipoprotein B levels in this fraction are also diminished[98] so that there must be a fall in the number of circulating particles and a change in the turnover rate of the lipoprotein. The earliest kinetic studies of LDL metabolism during cholestyramine therapy were published by Levy and Langer.[99] When they administered the resin to Type II hypercholesterolemic patients it increased the fractional catabolic rate of the lipoprotein protein moiety without affecting its synthesis. We investigated this effect in greater detail[100] by employing chemically modified LDL to discriminate between changes in the receptor and nonreceptor routes of LDL catabolism. FH heterozygotes exhibit half the normal clearance of LDL by the receptor-mediated pathway, but when given cholestyramine this parameter rose by 170% to values near to normal; and similar results were seen in bypass-treated hypercholesterolemics.[101] Subsequent animal studies[29,102] demonstrated that the resin selectively activated LDL receptors located on the surface membranes of liver cells. The increase in hepatic receptor-mediated uptake of LDL correlated strongly with its enhanced plasma catabolic rate. These findings indicate that the hepatocyte possesses a high affinity LDL receptor pathway and responds like other cells to a sterol deficit by synthesizing more receptors, which, when inserted into the membrane, promote increased cholesterol uptake from the environment. Cultured fibroblasts prefer to ingest preformed cholesterol in LDL rather than make it from acetate for themselves and so stimulation of receptors is probably one of the first responses the liver cell makes to the drain of the intracellular sterol pool. For some reason, however, it does not make maximum use of this facility, but also proceeds to replace a large part of the loss by activating HMG CoA reductase (as mentioned in Section III.A). A more complete expression of the potential receptor activity in the liver is seen in animals or patients given both cholestyramine and a reductase inhibitor like mevinolin. Here receptor activity is stimulated markedly and receptor-mediated LDL catabolism shows an increment over and above that observed with sequestrant resin alone.[103] Combined therapy of this kind reduces plasma LDL cholesterol levels by more than 50% compared to pretreatment values.[81,103] Thus, in these patients with a genetically determined inability to catabolize LDL the normal physiological pathway can be activated to levels which overcome virtually completely the inherited defect. This effect is not seen, of course, in those rare individuals homozygous for FH who have no receptors on their liver cells. They do respond to resin therapy or ileal bypass by activating cholesterol 7α-hydroxylase and depleting intracellular sterol levels. However, no receptors are made and so no plasma LDL lowering occurs. All of the replacement sterol which they require is supplied by *de novo* synthesis. The best therapy to date for true "receptor negative" individuals has been to transplant a normal liver with its receptor complement into a homozygous patient.[104] Following this maneuver, plasma cholesterol fell by 70% to near normal values.

There are obviously still many questions that remain to be answered regarding the metabolic response to interruption of the enterohepatic circulation. One intriguing finding is that the synthetic rate of LDL apo B in most subjects receiving resins or surgery remains at the pretreatment value[99,100] despite the channelling of hepatic cholesterol into bile acid synthesis. This contrasts with the effects of neomycin on LDL catabolism.[105] The antibiotic blocks cholesterol absorption from the gut, decreases the supply of sterol to the liver in chylomicrons, and so depletes the hepatic sterol pool. The fall in LDL cholesterol on neomycin treatment is 20 to 25%, similar to the drop observed during resin therapy, and yet this results from diminished synthesis with little or no change in the catabolic rate of the lipoprotein.[105] The answer to this paradox may lie in the purported compartmentalization of cholesterol within the liver.[106] It may be that lipoprotein synthesis derives its sterol from a pool that is separate from that feeding bile acid production. This hypothesis would help explain other observations such as that by Miller et al.[77] and Grundy et al.[63] who found that in patients given sequestrant resin during a cholesterol turnover study, biliary cholesterol had a lower specific activity than plasma sterol or biliary cholic acid which shared the same value for this parameter. Under these conditions biliary cholesterol, therefore, must be derived from a separate pool from that which is used for cholic acid synthesis. The former pool apparently obtains a higher percentage of its sterol from newly synthesized material, while the preferred substrate for 7α-hydroxylase appears to be cholesterol derived from plasma lipoprotein sources.

IV. IMPLICATIONS FOR THE PREVENTION OF ATHEROSCLEROSIS

Until the publication of the results of the Lipid Research Clinics (LRC) Primary Prevention Trial[107] there was no firm statistical evidence that lowering plasma cholesterol levels improved the risk of developing ischemic heart disease, although many clinical trials had obtained results that were suggestive of such an effect. This trial showed that cholestyramine is beneficial and in a concurrent secondary prevention study with the same compound, less progression of atherosclerosis was found in the drug-treated individuals than in those receiving the placebo.[108] Since the drug is a high molecular weight resin that is not absorbed or metabolized, its therapeutic benefits appear to derive from the changes described above in cholesterol and lipoprotein metabolism which center on the liver. This organ, under conditions in which bile acid production is stimulated, obtains cholesterol from plasma LDL via the receptor pathway. In our studies on FH heterozygotes[100] we were able to show that a direct consequence of this action was to alleviate the burden of cholesterol clearance into other, less efficient receptor-independent pathways. LDL cholesterol delivery to these metabolic routes fell by 25% during resin therapy. The particular cell types involved in degrading LDL via low affinity mechanisms are not known but there is some evidence to implicate macrophages in this process. These cells are replete with cholesterol in FH homozygotes,[109] who have to clear large amounts of LDL by the receptor-independent pathway. Furthermore, inhibition of reticuloendothelial activity in vivo suppresses receptor-independent LDL catabolism and leads to an increase in plasma cholesterol.[110] It is conceivable that during cholestyramine therapy the increased delivery of sterol to the liver and its reduced clearance by other pathways diminishes the amount of the lipid deposited in arterial extra- and intracellular sites. More specifically, it may limit the amount acquired by monocyte-macrophages during their passage through the intima, hence reducing their tendency to become foam-cells and participate in the initiation or progression of lesions.[59]

The LRC trial results showed a gradient of benefit in which there was a 2% drop in the risk of developing signs and symptoms of coronary heart disease for every 1% decrement in LDL cholesterol. Since there are now means of reducing LDL levels by more than 50% (see also Chapter 9), the prospects for curtailing the present epidemic of heart disease and improving the prognosis for hypercholesterolemic patients look promising.

ACKNOWLEDGMENTS

This work was performed during the tenure of grants from the Scottish Home and Health Department (K/MRS/50/C429), the Scottish Hospital Endowments Research Trust (HERT 673) and the Medical Research Council (G 8111558SA).

REFERENCES

1. **Turley, S. D. and Dietschy, J. M.,** Cholesterol metabolism and excretion, in *The Liver: Biology and Pathobiology,* Arias, I., Popper, H., Schachter, D., and Shafritz, D. A., Eds., Raven Press, New York, 1982, 467.
2. **Norum, K. R., Berg, T., Helgerud, P., and Drevon, C. A.,** Transport of cholesterol, *Physiol. Rev.,* 63, 1343, 1983.
3. **Havel, R. J., Goldstein, J. L., and Brown, M. S.,** Lipoproteins and lipid transport, in *Metabolic Control and Disease,* Bondy, P. K. and Rosenberg, L. E., Eds., W. B. Saunders, Philadelphia, 1980, 393.
4. **Danielson, H. and Sjovall, J.,** Bile acid metabolism, *Annu. Rev. Biochem.,* 44, 233, 1975.
5. **Rodwell, V. W., Nordstrom, J. L., and Mitschelen, J. J.,** Regulation of HMG CoA reductase, *Adv. Lipid Res.,* 14, 1, 1976.
6. **Brown, M. S. and Goldstein, J. L.,** Multivalent feedback regulation of HMG CoA reductase, a control mechanism coordinating isoprenoid synthesis and cell growth, *J. Lipid Res.,* 21, 505, 1980.
7. **Wilson, J. D.,** The role of bile acids in the overall regulation of steroid metabolism, *Ann. Intern. Med.,* 130, 493, 1972.
8. **Kane, J. P., Hardman, D. A., and Paulus, H. E.,** Heterogeneity of apolipoprotein B: isolation of a new species from human chylomicrons, *Proc. Natl. Acad. Sci. U.S.A.,* 77, 2465, 1980.
9. **Elovson, J., Huang, Y. O., Baker, N., and Kannan, R.,** Apolipoprotein B is structurally and metabolically heterogeneous in the rat, *Proc. Natl. Acad. Sci. U.S.A.,* 78, 157, 1981.
10. **Schaefer, E. J., Jarkins, L. L., and Brewer, H. B.,** Human chylomicron apolipoprotein metabolism, *Biochem. Biophys. Res. Commun.,* 80, 405, 1978.
11. **Morrisett, J. D., Jackson, R. L., and Gotto, A. M.,** Lipoproteins: structure and function, *Annu. Rev. Biochem.,* 44, 183, 1975.
12. **Biesbroeck, R., Oram, J. F., Albers, J. J., and Bierman, E. L.,** Specific high-affinity binding of high density lipoproteins to cultured human skin fibroblasts and arterial smooth muscle cells, *J. Clin. Invest.,* 71, 525, 1983.
13. **Oram, J. F., Brinton, E. A., and Bierman, E. L.,** Regulation of high density lipoprotein receptor activity in cultured human skin fibroblasts and human arterial smooth muscle cells, *J. Clin. Invest.,* 72, 1611, 1983.
14. **Hopkins, G. J. and Barter, P. J.,** Transfers of esterfied cholesterol and triglyceride between high density and every low density lipoproteins: In vitro studies of rabbits and humans, *Metabolism,* 29, 546, 1980.
15. **Albers, J. J., Tollefson, J. M., Chen, C-H., and Steinmetz, A.,** Isolation and characterisation of human plasma lipid transfer proteins, *Arteriosclerosis,* 4, 49, 1984.

16. **Carew, T. E., Pittman, R. C., and Steinberg, D.,** Tissue sites of native and reductively methylated [^{14}C] sucrose-labelled low density lipoprotein in rats, *J. Biol. Chem.*, 257, 8001, 1982.
17. **Langer, T., Strober, W., and Levy, R. I.,** The metabolism of low density lipoprotein in type II hyperlipoproteinemia, *J. Clin. Invest.*, 51, 1528, 1973.
18. **Packard, C. J. and Shepherd, J.,** Low density lipoprotein receptor pathway in man: its role in regulating plasma low density lipoprotein levels, *Atherosclerosis Rev.*, 11, 29, 1983.
19. **Bilheimer, D. W., Stone, N. J., and Grundy, S.,** Metabolic studies in familial hypercholesterolemia, *J. Clin. Invest.*, 64, 524, 1979.
20. **Goldstein, J. L. and Brown, M. S.,** The low density lipoprotein pathway and its relation to atherosclerosis, *Annu. Rev. Biochem.*, 46, 897, 1977.
21. **Innerarity, T. L., Kempner, E. S., Hui, D. Y., and Mahley, R. W.,** Functional unit of the low density lipoprotein receptor of fibroblasts. A 100,000-dalton structure with multiple binding sites, *Proc. Natl. Acad. Sci. U.S.A.*, 78, 4378, 1982.
22. **Schneider, W. J., Beisigel, U., Goldstein, J. L., and Brown, M. S.,** Purification of the low density lipoprotein receptor, an acidic glycoporotein of 164,000 molecular weight, *J. Biol. Chem.*, 257, 2664, 1982.
23. **Mahley, R. W., Innerarity, T. L., and Weisgraber, K. H.,** Alteration in metabolic activity of plasma lipoproteins following selective chemical modification of the apo-proteins, *Ann. N.Y. Acad. Sci.*, 348, 265, 1980.
24. **Brown, M. S., Anderson, R. G. W., and Goldstein, J. L.,** Recycling receptors: The round trip itinerary of migrant membrane proteins, Cell, 32, 663, 1983.
25. **Shepherd, J., Bicker, S., Lorimer, A. R., and Packard, C. J.,** Receptor-mediated low density lipoprotein catabolism in man, *J. Lipid Res.*, 20, 999, 1979.
26. **Slater, H., Packard, C. J., and Shepherd, J.,** Measurement of receptor-independent lipoprotein catabolism using 1,2 cyclohexanedione modified low density lipoprotein, *J. Lipid Res.*, 23, 92, 1982.
27. **Thomson, G. R., Soutar, A. K., Spengel, F. A., Jadhav, A., Gavigan, S. J. P., and Myant, N. B.,** Defects of receptor mediated low density lipoprotein catabolism in homozygous familial hypercholesterolemia and hypothyroidism in vivo, *Proc. Natl. Acad. Sci. U.S.A.*, 78, 2591, 1981.
28. **Kovanen, P. T., Basu, S. K., Goldstein, J. L., and Brown, M. S.,** Low density lipoprotein receptors in bovine adrenal cortex. II. Low density lipoprotein binding to membranes prepared from fresh tissue, *Endocrinology*, 104, 610, 1979.
29. **Slater, H., Packard, C. J., Bicker, S., and Shepherd, J.,** Effects of cholestyramine on receptor-mediated plasma clearance and tissue uptake of human low density lipoprotein in the rabbit, *J. Biol. Chem.*, 255, 10210, 1980.
30. **Spady, D. K., Bilheimer, D. W., and Dietschy, J. M.,** Rates of receptor-dependent and independent low density lipoprotein uptake in the hamster, *Proc. Natl. Acad. Sci. U.S.A.*, 80, 3499, 1983.
31. **Dietschy, J. M.,** Regulation of cholesterol metabolism in man and in other species, *Klin. Wochenschr.*, 62, 338, 1984.
32. **Pittman, R. C., Carew, T. E., Attie, A. D., Witztum, J. L., Watanabe, Y., and Steinberg, D.,** Receptor-dependent and receptor-independent degradation of low density lipoprotein in normal rabbits and in receptor-deficient mutant rabbits, *J. Biol. Chem.*, 257, 7994, 1982.
33. **Schwartz, K. B., Witztum, J. L., Schonfeld, G., Grundy, S. M., and Connor, W. E.,** Elevated cholesterol and bile acid synthesis in a young patient with homozygous familial hypercholesterolemia, *J. Clin. Invest.*, 64, 756, 1979.
34. **Attie, A. D., Pittman, R. C., and Steinberg, D.,** Hepatic catabolism of low density lipoprotein: mechanisms and metabolic consequences, *Hepatology*, 2, 269, 1982.
35. **Nagelkerke, J. F., Barto, K. P., and Berkel, T. J. C.,** In vitro and in vivo uptake and degradation of acetylated low density lipoprotein by rat liver endothelial, Kupffer and parenchymal cells, *J. Biol. Chem.*, 258, 12221, 1983.
36. **Blomhoff, R., Drevon, C. A., Eskild, W., Helgerud, P., Norum, K. R., and Berg, T.,** Clearance of acetyl low density lipoprotein by rat liver endothelial cells, *J. Biol. Chem.*, 259, 8898, 1984.
37. **Arnesjo, B., Nilsson, A., Barrowman, J., and Borgstrom, B.,** Intestinal digestion and absorption of cholesterol and lecithin in the human, *Scand. J. Gastroenterol.*, 4, 653, 1969.
38. **Sodhi, H. S., Kudhodkar, B. J., and Mason, D. T.,** Cholesterol metabolism in clinical hyperlipidemias, *Adv. Lipid Res.*, 17, 107, 1980.
39. **Kostner, G. and Holasek, A.,** Characterisation and quantitation of the apolipoproteins from human chyle chylomicrons, *Biochemistry*, 11, 1217, 1972.
40. **Redgrave, T. and Carlson, L. A.,** Changes in plasma very low density and low density lipoprotein content, composition and size after a fatty meal in normo- and hypertriglyceridemic man, *J. Lipid Res.*, 20, 217, 1979.
41. **Eisenberg, S.,** High density lipoprotein metabolism, *J. Lipid Res.*, 25, 1017, 1984.

42. **Sherrill, B. C.,** Kinetic characteristics of the hepatic transport of chylomicron remnants, in *Disturbances in Lipid and Lipoprotein Metabolism,* Dietschy, J. M., Gotto, A. M., and Ontko, J. A., Eds., American Physiological Society, Bethesda, Md., 1978, 99.
43. **Sherrill, B. C., Innerarity, T. L., and Mahley, R. W.,** Rapid hepatic clearance of the canine lipoproteins containing only the E apoprotein by a high affinity receptor, *J. Biol. Chem.,* 255, 1804, 1980.
44. **Kita, T., Goldstein, J. L., Brown, M. S., Watanabe, Y., Hornick, C. A., and Havel, R. J.,** Hepatic uptake of chylomicron remnants by WHHL rabbits. A mechanism genetically distinct from the low density lipoprotein receptors, *Proc. Natl. Acad. Sci. U.S.A.,* 79, 3623, 1982.
45. **Kane, J. P., Sata, T., Hamilton, R. L., and Havel, R. J.,** Apoporotein composition of very low density lipoproteins of human serum, *J. Clin. Invest.,* 56, 1622, 1975.
46. **Sigurdsson, G., Nicoll, A., and Lewis, B.,** Conversion of very low density lipoprotein to low density lipoprotein. A metabolic study of apolipoprotein B in human subjects, *J. Clin. Invest.,* 56, 1481, 1975.
47. **Packard, C. J., Munro, A., Lorimer, A. R., Gotto, A. M., and Shepherd, J.,** Metabolism of apolipoprotein B in large triglyceride-rich very low density lipoproteins of normal and hypertriglyceridemic subjects, *J. Clin. Invest.,* 74, in press, 1984.
48. **Drevon, C. A., Engelhorn, S. C., and Steinberg, D.,** Secretion of very low density lipoproteins enriched in cholesteryl esters by cultured rat hepatocytes during stimulation of intracellular cholesterol esterification, *J. Lipid Res.,* 21, 1065, 1980.
49. **Kuchinskiene, Z. and Carlson, L. A.,** Composition, concentration and size of low density lipoproteins and of subfractions of very low density lipoproteins from serum of normal men and women, *J. Lipid Res.,* 23, 762, 1982.
50. **Soutar, A. K., Myant, N. B., and Thomson, G. R.,** The metabolism of very low density and intermediate density lipoproteins in patients with familial hypercholesterolemia, *Atherosclerosis,* 43, 217, 1982.
51. **Packard, C. J. and Shepherd, J.,** unpublished data, 1984.
52. **Hamilton, R. L., Williams, M. C., Fielding, C. J., and Havel, R. J.,** Discoidal bilayer structure of nascent high density lipoprotein from perfused rat liver, *J. Clin. Invest.,* 58, 667, 1976.
53. **Green, P. H. R. and Glickman, R. M.,** Intestinal lipoprotein, *Metabolism,* 22, 1153, 1981.
54. **Shepherd, J., Packard, C. J., Patsch, J. R., Gotto, A. M., and Taunton, O. D.,** Metabolism of apolipoprotein AI and AII and its influence on the high density lipoprotein subfraction distribution in males and females, *Eur. J. Clin. Invest.,* 8, 115, 1978.
55. **Sigurdsson, G., Noel, S-P., and Havel, R. J.,** Quantification of the hepatic contribution to the catabolism of high density lipoproteins in the rat, *J. Lipid Res.,* 20, 316, 1979.
56. **Glass, C. K., Pittmann, R. C., Keller, G. A., and Steinberg, D.,** Tissue sites of degradation of apoprotein AI in the rat, *J. Biol. Chem.,* 258, 7161, 1983.
57. **Smith, E. B. and Slater, R. S.,** Relationship between low density lipoproteins in aortic intima and serum lipid levels, *Lancet,* 1, 463, 1972.
58. **Brown, M. S. and Goldstein, J. L.,** Lipoprotein metabolism in the macrophage, *Annu. Rev. Biochem.,* 52, 223, 1983.
59. **Faggioto, A., Ross, R., and Harker, L.,** Studies of hypercholesterolemia in the non-human primate 1. Changes that lead to fatty streak formation, *Arteriosclerosis,* 4, 323, 1984.
60. **Kesaniemi, Y. A., Vega, G. L., and Grundy, S. M.,** Kinetics of apolipoprotein B in normal and hyperlipidemic man: review of current data, in *Lipoproteins Kinetics and Modelling,* Berman, M., Grundy, S. M., and Howard, B. V., Eds., Academic Press, New York, 1982, 181.
61. **Goh, E. H. and Heimberg, M.,** Relationship between the activity of hepatic 3-hydroxy-3-methylglutaryl coenzyme A reductase and secretion of very low density lipoprotein cholesterol in the isolated perfused liver and in the intact rat, *Biochem. J.,* 184, 1, 1979.
62. **Schoenfield, L. S., Bonorris, G. G., and Ganz, P.,** Induced alterations in the rate limiting enzymes of hepatic and bile acid synthesis in the hamster, *J. Lab. Clin. Med.,* 82, 858, 1973.
63. **Grundy, S. M., Ahrens, E. H., and Salen, G.,** Interruption of the enterohepatic circulation of bile acids in man: comparative effects of cholestyramine and ileal exclusion on cholesterol metabolism, *J. Lab. Clin. Med.,* 78, 94, 1971.
64. **Miettinen, T. A. and Lempinen, M.,** Cholestyramine and ileal bypass in the treatment of familial hypercholesterolemia, *Eur. J. Clin. Invest.,* 7, 509, 1977.
65. **Angelin, B., Bjorkhem, I., and Einarsson, K.,** Influence of bile acids on soluble phosphatidic acid phosphatase in rat liver, *Biochem. Biophys. Res. Commun.,* 100, 606, 1981.
66. **Packard, C. J. and Shepherd, J.,** The hepatobiliary axis and lipoprotein metabolism: effects of bile acid sequestrants and ileal bypass surgery, *J. Lipid Res.,* 23, 1081, 1982.
67. **Hamprecht, B., Roscher, R., Waltinger, G., and Nussler, C.,** The influence of bile acids on the activity of rat liver HMG CoA reductase. II. Effect of cholic acid on lymph fistula rats, *Eur. J. Biochem.,* 18, 15, 1971.
68. **Mosbach, E. H.,** Regulation of bile acid synthesis, in *Bile Acids in Human Diseases,* Back, P. and Gerok, W., Eds., F. K. Schattauer Verlag, Stuttgart, 1972, 89.

69. **Nervi, F. O. and Dietschy, J. M.,** The mechanisms of and the interrelationship between bile acid and chylomicron-mediated regulation of hepatic cholesterol synthesis in the liver of the rat, *J. Clin. Invest.*, 61, 895, 1978.
70. **Angelin, B., Einarsson, K., Hellstrom, K., and Leijd, B.,** Bile acid kinetics in relation to endogenous triglyceride metabolism in various types of hyperlipoproteinemia, *J. Lipid Res.*, 19, 1004, 1978.
71. **Angelin, B. and Leijd, B.,** Effects of cholic acid on the metabolism of endogenous plasma triglyceride and on biliary lipid composition in hyperlipoproteinemia, *J. Lipid Res.*, 21, 1, 1980.
72. **Nestel, P. J. and Grundy, S. M.,** Changes in plasma triglyceride metabolism during withdrawal of bile, *Metabolism,* 125, 1259, 1976.
73. **Beil, U., Crouse, J. R., Einarsson, K., and Grundy, S. M.,** Effects of interruption of the enterohepatic circulation of bile acids on the transport of very low density lipoprotein triglycerides, *Metabolism,* 31, 438, 1982.
74. **McNamara, D. J., Davidson, N. O., Samuel, P., and Ahrens, E. H.,** Cholesterol absorption in man: effect of administration of clofibrate and/or cholestyramine, *J. Lipid Res.*, 21, 1058, 1980.
75. **Wood, P. D., Shiola, R., Estrich, D. L., and Splitter, S. D.,** Effects of cholestyramine on composition of duodenal bile in obese human subjects, *Metabolism,* 21, 107, 1972.
76. **Moutafis, C. D. and Myant, N. B.,** The metabolism of cholesterol in two hypercholesterolemic patients treated with cholestyramine, *Clin. Sci.*, 37, 443, 1969.
77. **Miller, N. E., Clifton-Bligh, P., and Nestel, P. J.,** Effects of colestipol, a new bile acid sequestering resin on cholesterol metabolism in man, *J. Lab. Clin. Med.*, 82, 876, 1973.
78. **Goodman, D. S., Noble, R. P., and Dell, R. B.,** The effects of colestipol resin and of colestipol plus clofibrate on the turnover of plasma cholesterol in man, *J. Clin. Invest.*, 52, 2646, 1973.
79. **Mitropoulos, K., Knight, B. L., and Reeves, B. E.,** 3-hydroxy-3-methylglutaryl coenzyme A reductase, *Biochem. J.*, 185, 435, 1980.
80. **Higgins, M. J. P., Brady, D., and Rudney, H.,** Rat liver 3-hydroxy-3-methylglutaryl coenzyme A reductase. A comparison and immunological study of purified solubilised preparations and alterations of enzyme levels by cholestyramine feeding, *Arch. Biochem. Biophys.*, 163, 272, 1974.
81. **Mabuchi, H., Sakai, T., Sakai, Y., Yoshimura, A., Watanabe, A., Wakasugi, T., Kaizumi, J., and Takedam, R.,** Reduction of serum cholesterol in heterozygous patients with familial hypercholesterolemia: additive effects of compactin and cholestyramine, *N. Engl. J. Med.*, 308, 609, 1983.
82. **Angelin, B., Einarsson, K., Hellstrom, K., and Leijd, B.,** Effects of cholestyramine and chenodeoxycholic acid on the metabolism of endogenous triglyceride in hyperlipoproteinaemia, *J. Lipid Res.*, 19, 1017, 1978.
83. **Turley, S. D. and Dietschy, S. M.,** Regulation of biliary cholesterol output in the rat: dissociation from the rate of hepatic cholesterol synthesis, the size of the hepatic cholesteryl ester pool and the hepatic uptake of chylomicron cholesterol, *J. Lipid Res.*, 20, 923, 1979.
84. **Redinger, R. N. and Passi, R. B.,** Effects of plasma cholesterol lowering agents on hepatobiliary lipid metabolism and cholesterol turnover in the Rhesus monkey, *Can. J. Physiol. Pharmacol.*, 57, 235, 1979.
85. **Dietschy, J. M. and Wilson, J. D.,** Regulation of cholesterol metabolism, *N. Engl. J. Med.*, 282, 1128, 1970.
86. **Dietschy, J. M. and Gamel, W. G.,** Cholesterol synthesis in the intestine of man: regional differences and control mechanisms, *J. Clin. Invest.*, 50, 872, 1971.
87. **Panini, S. R., Lehrer, G., Rogers, G. H., and Rudney, H.,** Distribution of 3-hydroxy-3-methylglutaryl coenzyme A reductase and alkaline phosphatase activities in isolated ileal epithelial cells of fed, fasted, cholestyramine fed, and 4-aminopyrazolo [3,4-d] pyrimidine-treated rats, *J. Lipid Res.*, 20, 879, 1979.
88. **Gangl, A. and Ockner, R. K.,** Intestinal metabolism of lipids and lipoproteins, *Gastroenterology,* 68, 167, 1975.
89. **Shepherd, J., Packard, C. J., Morgan, H. G., Third, J. L. H. C., Stewart, J. M., and Lawrie, T. D. V.,** The effects of cholestyramine on high density lipoprotein metabolism, *Atherosclerosis,* 33, 433, 1979.
90. **Shepherd, J. and Packard, C. J.,** High density lipoprotein apolipoprotein metabolism, in *Clinical and Metabolic Aspects of High Density Lipoproteins,* Miller, N. E. and Miller, G. J., Eds., Elsevier, Amsterdam, 1984, 246.
91. **Bearnot, H. R., Glickman, R. M., Weinberg, L. M., Green, P. H. R., and Tall, A. R.,** Effect of biliary diversion on rat mesenteric lymph apolipoprotein AI and high density lipoprotein, *J. Clin. Invest.*, 69, 210, 1982.
92. **Witztum, J. L., Schonfeld, G., and Weidman, S. M.,** The effects of colestipol on the metabolism of very low density lipoproteins in man, *J. Lab. Clin. Med.*, 88, 1008, 1976.
93. **Adler, R., Mangules, E., Motson, R., and Ockner, R. K.,** Increase in production of triglyceride-rich lipoproteins after partial biliary diversion in the rhesus monkey, *Metabolism,* 27, 607, 1978.
94. **Clifton-Bligh, P., Miller, N. E., and Nestel, P. J.,** Changes in lipoprotein lipids in hypercholesterolemic patients treated with the bile acid sequestering resin, colestipol, *Clin. Sci.*, 47, 547, 1974.

95. **Angelin, B., Raviola, C. A., Innerarity, T. L., and Mahley, R. W.,** Regulation of hepatic lipoprotein receptors in the dog, *J. Clin. Invest.*, 71, 817, 1983.
96. **Clegg, R. J., Shepherd, J., and Packard, C. J.,** unpublished data, 1984.
97. **Glueck, C. J., Ford, S. J., Scheel, D., and Steiner, P.,** Colestipol and cholestyramine resins, comparative effects in familial type II hyperlipoproteinemia, *JAMA*, 222, 676, 1972.
98. **Witztum, J. L., Schonfeld, G., Weidman, S. W., Giese, E., and Dillingham, M. A.,** Bile acid sequestrant therapy alters the composition of low density and high density lipoproteins, *Metabolism*, 28, 221, 1979.
99. **Levy, R. I. and Langer, T.,** Hypolipidemic drugs and lipoprotein metabolism, *Adv. Exp. Med. Biol.*, 26, 155, 1972.
100. **Shepherd, J., Packard, C. J., Bicker, S., Lawrie, T. D. V., and Morgan, H. G.,** Cholestyramine promotes receptor mediated low density lipoprotein catabolism, *N. Engl. J. Med.*, 302, 1219, 1980.
101. **Spengel, F. A., Jadhav, A., Duffield, R. G. M., Wood, C. B., and Thomson, G. R.,** Superiority of partial ileal bypass over cholestyramine in reducing cholesterol in familial hypercholesterolemia, *Lancet*, II, 768, 1981.
102. **Kovanen, P. T., Bilheimer, D. W., Goldstein, J. L., Jaramillo, J. J., and Brown, M. S.,** Regulatory role for hepatic low density lipoprotein receptors in vivo in the dog, *Proc. Natl. Acad. Sci. U.S.A.*, 78, 1194, 1981.
103. **Bilheimer, D. W., Grundy, S. M., Brown, M. S., and Goldstein, J. L.,** Mevinolin and colestipol stimulate receptor mediated clearance of low density lipoprotein from plasma in familial hypercholesterolemia heterozygotes, *Proc. Natl. Acad. Sci. U.S.A.*, 80, 4124, 1983.
104. **Starzl, T. E., Bahnson, H. T., Hardesty, R. L., Iwatsuki, S., Gartner, J. C., Bilheimer, D. W., Shaw, B. W., Griffith, B. P., Zitelli, B. J., Malatak, J. J., and Urbach, A. H.,** Heart-liver transplantation in a patient with familial hypercholesterolemia, *Lancet*, 1, 1382, 1984.
105. **Kesaniemi, Y. A. and Grundy, S. M.,** Turnover of low density lipoproteins during inhibition of cholesterol absorption by neomycin, *Arteriosclerosis*, 4, 41, 1984.
106. **Mitropoulos, K. A., Myant, N. B., Gibson, G. F., Balasubramaniam, S., and Reeves, B.,** Cholesterol precursor pools for the synthesis of cholic and chenodeoxycholic acids in rats, *J. Biol. Chem.*, 249, 6052, 1974.
107. The Lipid Research Clinics Program, The Lipid Research Clinics Coronary Primary Prevention Trial Results. 1. Reduction in incidence of coronary heart disease, *JAMA*, 251, 351, 1984.
108. **Brensike, J. F., Levy, R. I., Kelsey, S. F., Passamani, E. R., Richardson, J. M., Loh, I. K., Stone, N. J., Aldrich, R. F., Battaglini, J. W., Moriarty, D. J., Fisher, M. R., Friedman, L., Friedewald, W., Detre, K. M., and Epstein, S. E.,** Effects of therapy with cholestyramine on progression of coronary arteriosclerosis: results of the NHLBI type II coronary intervention study, *Circulation*, 69, 313, 1984.
109. **Goldstein, J. L. and Brown, M. S.,** Familial hypercholesterolemia: Pathogenesis of a receptor disease, *Johns Hopkins Med. J.*, 143, 8, 1978.
110. **Slater, H. R., Packard, C. J., and Shepherd, J.,** Receptor independent catabolism of low density lipoprotein: involvement of the reticuloendothelial system, *J. Biol. Chem.*, 257, 307, 1982.

Chapter 12

CHOLESTEROL 7α-HYDROXYLASE AND BILE ACID SYNTHESIS IN RELATION TO TRIGLYCERIDE AND LIPOPROTEIN METABOLISM

Bo Angelin, Ingemar Björkhem, and Kurt Einarsson

TABLE OF CONTENTS

I. INTRODUCTION

The conversion of cholesterol into bile acids by the liver is a major pathway for cholesterol excretion from the body.[1,2] It is evident that the synthesis of bile acids will be influenced by the inflow of cholesterol to the liver, but there are also several lines of evidence indicating that perturbations of bile acid synthesis will reciprocally influence on plasma lipoprotein turnover.

In this brief review, we will summarize some recent studies regarding the metabolic relationship(s) between cholesterol 7α-hydroxylase activity and bile acid synthesis on the one hand and triglyceride and lipoprotein metabolism on the other (other aspects of these relationships are discussed in Chapter 11).

II. HEPATIC LIPOPROTEIN METABOLISM

The liver has a central role in the regulation of lipoprotein metabolism (Figure 1). It is the major site of very low density lipoprotein (VLDL) assimilation and secretion.[4,5] Furthermore, the liver is presently recognized as the dominant site for low density lipoprotein (LDL) uptake and degradation.[3,5,6] Chylomicrons, which are synthesized from dietary fat and cholesterol in the intestinal wall and secreted into the lymphatics, are converted into chylomicron remnants through the action of tissue lipoprotein lipases. These remnants are rapidly cleared from the plasma by receptor-mediated endocytosis in the liver.[3,5,6] The role of the liver in the synthesis and elimination of high density lipoprotein (HDL) is less well established: the current view is, however, that delivery of cholesterol to the liver can occur by HDL particles.[5]

The synthesis and assimilation of VLDL occurs in the hepatocyte endoplasmic reticulum and Golgi apparatus.[4] The synthesis of triglycerides is probably a major driving force in the production of VLDL. According to current concepts, availability of substrate is the most important determinant of the rate of biosynthesis of triglycerides.[4,7] The relative activity of the different enzymes involved in the sequence of reactions leading to triglyceride are probably less important from a regulatory point of view. Nevertheless, it has been well documented that the phosphatidic acid phosphatase, the enzyme catalyzing the conversion of phosphatidic acid into diglyceride, rapidly responds to physiological changes in the rate of triglyceride biosynthesis.[7] As this enzyme activity is lower than that of other enzymes involved, it is assumed to have a rate-limiting function in the biosynthesis of triglyceride.[7] Also the other constituents of VLDL may be of critical importance in the production of the lipoprotein particle. Lack of apolipoprotein B, as occurs in abetalipoproteinemia, results in deficient VLDL synthesis and hepatic triglyceride accumulation.[8]

The elimination of VLDL triglycerides occurs mainly in peripheral tissues, particularly adipose tissue and skeletal muscle.[4,5] Although some remnants of VLDL catabolism, intermediate density lipoproteins (IDL), may be taken up as such by the liver, a major fraction of VLDL is normally converted into the cholesterol-rich LDL.[4,5] At least in some situations, direct production of LDL by the liver may also occur. The elimination of plasma LDL occurs mainly via specific LDL receptors, recognizing apolipoprotein (apo-) B as well as apo-E.[3,5,6] Such receptors may be found on any cell type, but the major sites of LDL catabolism in vivo are the steroid-synthesizing endocrine organs and the liver.[3,5,6] In all tissues studied, the expression of the LDL (apo-B,E) receptor-binding activity is regulated by the cellular demand for cholesterol. The complex role of the hepatocyte in the regulation of cholesterol metabolism (Figure 1) results in a more complex picture of lipoprotein receptor modulation than generally encountered. As reviewed elsewhere,[3] several situations exist where the LDL receptor (but not the chylomicron remnant or apo-E receptor) is metabolically regulated in animals: LDL receptor binding capacity is diminished with increasing age, with feeding of

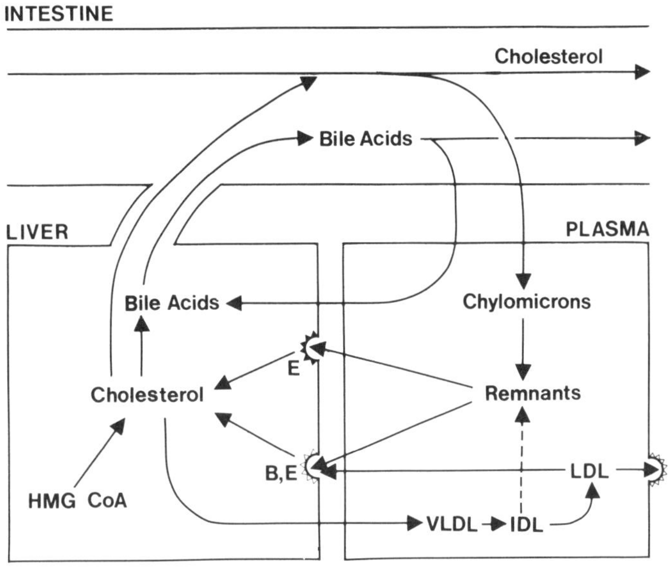

FIGURE 1. Simplified scheme of cholesterol, bile acid, and lipoprotein metabolism. (From Angelin, B., in *Liver and Lipid Metabolism*, Calandra, S., Carulli, N., and Salvioli, G., Eds., Elsevier, Amsterdam, 1984, 187. With permission.)

a cholesterol-rich diet, and with bile acid infusion. Receptor binding is increased by estrogen treatment, by starvation in some species, by cholestyramine treatment, and by blockage of cholesterol biosynthesis. It is evident that the efficiency of the hepatic uptake is of major importance in the determination of plasma LDL concentration in a given situation.

The uptake of chylomicrons by the liver appears to be much less regulated.[3,5,6] Thus, in the species studied, the specific receptor for chylomicron remnants (recognizing apo-E) appears to be very refractory to metabolic regulation, and the major determinant of chylomicron cholesterol inflow to the liver will instead be the amount of dietary cholesterol absorbed. A specific situation exists in hyperlipoproteinemia type III, where a defective apo-E occurs.[9] This apo-E cannot be recognized by the remnant receptor, and chylomicron remnants accumulate in plasma.[9]

The metabolic fate of HDL is less well known. The liver appears to be very active in converting HDL_2 into HDL_3, presumably via the hepatic triglyceride lipase.[10] Some indirect evidence exists that HDL free cholesterol may be a preferential precursor for biliary cholesterol and bile acid biosynthesis.[11] The validity of such studies is difficult to assess, however, as considerable exchange of free cholesterol between different lipoproteins is known to occur in vivo.

III. BILE ACID SYNTHESIS IN RELATION TO LIPOPROTEIN METABOLISM — IN VIVO ASPECTS

A. Bile Acid Metabolism in Various Forms of Hyperlipoproteinemia

Disturbances of lipoprotein metabolism, either primary (and in general inborn errors of metabolism) or secondary to other metabolic disorders, have been extensively studied with regard to bile acid metabolism during recent years. In early studies, a slight reduction in bile acid synthesis, and particularly a reduced synthesis of cholic acid, was observed in many patients with hypercholesterolemia (hyperlipoproteinemia type IIa).[12] Since the abnormality could be completely corrected by treatment with cholestyramine,[12] this was not a primary defect in these patients. Later studies with carefully selected patients have clearly demonstrated that heterozygous familial hypercholesterolemia (in which there is a reduction of hepatic LDL receptor expression) is not associated with any quantitative or qualitative abnormality of bile acid biosynthesis.[13] Furthermore, hypothyroidism, which appears to be linked to a reduced hepatic LDL receptor-mediated uptake,[14] is not associated with a reduced synthesis of bile acid.[15] Finally, patients with a total lack of LDL receptor activity (homozygous, receptor-negative) can still synthesize normal amounts of bile acids and increase their synthetic rate in response to interruption of the enterohepatic circulation.[16] Thus, taken together, the evidence indicates that partial or total lack of LDL receptor activity in the liver is not associated with a reduced production rate of bile acids.

Patients with very low HDL levels have also been studied with regard to cholesterol metabolism. In one patient with fish-eye disease, with HDL levels less than 10% of normal, we were unable to find any quantitative abnormality of bile acid metabolism. The ratio between cholic acid and chenodeoxycholic acid production rates was greatly reduced, however.[17] Similar findings were obtained in two siblings with another genetic HDL deficiency state.[18] Patients lacking the B apoprotein (abetalipoproteinemia) also appear to have normal rates of bile acid biosynthesis.[19] Thus, difficulties in delivery of HDL or LDL to the liver do not appear to be linked to major disturbances of the biosynthesis of bile acids in man. This is probably due to the fact that the hepatic cholesterogenesis is compensatorily increased in these situations, and that more bile acid than usual is derived directly from endogenous hepatic cholesterol. Whether this is of importance for the abnormal ratio between the synthesis of cholic acid and chenodeoxycholic acid sometimes observed remains to be studied.

An interesting situation occurs in type III hyperlipoproteinemia, where (as discussed above) a defective apo-E results in markedly reduced receptor recognition of chylomicron remnants. In these patients, the synthesis rate of bile acids is elevated.[20] This increase may be due to the reduced inflow of dietary cholesterol to the liver, resulting in a reduced feedback inhibition of hepatic cholesterol synthesis. The increased rate of cholesterol synthesis has to be counterbalanced by an increased bile acid production. It is of particular importance to note that the abnormality appeared to be present also in subjects with the apoprotein abnormality without manifest hyperlipoproteinemia.[20]

A particularly important finding is the connection between bile acid biosynthesis and hypertriglyceridemia. It has now been demonstrated in several laboratories that the rate of bile acid formation is frequently enhanced in patients with endogenous hypertriglyceridemia (type IV hyperlipoproteinemia).[12,21] In all published studies, there is a marked heterogeneity, however, and no correlation between plasma VLDL levels and bile acid biosynthesis is generally observed. In more detailed studies from our laboratory,[22,23] we have found a positive correlation between VLDL triglyceride production rate and bile acid biosynthesis in patients with isolated hypertriglyceridemia. About one half of our patients displayed definitely increased levels of bile acid biosynthesis, whereas the other half had levels within the normal range. In these studies, little or no information was available with regard to the underlying genetic mechanism for development of hypertriglyceridemia.

Primary type IV hyperlipoproteinemia may occur as the result of at least two distinct genetic disorders, familial combined hyperlipoproteinemia and familial hypertriglyceridemia.[24] In two separate studies of genetically well-characterized patients with hypertriglyceridemia, we have been able to define the genetic basis for the presence of an abnormal bile acid metabolism in hypertriglyceridemia.[25,26] Thus, patients with familial hypertriglyceridemia (in some cases even without being overtly hypertriglyceridemic) display grossly elevated bile acid production rates, particularly of cholic acid.[25,26] There is also some evidence of a defective intestinal handling of the bile acid pool in these individuals.[25,27] On the other hand, patients with hypertriglyceridemia due to familial combined hyperlipidemia generally have bile acid production rates within the normal range.[25,26,28] As the VLDL triglyceride production rate is generally more elevated in patients with familial hypertriglyceridemia than in patients with familial combined hyperlipidemia,[29,30] these results may explain both the heterogeneity of bile acid synthetic rates and the correlation with VLDL synthesis in a mixed hypertriglyceridemic population.

B. Bile Acid Metabolism During Perturbations of Lipoprotein Metabolism

Changing the composition of chylomicron remnants reaching the liver by dietary manipulations is known to affect bile acid biosynthesis in animals.[31] Also in man, some increase in bile acid production rate may occur during cholesterol feeding, but this is not a consistent finding.[2] A slight increase in bile acid production rate was also seen when carbohydrate was substituted for fat in humans, and it was hypothesized that the latter change was due to an increase in VLDL production rate.[32] Treatment with drugs primarily affecting lipoprotein metabolism, such as nicotinic acid and clofibrate, do not appear to affect bile acid biosynthesis in patients with initially normal production rates.[33,34] On the other hand, patients with hypertriglyceridemia and elevated bile acid production rates frequently reduce their bile acid synthesis when such drugs are given.[33,34] Similar, although less marked, changes have also been observed during weight reduction of hypertriglyceridemic, obese patients.[35] The exact mechanism(s) of these parallel reductions is difficult to define, however, as both drugs and hypocaloric treatment regimens may have primary effects also on bile acid and cholesterol metabolism. At any rate, it is clear that parallel changes in VLDL production rate and bile acid biosynthesis may occur in patients with hypertriglyceridemia.

C. Effects of Interaction with Bile Acid Metabolism on Lipoprotein Kinetics

Interruption of bile acid enterohepatic circulation by cholestyramine treatment or by biliary diversion results in an increased VLDL-triglyceride production rate.[36-39] This is frequently counterbalanced by an increased catabolic rate of the VLDL particle, resulting in transient or only mildly elevated VLDL triglyceride levels. Suppression of bile acid biosynthesis by treatment with chenodeoxycholic acid, on the other hand, leads to a reduced VLDL triglyceride production rate and slightly decreased VLDL triglyceride levels.[38,40] Treatment with cholic acid[41] and ursodeoxycholic acid[42] have not produced as marked effects on triglyceride metabolism as chenodeoxycholic acid. Interestingly, cholic acid and ursodeoxycholic acid also do not affect hepatic cholesterol synthesis to the same degree as chenodeoxycholic acid.[43] A more detailed discussion of possible mechanisms on the molecular level is given below.

As mentioned, the capacity for LDL receptor-mediated uptake of cholesterol in most tissues appears to be regulated by the metabolic demands of the cell.[3,5] Situations where an increased biosynthesis of cholesterol is induced (biliary diversion, cholestyramine treatment, ileal resection) result in an increased hepatic catabolism of LDL,[44,45] and an increased expression of LDL receptors.[3,5,21] On the other hand, reinfusion of bile acids, causing a normalization of the induced state, results in down-regulation of hepatic LDL receptors in the dog.[3,46] Accordingly, treatment of humans with chenodeoxycholic acid — but not with cholic acid or ursodeoxycholic acid — results in elevated plasma LDL levels.[3,46-50] An interesting clinical observation is that serum cholesterol levels were markedly increased after treatment with disulfiram.[51] This compound has recently been shown to exert an inhibitory effect on the cholesterol 7α-hydroxylase in rat liver.[52]

Blockage of hepatic cholesterol synthesis makes the liver dependent on lipoprotein receptors for its cholesterol demand. This leads to an increased expression of LDL receptors, in particular during simultaneous stimulation of bile acid biosynthesis.[3,5] Thus, perturbations of bile acid kinetics may influence LDL catabolism in several clinical situations. A reduction in bile acid production occurs with increasing age in humans;[53] whether this is of importance for the well-known increase in LDL levels with ageing remains to be studied.

In summary, strong interrelationships exist between the cholesterol-bile acid metabolism and the lipoprotein-cholesterol axis. Due to an efficient hepatic cholesterol synthesis, the synthesis of bile acids is not easily influenced by disturbances of hepatic LDL receptors; blockage of chylomicron remnant uptake (as in type III hyperlipoproteinemia) may actually even lead to an overproduction of bile acids. On the other hand, disturbances of bile acid metabolism may trigger an increased VLDL triglyceride production as occurs in some forms of hypertriglyceridemia and during cholestyramine treatment. Furthermore, suppression of bile acid biosynthesis may induce hypercholesterolemia due to down-regulation of hepatic LDL receptors; this is seen at the same time as hepatic VLDL production rate is reduced.

IV. BILE ACID SYNTHESIS IN RELATION TO LIPOPROTEIN METABOLISM — IN VITRO ASPECTS

A. 7α-Hydroxylase in Various Forms of Hyperlipoproteinemia

Very little information is presently available as regards the cholesterol 7α-hydroxylase activity in hyperlipoproteinemia. In a preliminary study by Salen et al.,[54] a relatively low activity of cholesterol 7α-hydroxylase was found in a patient with type IIa hyperlipoproteinemia (hypercholesterolemia), whereas a high activity was demonstrated in a few patients with hypertriglyceridemia. Although these findings are in agreement with the known in vivo data discussed above, they need to be confirmed and extended before any definite conclusions with regard to cholesterol 7α-hydroxylase in situations with an increased or decreased bile acid biosynthesis may be drawn. It has been established that HMG CoA reductase is normal in patients with hypercholesterolemia (type IIa or IIb hyperlipoproteinemia), whereas it is increased about twofold in patients with pure hypertriglyceridemia (type IV hyperlipoproteinemia).[55] In these studies, no determination of cholesterol 7α-hydroxylase was possible, but it could be established that the low synthetic rate of cholic acid frequently seen in type II hyperlipoproteinemia was not due to a deficiency of the 12α-hydroxylase.[55] The relative unimportance of this enzyme in the determination of the ratio between cholic acid and chenodeoxycholic acid synthesis has later been indirectly demonstrated by using labeled precursors.[56]

B. Perturbations of Lipoprotein Metabolism

In several animal studies, an increased amount of cholesterol in the diet has been shown to stimulate cholesterol 7α-hydroxylase activity; in some experiments the results have been less convincing.[57,58] Variations in the amount of dietary cholesterol and the period of feeding may explain some of the differences. Starvation is known to reduce cholesterol 7α-hydroxylase activity.[57] On the other hand, diversion of intestinal lipoproteins from the circulation by a lymph fistula results in a twofold increase in 7α-hydroxylase activity in rats.[59] Treatments with agents blocking cholesterol reabsorption (such as tomatine or beta-sitosterol) do not appreciably change cholesterol 7α-hydroxylase, whereas they result in an increased hepatic cholesterol synthesis (measured as HMG CoA reductase activity).[58]

The varying effects of these different treatments on 7α-hydroxylase may indicate that the delivery of intact chylomicron remnants to the liver could be one important factor in the regulation of enzyme activity. Further studies are needed to test this possibility, however. The above dietary perturbations also represent the rare situations in which there are differing effects on HMG CoA reductase activity and cholesterol 7α-hydroxylase activity.[58] It has to be kept in mind that with an increased hepatic cholesterol synthesis, bile acid biosynthesis may well be increased as a consequence of an increased precursor pool for the cholesterol 7α-hydroxylase. Under such conditions, depending upon the type of assay used, the apparent activity of the cholesterol 7α-hydroxylase may be unchanged. For understanding the delicate modulation of cholesterol 7α-hydroxylase activity and bile acid production by the influx of

Table 1
EFFECTS OF VARIOUS TREATMENTS ON PHOSPHATIDIC ACID PHOSPHATASE, CHOLESTEROL 7α-HYDROXYLASE, AND HMG CoA REDUCTASE ACTIVITIES IN RAT LIVER

Situation	Phosphatidic acid phosphatase	Cholesterol 7α-hydroxylase	HMG CoA reductase
Bile fistula	165 ± 17	300 ± 36	294 ± 10
Cholestyramine treatment	168 ± 12	693 ± 39	685 ± 61
Lymph fistula	203 ± 12	231 ± 13	260 ± 26
Cholesterol feeding	126 ± 11	157 ± 25	21 ± 6

Note: Values for phosphatidic acid phosphatase represent assay of cytosolic enzyme using micellar substrate; 7α-hydroxylase data represent assay using endogenous substrate.[62] Results are given as % of simultaneous control animals for each enzyme; values are mean ± SEM.

Data from Björkhem, I., Angelin, B., Berglund, L., Einarsson, K., Ewerth, S., and Lewenhaupt, A., in *Coordinate Regulation of Cholesterol Metabolism,* Galli, G., Sanghvi, A., and Scallen, T. J., Eds., Raven Press, New York, in press, 1985.

intestinal lipoproteins, further studies are needed in which both net bile acid production rate as well as microsomal cholesterol and cholesterol 7α-hydroxylase activity are measured.

An increased inflow of triglyceride-rich lipoproteins in the liver has been demonstrated to increase hepatic HMG CoA reductase activity; this increase was not seen with cholesterol-enriched remnants.[60] Whether this is also coupled to an increased cholesterol 7α-hydroxylase activity is not known. Stimulation of cholesterol and bile acid production rates by inflow of triglyceride may be one of the mechanisms for the well-known diurnal variation in cholesterol 7α-hydroxylase and HMG CoA reductase activities, although other factors are definitely of importance, too.[57,58]

C. Effects on Triglyceride Metabolism of Perturbations of Bile Acid Turnover

In a series of experiments,[42,61,62] we have determined in parallel activities of the supposedly rate limiting enzyme in hepatic triglyceride biosynthesis, the phosphatidic acid phosphatase, the cholesterol 7α-hydroxylase, and the HMG CoA reductase. In these experiments, rats have been treated by various procedures disturbing the normal synthesis of bile acids and cholesterol. As summarized in Table 1, parallel changes occurred in phosphatidic acid phosphatase activity (both of microsomal and cytosolic enzyme) and cholesterol 7α-hydroxylase in bile fistula animals, during treatment with cholestyramine, during lymph diversion, and during cholesterol feeding. The magnitude of the changes in phosphatidic acid activity observed was in good agreement with the degree of stimulation of VLDL triglyceride production rate seen during cholestyramine feeding (Table 2).

Thus, we were able to show parallel changes in the three enzyme systems during a variety of situations in which hepatic cholesterol and bile acid metabolism was perturbed. The common mediator(s) responsible for the parallel regulation of these three systems have however not yet been defined.

Table 2
PARALLEL INCREASES IN VLDL TRIGLYCERIDE PRODUCTION RATE AND HEPATIC PHOSPHATIDIC ACID PHOSPHATASE ACTIVITIES IN CHOLESTYRAMINE-FED RATS

	Increase relative to control animals
VLDL triglyceride production rate	+40%
Phosphatidic acid phosphatase activity	
Soluble enzyme, micellar substrate	+68%
Soluble enzyme, microsomal substrate	+29%
Microsomal enzyme, micellar substrate	+46%

Note: VLDL triglyceride production rate was measured using the Triton procedure.[63]

Data from References 42 and 62.

V. CONCLUDING REMARKS

There are several points of connection between the metabolism of lipoproteins and the metabolism of cholesterol and bile acids. It is evident that perturbations of one of the systems may lead to changes in the other. Of particular importance is the fact, that interaction with bile acid biosynthesis (and thus regulation of cholesterol 7α-hydroxylase activity) results in clear-cut adaptive responses both in VLDL triglyceride production rate and LDL catabolic rate. Further studies of the mechanisms of these apparent relationships should provide important information not only with regard to the regulation of lipoprotein metabolism, but also to our possibilities of influencing lipoprotein metabolism and cholesterol net excretion in therapeutic situations.

ACKNOWLEDGMENTS

The skillful secretarial assistance of Ms. Lena Ericsson in gratefully acknowledged. The authors' work is supported by grants from the Swedish Medical Research Council (03X-3141 and 03X-4793).

REFERENCES

1. **Hofmann, A. F.,** The enterohepatic circulation of bile acids in man, *Adv. Intern. Med.,* 21, 501, 1976.
2. **Grundy, S. M.,** Cholesterol metabolism in man, *West. J. Med.,* 128, 13, 1978.
3. **Angelin, B.,** Regulation of hepatic lipoprotein receptor expression, in *Liver and Lipid Metabolism,* Calandra, S., Carulli, N., and Salvioli, G., Eds., Elsevier, Amsterdam, 1984, 187.
4. **Jackson, R. L., Morrisett, J. D., and Gotto, A. M., Jr.,** Lipoprotein structure and metabolism, *Physiol. Rev.,* 56, 259, 1976.
5. **Brown, M. S., Kovanen, P. T., and Goldstein, J. L.,** Regulation of plasma cholesterol by lipoprotein receptors, *Science,* 212, 628, 1981.
6. **Mahley, R. W. and Innerarity, T. L.,** Lipoprotein receptors and cholesterol homeostasis, *Biochim. Biophys. Acta,* 737, 197, 1983.
7. **Bell, R. M. and Coleman, R. A.,** Enzymes of glycerolipid synthesis in eukaryotes, *Annu. Rev. Biochem.,* 49, 459, 1980.

8. **Isselbacher, K. J., Scheig, R., Plotkin, G. R., and Caulfield, J. B.,** Congenital β-lipoprotein deficiency: an hereditary disorder involving a defect in the absorption and transport of lipids, *Medicine,* 43, 347, 1964.
9. **Mahley, R. W. and Angelin, B.,** Type III hyperlipoproteinemia: recent insights into the genetic defect of familial dysbetalipoproteinemia, *Adv. Intern. Med.,* 29, 385, 1984.
10. **Nilsson-Ehle, P., Garfinkel, A. S., and Schotz, M. C.,** Lipolytic enzymes and plasma lipoprotein metabolism, *Annu. Rev. Biochem.,* 49, 667, 1980.
11. **Schwartz, C. C., Halloran, L. G., Vlahcevic, Z. R., Gregory, D. H., and Swell, L.,** Preferential utilization of free cholesterol from high-density lipoproteins for biliary cholesterol secretion, *Science,* 200, 62, 1978.
12. **Hellström, K. and Einarsson, K.,** Bile acid metabolism in hyperlipoproteineaemia, *Clin. Gastroenterol.,* 6, 103, 1977.
13. **Angelin, B.,** Bile acid metabolism in heterozygous familial hypercholesterolemia: a study comparing affected and unaffected siblings of four kindreds, submitted for publication, 1985.
14. **Thompson, G. R., Soutar, A. K., Spengel, F. A., Jadhav, A., Gavigan, S. J. P., and Myant, N. B.,** Defects of receptor-mediated low density lipoprotein catabolism in homozygous familial hypercholesterolemia and hypothyroidism in vivo, *Proc. Natl. Acad. Sci. U.S.A.,* 78, 2591, 1981.
15. **Angelin, B., Einarsson, K., and Leijd, B.,** Bile acid metabolism in hypothyroid subjects: response to substitution therapy, *Eur. J. Clin. Invest.,* 13, 99, 1983.
16. **Deckelbaum, R. J., Lees, R. S., Small, D. M., Hedberg, S. E., and Grundy, S. M.,** Failure of complete biliary diversion and oral bile acid therapy in the treatment of homozygous familial hypercholesterolemia, *N. Engl. J. Med.,* 296, 465, 1977.
17. **Angelin, B. and Carlson, L. A.,** Bile acids and high density lipoproteins: biliary lipid metabolism in fish eye disease, *Eur. J. Clin. Invest.,* in press, 1980.
18. **Beher, W. T., Gabbard, A., Norum, R. A., and Stradnieks, S.,** Effect of blood high density lipoprotein cholesterol concentration on fecal steroid excretion in humans, *Life Sci.,* 32, 2933, 1983.
19. **Illingworth, D. R., Connor, W. E., Lin, D. S., and Diliberti, J.,** Lipid metabolism in abetalipoproteinemia: a study of cholesterol absorption and sterol balance in two patients, *Gastroenterology,* 78, 68, 1980.
20. **Angelin, B., Einarsson, K., Holmquist, L., and Leijd, B.,** Bile acid metabolism in primary dysbetalipoproteinemia, submitted for publication, 1985.
21. **Angelin, B., Rudling, M., Einarsson, K., Ewerth, S., Leijd, B., Hershon, K., and Brunzell, J.,** Enterohepatic circulation of bile acids and lipoprotein metabolism, in *Enterohepatic Circulation of Bile Acids and Sterol Metabolism,* Paumgartner, G., Stiehl, A., and Gerok, W., Eds., MTP Press, Lancaster, 1985, 159.
22. **Angelin, B., Einarsson, K., Hellström, K., and Leijd, B.,** Bile acid kinetics in relation to endogenous triglyceride metabolism in various types of hyperlipoproteinemia, *J. Lipid Res.,* 19, 1004, 1978.
23. **Angelin, B., Einarsson, K., Leijd, B., and Wallentin, L.,** Plasma cholesterol esterification rate in type IV hyperlipoproteinemia. Relation to bile acid kinetics and triglyceride metabolism, *J. Lab. Clin. Med.,* 97, 502, 1981.
24. **Motulsky, A. G.,** Current concepts in genetics. The genetic hyperlipidemias, *N. Engl. J. Med.,* 294, 823, 1976.
25. **Angelin, B., Hershon, K. C., and Brunzell, J. D.,** Bile acid metabolism in hereditary forms of hypertriglyceridemia: evidence for an increased synthesis rate in monogenic familial hypertriglyceridemia, submitted for publication, 1985.
26. **Angelin, B., Einarsson, K., and Leijd, B.,** Biliary lipid metabolism in primary hypertriglyceridemia with special reference to hereditary forms, submitted for publication, 1985.
27. **Angelin, B., Björkhem, I., and Einarsson, K.,** Individual serum bile acid concentrations in normo- and hyperlipoproteinemia as determined by mass fragmentography: relation to bile acid pool size, *J. Lipid Res.,* 19, 527, 1978.
28. **Beil, U., Grundy, S. M., Crouse, J. R., and Zech, L.,** Triglyceride and cholesterol metabolism in primary hypertriglyceridemia, *Arteriosclerosis,* 2, 44, 1982.
29. **Chait, A., Albers, J. J., and Brunzell, J. D.,** Very low density lipoprotein overproduction in genetic forms of hypertriglyceridaemia, *Eur. J. Clin. Invest.,* 10, 17, 1980.
30. **Janus, E. D., Nicoll, A. M., Turner, P. R., Magill, P., and Lewis, B.,** Kinetic bases of the primary hyperlipidaemias: studies of apolipoprotein B turnover in genetically defined subjects, *Eur. J. Clin. Invest.,* 10, 161, 1980.
31. **Botham, K. M. and Boyd, G. S.,** The effect of dietary fat on bile salt synthesis in rat liver, *Biochim. Biophys. Acta,* 752, 307, 1983.
32. **Andersén, E.,** On the metabolism of cholesterol and bile acids in normo- and hyperlipidaemic subjects, *Acta Med. Scand.,* Suppl. 643, 1, 1980.
33. **Einarsson, K., Hellström, K., and Kallner, M.,** The effect of clofibrate on the elimination of cholesterol as bile acids in patients with hyperlipoproteinaemia type II and IV, *Eur. J. Clin. Invest.,* 3, 345, 1973.

34. **Einarsson, K., Hellström, K., and Leijd, B.,** Bile acid kinetics and steroid balance during nicotinic acid therapy in patients with hyperlipoproteinemia type II and IV, *J. Lab. Clin. Med.*, 90, 613, 1977.
35. **Leijd, B.,** The effect of weight reduction on biliary lipid metabolism in patients with hyperlipoproteinemia type IV, *Metabolism*, 30, 1185, 1981.
36. **Nestel, P. J. and Grundy, S. M.,** Changes in plasma triglyceride metabolism during withdrawal of bile, *Metabolism*, 25, 1259, 1976.
37. **Adler, R., Margules, E., Motson, R., Way, L., and Ockner, R.,** Increased production of triglyceride-rich lipoproteins after partial biliary diversion in the rhesus monkey, *Metabolism*, 27, 607, 1978.
38. **Angelin, B., Einarsson, K., Hellström, K., and Leijd, B.,** Effects of cholestyramine and chenodeoxycholic acid on the metabolism of endogenous triglyceride in hyperlipoproteinemia, *J. Lipid Res.*, 19, 1017, 1978.
39. **Beil, U., Crouse, J. R., Einarsson, K., and Grundy, S. M.,** Effects of interruption of the enterohepatic circulation of bile acids on the transport of very low density lipoprotein triglycerides, *Metabolism*, 31, 438, 1982.
40. **Begemann, F.,** Influence of chenodeoxycholic acid on the kinetics of endogenous triglyceride transport in man, *Eur. J. Clin. Invest.*, 8, 283, 1978.
41. **Angelin, B. and Leijd, B.,** Effects of cholic acid on the metabolism of endogenous plasma triglyceride and on biliary lipid composition in hyperlipoproteinemia, *J. Lipid Res.*, 21, 1, 1980.
42. **Angelin, B., Björkhem, I., and Einarsson, K.,** The influence of perturbations of bile acid metabolism on the synthesis and catabolism of very low density lipoproteins, in *Recent Advances in Bile Acid Research*, Barbara, L., Ed., Raven Press, New York, in press, 1985.
43. **Angelin, B. and Einarsson, K.,** Regulation of HMG CoA reductase in human liver, in *Regulation of HMG CoA Reductase*, Preiss, B., Ed., Academic Press, New York, in press, 1985.
44. **Shepherd, J., Packard, C. J., Bicker, S., Lawrie, T. D., and Morgan, H. G.,** Cholestyramine promotes receptor-mediated low-density-lipoprotein catabolism, *N. Engl. J. Med.*, 302, 1219, 1980.
45. **Spengel, F. A., Jadhav, A., Duffield, R. G. M., Wood, C. B., and Thompson, G. R.,** Superiority of partial ileal bypass over cholestyramine in reducing cholesterol in familial hypercholesterolaemia, *Lancet*, 2, 768, 1981.
46. **Angelin, B., Raviola, C. A., Innerarity, T. L., and Mahley, R. W.,** Regulation of hepatic lipoprotein receptors in the dog. Rapid regulation of apolipoprotein B, E receptors, but not of apolipoprotein E receptors, by intestinal lipoproteins and bile acids, *J. Clin. Invest.*, 71, 816, 1983.
47. **Angelin, B., Einarsson, K., and Leijd, B.,** Clofibrate treatment and bile cholesterol saturation: short-term and long-term effects and influence of combination with chenodeoxycholic acid, *Eur. J. Clin. Invest.*, 11, 185, 1981.
48. **Albers, J. J., Grundy, S. M., Cleary, P. A., Small, D. M., Lachin, J. M., and Schoenfield, L. J.,** National Cooperative Gallstone Study. The effect of chenodeoxycholic acid on lipoproteins and apolipoproteins, *Gastroenterology*, 82, 638, 1982.
49. **Leijd, B. and Angelin, B.,** Effects of chenodeoxycholic acid and cholic acid feeding on plasma lipoproteins and cholesterol absorption in hyperlipoproteinemia: a cross-over study, in *Bile Acids and Cholesterol in Health and Disease*, Paumgartner, G., Stiehl, A., and Gerok, W., Eds., MTP Press, Lancaster, 1983, 191.
50. **Angelin, B., Ewerth, S., and Einarsson, K.,** Ursodeoxycholic acid treatment in cholesterol gallstone disease: effects on hepatic 3-hydroxy-3-methylglutaryl coenzyme A reductase activity, biliary lipid composition, and plasma lipid levels, *J. Lipid Res.*, 24, 461, 1983.
51. **Major, L. F. and Goyer, P. F.,** Effects of disulfiram and pyridoxine on serum cholesterol, *Ann. Intern. Med.*, 88, 53, 1978.
52. **Andersson, S. and Boström, H.,** Effect of disulfiram on rat liver cholesterol 7α-hydroxylase, *Biochem. Pharmacol.*, 33, 2930, 1984.
53. **Einarsson, K., Nilsell, K., Leijd, B., and Angelin, B.,** Influence of age on secretion of cholesterol and synthesisof bile acids by the liver, *N. Engl. J. Med.*, 313, 277, 1985.
54. **Nicolau, G., Shefer, S., Salen, G., and Mosbach, E. H.,** Determination of hepatic cholesterol 7α-hydroxylase activity in man, *J. Lipid Res.*, 15, 146, 1974.
55. **Ahlberg, J., Angelin, B., Björkhem, I., Einarsson, K., and Leijd, B.,** Hepatic cholesterol metabolism in normo- and hyperlipidemic patients with cholesterol gallstones, *J. Lipid Res.*, 20, 107, 1979.
56. **Björkhem, I., Eriksson, M., and Einarsson, K.,** Evidence for a lack of regulatory importance of the 12α-hydroxylase in formation of bile acids in man: an in vivo study, *J. Lipid Res.*, 24, 1451, 1983.
57. **Myant, N. B. and Mitropoulos, K. A.,** Cholesterol 7α-hydroxylase, *J. Lipid Res.*, 18, 135, 1977.
58. **Björkhem, I.,** Mechanism of bile acid biosynthesis in mammalian liver, in *New Comprehensive Biochemistry*, Danielsson, H. and Sjövall, J., Eds., Elsevier, Amsterdam, in press, 1985.
59. **Björkhem, I., Blomstrand, R., Lewenhaupt, A., and Svensson, L.,** Effect of lymphatic drainage on 7α-hydroxylation of cholesterol in rat liver, *Biochem. Biophys. Res. Commun.*, 85, 532, 1978.

60. **VanZuiden, P. A., Erickson, S. K., and Cooper, A. D.,** Effect of removal of lipoproteins of different composition on hepatic 3-hydroxy-3-methylglutaryl coenzyme A reductase activity and hepatic very low density lipoprotein secretion, *J. Lipid Res.*, 24, 418, 1983.
61. **Angelin, B., Björkhem, I., and Einarsson, K.,** Influence of bile acids on the soluble phosphatidic acid phosphatase in rat liver, *Biochem. Biophys. Res. Commun.*, 100, 606, 1981.
62. **Björkhem, I., Angelin, B., Berglund, L., Einarsson, K., Ewerth, S., and Lewenhaupt, A.,** Effect of flux of lymph and bile acids on the hepatic phosphatidic acid phosphatase in rat liver, in *Coordinate Regulation of Cholesterol Metabolism,* Galli, G., Sanghvi, A., and Scallen, T. J., Eds., Raven Press, New York, in press, 1985.
63. **Otway, S. and Robinson, D. S.,** The use of a non-ionic detergent (Triton WR 1339) to determine rates of triglyceride entry into the circulation of the rat under different physiological conditions, *J. Physiol.*, 190, 321, 1967.

Chapter 13

ALTERATIONS OF BILE ACID METABOLISM IN VARIOUS DISEASE STATES*

Z. Reno Vlahcevic, William Kubaska, and Phillip Hylemon

TABLE OF CONTENTS

* In part supported by NIH grant 1 R01 AM34985-01, and Veterans Administration Research Grant.

I. BILE ACID METABOLISM IN NORMAL SUBJECTS

For the last 15 years, a great deal of progress has been made in our understanding of the metabolism of bile acids in health and disease. This significant progress was heralded by a series of studies defining the physical chemical properties of bile acids, along with an understanding of the role of bile acids in the solubilization of biliary cholesterol and the digestion of fat. Also, the importance of bile acid synthesis as a major excretory pathway for the elimination of cholesterol from the body was documented. In view of these, and many other related developments, a summary of current concepts of bile acid metabolism is timely. In this chapter, special emphasis will be given to the alterations of bile acid metabolism in various diseases, with an insight into how accrued knowledge of bile acid metabolism may influence the approach to diagnosis and therapy.

Cholesterol is the obligatory precursor of bile acids in all species studied.[1,2] Cholesterol degradation to bile acids occurs exclusively in the liver. The generally accepted view of how cholesterol is metabolized into primary bile acids is based primarily on investigations carried out in the rat. Bile acid biosynthetic pathways in man may differ somewhat from those in rat. For example, several alternate pathways of cholesterol degradation to bile acids have been reported in man.[3,4] These include a pathway via the intermediate formation of 26-hydroxycholesterol which favors chenodeoxycholic acid (CDCA) synthesis,[4] and a 25-hydroxycholesterol pathway with almost exclusive formation of cholic acid (CA).[5] The relative contributions of these alternate pathways to bile acid synthesis in man remain to be elucidated. The general role of cholesterol 7α-hydroxylase in the control of bile acid synthesis is reviewed in detail in Chapters 2 and 3.

In man, cholic and chenodeoxycholic acids are the two ''primary'' bile acids which are formed in the process of degradation of cholesterol in the liver. These two bile acids are conjugated with glycine and taurine and then excreted in bile. The ratio of glycine to taurine conjugates differs from species to species. In man, the ratio of glycine to taurine conjugates is about 3:1.[1,2] Under normal circumstances very little bile acid is absorbed in the proximal intestine allowing for optimal micellar solubilization and absorption of ingested fat. About 95% of bile acids are absorbed in the terminal ileum by an active transport mechanism.[6] The absorbed bile acids (in free and conjugated form) are returned to the liver via portal circulation where the free bile acids are reconjugated and again secreted into the bile. Hepatic uptake of conjugated and free bile acids is dependent on a carrier-mediated saturable mechanism.[7] There is evidence for a bile acid receptor protein in liver surface membranes which may be responsible for the initial binding of bile acids for transport across the hepatocyte.[8] Under normal physiological circumstances, bile acid uptake by the liver is so efficient that very low concentrations of bile acids are found in the peripheral circulation. However, in liver disease, plasma bile acids are often elevated.[9,10] It has been estimated that the bile acid pool circulates 6 to 8 times per day; about 20% of the total bile acid pool is lost per day into the feces and must be replaced by new synthesis. Thus, under physiological circumstances, the preservation of bile acids is very effective allowing for the continuous efficient absorption of ingested fat. In the colon, the primary bile acids are subjected to deconjugation and/or 7α-dehydroxylation by the intestinal microbial flora. The later process results in the formation of two ''secondary'' bile acids, deoxycholic acid (DCA) and lithocholic acid (LCA), which are formed from cholic and chenodeoxycholic acid, respectively.[11] About one-third to one-half of the newly formed DCA is absorbed in the colon and transported to the liver. Newly formed LCA is also absorbed, but to a lesser extent than DCA. LCA is conjugated with either glycine or taurine. In man, LCA is also sulfated at the C-3 position to form sulfolithocholylglycine and sulfolithocholyltaurine. These two conjugates of LCA are not rapidly absorbed in the intestines and are readily excreted in the feces. Thus, very little LCA is found in bile. The sulfation probably represents a protective mechanism against known hepatotoxic effects of this bile acid.[12]

The total amount of bile acids circulating in the enterohepatic circulation (liver, biliary radicles, gallbladder, intestines, and portal circulation) is defined as the "bile acid pool". The size of cholic and chenodeoxycholic acid pools can be determined by isotope dilution techniques.[13,14] On an average, the bile acid pool size in normal subjects ranges between 2 to 3 g. Cholic acid synthesis is higher than chenodeoxycholic acid synthesis, while fractional daily turnover of chenodeoxycholic acid is slower than turnover of cholic acid. As a result, about equal proportions of these two bile acids are found in bile.[13,14] Deoxycholic acid comprises about 20% of total bile acids while lithocholic acid is usually less than 2%.

Bile acid synthesis is thought to be regulated by the composition and magnitude of the circulating bile acid pool. Bile acids returning to the liver via portal circulation regulate their own synthesis by acting at the level of cholesterol 7α-hydroxylase, which appears to be the rate limiting enzyme in the bile acid biosynthetic pathway.[15,16] Bile acid synthesis increases severalfold after complete interruption of the enterohepatic circulation, such as in total bile fistula or after ileal resection.[17] Administration of cholestyramine, a bile acid binding resin, also results in an increase in bile acid synthesis. Conversely, oral administration of bile acids results in the suppression of bile acid synthesis.[18,19] The exact mechanism by which bile acid synthesis is regulated at the molecular level is not known. However, recent studies[20] raise some questions about the role of bile acids as the sole effector molecules regulating bile acid synthesis in vivo. The action of bile acids in the feedback regulation of cholesterol 7α-hydroxylase is also discussed in Chapter 3.

Bile is an aqueous solution consisting of 85% water and 15% solids. Most of the solids in bile are cholesterol, bile acids, and phospholipids (lecithin), which are frequently referred to as biliary lipids. Cholesterol, a water-insoluble compound, is solubilized in bile in the form of mixed micelles containing bile acids and lecithin. The ability of bile acids to solubilize cholesterol increases approximately threefold in the presence of lecithin.[21] The interrelationship of these three lipids in bile may play an important role in the pathogenesis of cholesterol gallstone formation in man.[21,22]

Cholesterol is implicated in several disease states (atherosclerosis and cholesterol cholelithiasis), and cholesterol conversion to bile acids represents approximately 50% of the net removal of cholesterol from the body. Hence, the interrelationship between the metabolism of bile acids and cholesterol may be extremely important in the maintenance of cholesterol homeostasis. The body acquires cholesterol either through the diet or via endogenous synthesis (liver, intestines, other tissues). HMG-CoA reductase is the rate-limiting enzyme in the cholesterol biosynthetic pathway. This enzyme is strongly repressed by cholesterol associated with chylomicron remnants.[23] Considerable controversy exists as to whether bile acids regulate cholesterol synthesis via a direct action on HMG-CoA reductase,[24] by their effect on intestinal cholesterol absorption,[25] or by modulation of lipoprotein cholesterol uptake.[26] Under normal physiological conditions, most extrahepatic tissues do not synthesize significant amounts of cholesterol, but depend mostly on plasma low density lipoprotein (LDL)-cholesterol.[27-29] The exact mechanism of mobilization of cholesterol from peripheral tissue has not been elucidated but it appears that high density lipoproteins (HDL) play an important role.[30] Ultimately, all cellular cholesterol must be transported to the liver for removal from the body. Recent studies indicate that separate cholesterol precursor compartments, one for bile acid synthesis and one for biliary cholesterol exist in the liver.[26] Plasma-free cholesterol and, more specifically, HDL-cholesterol appear to be the major contributors of cholesterol to these hepatic precursor compartments.[26,31-33] However, the prevalent view is that the liver removes cholesterol from the plasma via two high affinity apo E and apo B,E receptors.[34,35] The apo E receptor is specific for chylomicron remnants.[24-36] The apo B,E receptor reacts with LDL apo B-containing lipoprotein and HDLs containing apo E. The apo B,E receptors resemble LDL receptors of extrahepatic tissue.[37-41] See Chapters 9, 11, and 12 for extensive further comment on the role of cholesterol 7α-hydroxylase in the control of lipoprotein turnover.

From these considerations, it is apparent that diseases affecting the liver (major site of synthesis of bile acids and cholesterol), gallbladder (reservoir for bile acid pool during interdigestive periods), and intestines (site of bile acid absorption, bacterial biotransformation, and cholesterol absorption) may result in marked abnormalities of bile acid and cholesterol metabolism.

II. ALTERATION OF BILE ACID METABOLISM IN VARIOUS DISEASES

A. Bile Acid Metabolism in Liver Disease

In considering the disturbances of bile acid metabolism produced by liver disease, the central role of the liver in almost all phases of bile acid metabolism must be appreciated. Synthesis, conjugation, uptake, and secretion of bile acids all take place in the liver. All of these processes are responsible for the maintenance of integrity of the enterhepatic circulation, which in turn allows for the normal physiologic action of bile acids to proceed. Therefore, it is not surprising that diseases of the liver and/or biliary tract are associated with profound disturbances of bile acid and, to a lesser extent, cholesterol metabolism.

1. Acute Liver Cell Diseases (Hepatitis)

In acute diseases of the liver, such as hepatitis (regardless of etiology), the major pathologic feature is a variable degree of liver cell necrosis and liver cell death. In hepatitis, the anatomic alterations associated with chronic liver disease are not present (fibrosis, alteration of hepatic vasculature, etc). There is a paucity of the abnormalities of bile acid metabolism in acute liver disease. In acute hepatitis, there is a delayed clearance of intravenously administered cholic acid, and a decrease in cholic acid biosynthesis, cholic acid pool size, bile acid conjugation, and biliary bile acid concentrations.[43] As in chronic liver disease, plasma bile acids are elevated and the excretion of sulfated bile acids in urine is increased.[10]

In chronic hepatitis, plasma bile acid concentration is increased.[10,44] Several publications report that plasma bile acid levels could be used as a sensitive test for the improvement (or worsening) of liver disease.[44-47] Apparently, elevation of plasma bile acid levels may persist in patients with chronic hepatitis in the face of normalization of liver function. Also, during the relapse of chronic active hepatitis, the elevation of plasma bile acids occurs earlier than the elevation of routine liver enzymes (SGOT, SGPT).[44] A bile acid tolerance test has been reported to be a more sensitive index of active liver disease than liver enzyme levels.[47] With the wide availability of bile acid radioimmunoassay, it is now practical and relatively inexpensive to monitor bile acid levels routinely in patients with acute or chronic liver diseases. However, the advantages of this approach over routine assessment of liver enzyme levels are not great, which is probably the reason why this test has not gained wide acceptance.

2. Chronic Liver Disease

In patients with chronic liver disease (cirrhosis), several major anatomic and functional abnormalities occur, such as extensive porto-systemic shunts, reduction in functional liver cell mass, various degrees of intracellular cholestasis, and abnormal cholesterol and lipoprotein metabolism. The combination of all these factors account for many observed abnormalities of bile acid metabolism. The presence of extensive porto-systemic shunts results in an impaired bile acid uptake by the liver and a high concentration of bile acids in the peripheral blood and ascites.[48,49] Bile acid distribution is altered since bile acids are no longer confined solely to the enterohepatic circulation. As a consequence, there is a higher urinary concentration of bile acids and deposition of bile acids into the skin.[50] Several studies have demonstrated that a significant proportion of serum and urinary bile acids are sulfated in cirrhosis.[9,10] Renal clearance of sulfated bile acids is 20 to 200 times greater than the clearance of nonsulfated bile acids.[10]

In patients with cirrhosis, there is a striking reduction in the percentage of cholic acid, coupled with reduction or complete absence of deoxycholic acid and a relative increase in the percentage of chenodeoxycholic acid.[48,49] The total bile acid pool in these patients is markedly diminished when compared to normals (1.2 vs. 2.5 g). Diminution of the bile acid pool size results from a marked reduction in the size of cholic and deoxycholic acid pools, while the chenodeoxycholic acid pool is not significantly altered. The most remarkable observation, however, was the finding of a marked reduction of cholic acid synthesis in cirrhosis, while chenodeoxycholic acid synthesis was not changed remarkably. The mechanism responsible for the selective decrease in cholic acid synthesis is not well understood. The most likely defect was thought to be a change in 12α-hydroxylase, which is specific for the cholic acid biosynthetic pathway. However, after administration of cholestyramine,[51] or after institution of bile fistula,[52] patients with cirrhosis substantially increased bile acid synthesis suggesting relatively normal 12α-hydroxylase activity. In vivo studies in cirrhotic patients do not clearly demonstrate a definitive block in the cholic acid biosynthetic pathway.[53] Alternatively, these patients may be lacking cholesterol substrate for bile acid synthesis either due to reduced cholesterol synthesis or reduced hepatic uptake of compositionally abnormal plasma lipoproteins, or both.[53] The fractional turnover rate of cholic acid in patients with cirrhosis is markedly decreased while the turnover of chenodeoxycholic acid is normal.[48,49] There is not a good explanation for the differences in the turnover rates of these two bile acids. Extensive studies of secondary bile acid metabolism in patients with cirrhosis suggest that conversion of cholic acid to deoxycholic acid is selectively impaired, presumably secondary to abnormal colonic flora.[54,55] McCormick et al.[56] correlated cholic and chenodeoxycholic acid synthesis and fractional turnovers with the severity of liver disease. They found an inverse correlation between cholic acid synthesis and the severity and progression of liver disease. By contrast, there was no correlation between chenodeoxycholic acid synthesis and the severity of liver disease.

The reduction of bile acid pool size may play a role in the pathogenesis of cholesterol gallstones; therefore, one might expect that patients with cirrhosis should have a higher incidence of cholesterol gallstones.[13] Indeed, several authors have reported that the incidence of gallstones in liver disease patients is about threefold higher than in the general population.[57,58] However, the majority of gallstones in cirrhosis appear to be calcium bilirubinate stones. The most likely explanation for the low incidence of cholesterol cholelithiasis in patients with cirrhosis is marked reduction of biliary cholesterol secretion.[59] Reduced cholesterol synthesis, inadequate absorption of dietary cholesterol, or a combination of both could be responsible for the marked reduction of biliary cholesterol secretion observed in these patients. As a result of decreased biliary cholesterol secretion, the bile of patients with cirrhosis is undersaturated with cholesterol in spite of significant diminution of the bile acid pool size.[59]

Several studies have shown that mild to moderate steatorrhea is common in patients with cirrhosis.[60,61] This observation is best explained by the reduction of the duodenal concentration of bile acids secondary to a diminished bile acid pool and/or redistribution of bile acid pool between enterohepatic circulation and plasma.

3. Bile Acid Metabolism in Cholestasis

In cholestasis, bile acid composition is also altered. In addition to naturally occurring bile acids several less common bile acids are encountered such as 3β-hydroxy-5-cholenoic acid, ursodeoxycholic acid, hyocholic, hyodeoxycholic acid, etc. In cholestasis, sulfation and glucuronidation of bile acids become an important bile acid excretory pathway.[62-64] Following the administration of ^{14}C cholic and ^{14}C chenodeoxycholic acid to patients with biliary atresia, more than 75% of the administered bile acids were secreted in urine as sulfated metabolites.[63] No significant differences in the sulfation of bile acids were detected between patients with extrahepatic obstruction, hepatitis, and cirrhosis. Patients with primary biliary cirrhosis often suffer from pruritis and at the same time have a large increase in plasma bile acids.[50] However, the correlation between bile acids in plasma and presence/absence of pruritis is low. Several other cholestatic entities have been described such as cholestasis of pregnancy, intrahepatic and extrahepatic cholestasis of infancy, etc.[66] The role of bile acids, if any, in the pathogenesis of these disorders is not clear; the activity of cholesterol 7α-hydroxylase in drug-induced cholestasis is discussed in Chapter 9.

B. Abnormalities of Bile Acid Metabolism in Cholesterol Gallstone Disease

The incidence of cholelithiasis is about 15% in the U.S. and 75% among American Indians.[1,2] The majority of gallstones in developed countries are predominantly cholesterol stones. Several recent observations dramatically increased our knowledge of this disease. First, we now recognize that biliary cholesterol is solubilized in bile in the form of mixed micelles containing bile salts and phospholipids (lecithin).[21] Second, supersaturation of bile with cholesterol appears to be an important risk factor in the formation of cholesterol gallstones.[21] Third, supersaturated bile could be caused by an increase of biliary cholesterol excretion, a decrease of cholesterol-solubilizing substances in bile, i.e., bile acid and lecithin, or a combination of both factors.[13]

Several studies have shown that Caucasian males and American Indians (males and females) with cholesterol gallstones have smaller bile acid pool sizes than subjects without gallstones.[13,67,68] In this population, the major pathogenetic factor responsible for cholesterol gallstone formation may be the reduction of bile acid and phospholipid secretion, both of which could result in formation of bile supersaturated with cholesterol. The most plausible explanation for the decrease in bile acid pool size in these patients is an increased intestinal loss of bile acids along with a decreasae of *de novo* hepatic synthesis of bile acids.[13] The formation of lithogenic bile in Caucasian females, American Indians, and in obese patients was linked to an increase in biliary cholesterol secretion.[69,70] The mechanism responsible for increase of biliary cholesterol secretion is not known, but may be associated with an increase in dietary cholesterol,[71] effects of hormonal factors,[72] increase in endogenous cholesterol synthesis,[73] or an increase in fluxes of plasma free cholesterol and/or newly synthesized cholesterol into a hepatic cholesterol precursor compartment associated with the excretion of biliary cholesterol.[74]

The recognition that the bile acid pool is reduced in these patients provided the rationale for oral administration of bile acids to dissolve gallstones. The hypothesis was that administered bile acids will augment bile acid pool size which should lead to increased secretion of bile acids and phospholipids and, consequently, result in desaturation of bile. It was known for some time that cholesterol stones can be dissolved if they are bathed by bile undersaturated with cholesterol.[75] This hypothesis was tested in numerous clinical trials in which bile acids were administered orally for a period of 6 months to two years. The results of these studies show that gallstone dissolution can be accomplished by the prolonged administration of CDCA or UDCA, but not CA.[76-80] In these studies, all exogenously administered bile acids augment bile acid pool size but only those reducing biliary cholesterol secretion (CDCA and UDCA) were proven effective in dissolving cholesterol gallstones. UDCA has been shown to be equally as effective as CDCA, but in lower doses. UDCA also induces diarrhea less frequently than CDCA and has not been associated with an elevation of serum cholesterol levels.[82,83] Gallstone calcification, however, may be accelerated upon treatment with UDCA.[81-85] The exact mechanisms by which these two bile acids decrease biliary cholesterol excretion still remain obscure. One possible explanation is that they reduce intestinal cholesterol absorption, but these data are equivocal.[86-93] An alternative hypothesis is that chenodeoxycholic acid reduces hepatic cholesterol synthesis, which in turn results in a diminished rate of biliary cholesterol secretion and bile acid synthesis.[73,94-96] The effects of UDCA on cholesterol synthesis is also unclear.[85,95-98] The possible effects of these bile acids on rates of lipoprotein cholesterol uptake have not been adequately studied. Further experimentation is needed to elucidate the mechanism of action of these two cholelitholytic bile acids. The effect of the bile acids on cholesterol 7α-hydroxylase activity is also discussed in Chapter 9.

C. Bile Acids in Hyperlipoproteinemia

It has been recognized that different hyperlipidemias may be associated with certain abnormalities in bile acid metabolism. Type II hyperlipoproteinemia is characterized by a below normal synthesis rate of cholic acid, a defect compensated by the higher CDCA synthesis; total bile acid synthesis is normal. The reason for a specific reduction of cholic acid synthesis is not known.[99] In Type IV hyperlipoproteinemia, an abnormally high incidence of cholelithiasis was observed. Despite an increased incidence of cholesterol gallstones in Type IV, the bile acid pool size in these patients is similar to normal subjects.[99] Thus, the major underlying reason for the development of gallstones in these patients is probably not altered bile acid metabolism but more likely a poorly defined abnormality of cholesterol metabolism resulting in enhanced biliary cholesterol secretion. The regulation of cholesterol 7α-hydroxylase in hyperlipidemia is discussed in detail in Chapters 11 and 12.

D. Bile Acid Metabolism in Intestinal Diseases

The small intestine plays a central role in the conservation of bile acids, i.e., the maintenance of enterohepatic cycle. Therefore, multiple abnormalities of bile acid metabolism are expected to occur in the diseases of the small intestines, particularly the terminal ileum, where most of the absorption of bile acids takes place. In Crohn's disease involving the terminal ileum, or after ileal resection, the absorption of bile acid is either impaired or abolished, resulting in varying degrees of interruption of the EHC. Depending upon the extent of bile acid loss, and/or bacterial overgrowth, clinical symptoms of cholerheic enteropathy (bile acid diarrhea) or steatorrhea may be present. The appreciation of pathophysiology of these disorders, as it pertains to changes of bile acid metabolism, is necessary for a better understanding of clinical symptoms.

1. Contaminated Small Bowel Syndrome (CSBS)

In healthy subjects, a significant proportion of bile acids is deconjugated prior to absorption from the terminal ileum thus allowing for continuous intraluminal fat digestion.[1,2] The rate of deconjugation of bile acids is markedly increased if bacteria have access to the circulating bile acid pool more proximally in the intestine and/or if a larger than normal fraction of the bile acid pool is delivered into the colon, where an abundance of bacteria is normally present. Anaerobic bacteria are primarily responsible for the deconjugation of bile acids. Several studies have shown that malabsorption in CSBS may be due to the combination of a direct toxic effects of free bile acids on the jejunal mucosa[100,101] and a significant reduction of the concentration of conjugated bile acids in the intestines.[102] The clinical conditions in which CSBS is most likely to occur include diseases of small bowel motility (scleroderma, pseudo-obstruction, small bowel diverticulosis), diseases with multiple areas of small bowel narrowing and partial obstruction (most commonly Crohn's disease), jejuno-ileal bypass, small bowel resection, etc. The feature common to all of those disorders is the presence of bacteria in the upper intestine capable of deconjugating bile acids. Deconjugated bile acids have higher pKas, which in turn facilitates passive absorption of free bile acid in the jejunum ("jejunohepatic circulation"). Such premature absorption of bile acids may lead to an inadequate luminal micellar solubilization of digested fat, fat maldigestion and steatorrhea.

The objective clinical diagnosis of this condition is possible but not readily available in many medical centers. An increase in bacterial count is not reliable and requires small intestinal intubation.[103] The simpler tests are cholylglycine [^{14}C] breath test[104,105] and the more recently developed [^{14}C] xylose test.[106] The former test is based on the assumption that only bacteria can deconjugate bile acids and that the amino acid released after the cleavage of the amide bond is partially exhaled in breath air as [^{14}C] CO_2. Several investigators have shown the usefulness of this test in the detection of contaminated bowel syndrome.[105,106] The cholylglycine breath test does not provide information on the site of bile acid deconjugation (small intestine or colon); but if 24 hr fecal ^{14}C-excretion is determined simultaneously, separation between small intestinal bacterial overgrowth and bile acid malabsorption may be possible. The [^{14}C] xylose test appears to have a great degree of sensitivity and specificity than the cholylglycine [^{14}C] breath test in detecting the presence of small intestinal bacterial overgrowth.

The aim of therapy in CSBS is to reduce the bacterial overgrowth. This may be achieved by the administration of antibiotics and/or correction (usually surgical) of the small bowel abnormality leading to stasis of microbial overgrowth.[105] If antibiotics are effective, the diarrhea and steatorrhea should improve within a week after the beginning of therapy.

2. Bile Acid Malabsorption with Cholerheic Enteropathy (Regional Ileitis)

This complication is most commonly encountered in Crohn's disease involving the terminal ileum or in patients who have undergone an ileal resection. In the case of a poorly functioning (or resected) ileum, a significant portion of bile acids is not absorbed and appears in high concentration in the colon.[107] Mekhjian et al.[108] have shown that deoxycholic and chenodeoxycholic acids inhibit absorption and induce reversible secretion of water and sodium in the human perfused colon; these effects are apparently dose dependent. By contrast, cholic acid, even in high concentrations, did not result in excessive sodium and water secretion. Identical results have been reported using in vitro models.[109] This and subsequent studies provide evidence that secretory diarrhea produced by certain bile acids is probably mediated by cyclic AMP-dependent mechanisms.[110] Because of an increased intestinal bile acid loss, these patients have a reduction in portal blood concentrations of bile acids, which is then coupled with a compensatory increase of bile acid synthesis (synthesis = loss). Consequently, there is a mechanism for the bile acid pool size to remain constant. This in turn perpetuates diarrhea, as bile acid concentrations in the lumen of the colon remain abnormally high. The syndrome of cholerheic enteropathy is particularly evident in Crohn's disease of moderate intensity and ileal resection of less than 100 cm.[111]

The cholerheic enteropathy presents with a watery diarrhea of 5 to 15 bowel movements per day. Stool fat content is usually less than 20 g · day^{-1} on a 100 g fat diet.[111] Because there is no significant degree of steatorrhea, patients usually do not suffer fron extensive weight loss. The objective documentation of this disease entity is difficult. Excessive watery diarrhea with stool fat $<$20 g · day^{-1}, in the setting of Crohn's disease or ileal resection, strongly suggest that the patient has cholerheic enteropathy. Also, a [^{14}C] cholylglycine breath test will be abnormal. Once this entity is suspected the patients should be treated with bile acid-binding resins (cholestyramine, cholestipol). Cholestyramine binds bile acids intraluminally and prevents them from causing mucosal damage and secretory diarrhea.[112] The patient's response to the administration of cholestyramine is usually dramatic, with a marked reduction of the number of bowel movements.

3. Bile Acid Malabsorption with Steatorrhea (Regional Ileitis)

As previously mentioned, steatorrhea is frequently present in patients with CSBS. However, it is not unusual that steatorrhea also complicates advanced cases of regional ileitis or after ileal resection of more than 100 cm ileum without evidence of bacterial overgrowth.[111] Under these conditions, bile acid loss is extensive, so it may overcome the synthesizing capacity of the liver (loss $>$ synthesis) leading to eventual shrinkage of the bile acid pool.

In extreme cases, the bile acid pool is practically nonexistent as the EHC may be completely interrupted. In this new steady state, bile acid synthesis is maximal, because bile acids are almost totally lost on the first pass through the intestines. In this situation the intestinal concentration of bile acids is markedly reduced. The latter results in an inadequate micellar solubilization of ingested fat and consequently steatorrhea. Twenty-four hr stool fat in these patients is always above 20 g $\cdot$ day^{-1} on a 100 g fat diet.

Hofmann et al.[112] concluded that with large ileal resection, diarrhea might be due to the secretory effects of fatty acids rather than bile acids (''fatty acid diarrhea''). Subsequently, McJunkin et al.[113] reported that fecal pH may also be an important factor in diarrhea. In small ileal resections, high fecal dihydroxy bile acid concentration was present together with an alkaline pH and mild steatorrhea. In contrast, large ileal resections or jejunoileal bypass, the fecal pH was acidic and bile acid concentration was low (below 1.5 m*M*, i.e., below ''secretory'' levels). The authors concluded that dilution of colonic contents by large amounts of fluid delivered from the small intestine plays a role in diarrhea, secondary to large ileal resections or jejunoileal bypass.

The patient's ability to solubilize ingested fat is likely to be more effective in the morning, as bile acids synthesized overnight are sequestered in the gallbladder and released into the intestine at the time of the first meal. During the day, progressively smaller amounts of bile acids will be available for digestion of fat thus worsening maldigestion. An even worse scenario might be present if the patient has had a cholecystectomy (no reservoir for bile). This is not an unusual situation as the incidence of cholelithiasis in patients with Crohn's disease is three times higher than in the general population presumably because of a significant reduction of the bile acid pool size.[13] Cholestyramine therapy is not indicated in these patients as it worsens the steatorrhea. Therefore, the clinical approach is limited to the low fat diet, medium chain triglycerides (MCT), and substitution of calories with elemental diet; neither elemental diet nor MCT require bile acids for absorption.

4. Sprue

The steatorrhea in the topical and nontopical sprue is predominantly due to malabsorption of fat secondary to involvement of small intestinal mucosa. However, the available evidence suggests that maldigestion secondary to abnormalities of bile acid metabolism may also exist. Low-Beer et al.[114] have shown that the bile acid pool in sprue is enlarged presumably due to the infrequent contraction of the gallbladder, possibly due to impaired CCK release. It has been reported that the bile acid pool in sprue circulates 1 to 2 times per day rather than 6 to 7 times per day. Thus, bile acid excretion is reduced, but synthesis is unchanged resulting in an enlargement of the bile acid pool. The sequestration of the bile acid pool in the gallbladder combined with the sluggish EHC may result in the reduction of bile acid concentration in the intestinal lumen and consequently fat maldigestion. It is not certain to what extent, if any, bacterial overgrowth contributes to steatorrhea observed in patients with sprue.[115,116]

5. Idiopathic Bile Acid Diarrhea

A common problem in the practice of gastroenterology is chronic diarrhea without an obvious organic cause. Diarrhea in at least some of these patients may be related to selective malabsorption of bile acids in the ileum without any radiological evidence of ileal disease.[117] This syndrome was labeled as ''idiopathic bile acid catharsis'' and is frequently diagnosed as functional bowel disease. The objective evidence for the existence of idiopathic bile acid malabsorption is severalfold: (1) increased fecal [^{14}C] radioactivity after oral administration of [^{14}C] cholylglycine, (2) increase in fractional turnover rates of bile acids and (3) improvement of diarrhea after administration of cholestyramine.[118]

6. Bile Acid Metabolism in Jejunoileal Bypass

In the past, the jejunoileal bypass was a rather popular treatment for morbid obesity. This rather drastic surgery has been associated with many complications such as diarrhea, bile salt loss, liver disease, electrolyte imbalance, trace metal deficiency, and hyperabsorption of oxalate and hyperoxaluria.[119] Malabsorption of fat was attributed mainly to exclusion of major portions of absorptive surface of the jejunum. However, bacterial colonization of the jejunum could also play a contributory role. Fromm et al.[120] have shown that fecal excretion of bile acids was increased in these patients but was still below the levels required to induce secretory diarrhea. The bile acid breath test was positive in all patients studied. However, this increase was probably due to more frequent enterohepatic cycling rather than bacterial overgrowth, as the bacterial contamination of the jejunum could not be documented. The authors concluded that diarrhea and malabsorption syndrome after jejunoileal bypass are mainly due to a reduction in the absorptive surface of the small intestine (''overload diarrhea'') rather than due to secretory effects of bile acids or bacterial overgrowth.

E. Bile Acids in Gastric Ulcer, Gastritis, and Esophageal Reflux

It has been postulated that the bile reflux into the stomach is a major factor in the pathogenesis of gastritis and gastric ulcer.[121] Miranda et al.[122] have recently described several abnormalities in gastric ulcer patients: (1) the mean bile acid concentration in the stomach of patients with gastric ulcer was higher than in normal subjects, (2) antral motility was significantly decreased and (3) phase III of the interdigestive complex was not initiated in the antrum. Rhodes et al.[123] postulated that because of disturbance of antroduodenal motility coupled with the incompetency of the pyloric sphincter, there is an increase in the reflux of bile rich duodenal juice into the stomach. Several studies have demonstrated that bile acids and lysolecithin in an acid environment could lead to the breakage of the gastric mucosal barrier. Damaged gastric mucosa is thought to be susceptible to peptic ulceration. Evidence favoring this hypothesis include studies which demonstrate an increase in the gastric concentration of bile acids in patients with gastric ulcer,[122] incompetence of pyloric sphincter,[124] and an increase in bile regurgitation during the active phase of the ulcer.[122] Similar interactions between bile acids, lysolecithin, and pepsin in the acid environment could play a role in the pathogenesis of peptic esophagitis.[125]

Distinct entities attributed to bile reflux are alkalyne gastritis and esophagitis in the setting of partial gastrectomy. The treatment of this relatively common entity is cholestyramine and biliary diversion.

F. Inborn Errors of Bile Acid Metabolism

Cerebrotendious xanthomatosis (CTX) is a rare inherited lipid storage disease. The clinical symptoms include premature arteriosclerosis, cataracts, dementia, and large accumulations of cholesterol and cholestanol in the brain, lungs, and Achilles tendons.[126] The accumulation of these two sterols is presumably secondary to augmented synthesis. Another major metabolic abnormality in this disease is a major block in the bile acid biosynthetic pathway, resulting in deficient bile acid formation, reduction in relative proportion of chenodeoxycholic acid and increased excretion of bile acid precursors in bile, urine, and feces.[127,128] Lack of mitochondrial 26β-hydroxylase appears to be the major defect. This results in the exposure of large amounts of 5β cholestane-3α, 7α, 12α-triol, a precursor of cholic acid, to microsomal 23-, 24-, 25-hydroxylases with subsequent preferential formation of cholic acid.[129] Chenodeoxycholic acid administered to these patients for a prolonged period of time resulted in an improvement of neurologic functions.[130] The mode of action of CDCA in this setting is not well understood, but it is conceivable that CDCA reduces cholesterol and cholestanol synthesis with resultant decrease in plasma cholestanol levels.[130]

Another inborn error of bile acid synthesis has been described in two children with severe intrahepatic cholestasis.[131] A block was found in the side chain oxidation of CA resulting in increased levels of trihydroxy cholestanoic acid in bile.[132] This defect is specific for CA synthesis, since CDCA synthesis appears to be normal.

G. Bile Acids and Colon Carcinogenesis

Epidemiological studies have shown that large bowel cancer is much more prevalent in industrialized countries such as those in North America and Western Europe, as compared to relatively underdeveloped countries such as those in South America, Africa, and Asia.[133]

One essential difference between industrialized and underdeveloped countries is reflected in dietary habits. "Western type" diets are characterized by high intake of animal protein, fats, refined carbohydrates, and total calories as well as a low fiber content.[134] In contrast, non-Western societies consume much less meat with the bulk of their protein coming in the form of vegetable protein. Their diets also include relatively large amounts of fiber and unrefined cereals.

Experiments have shown that animals fed a diet relatively high (20%) in total fat are more susceptible to chemical mutagens that initiate colon cancer than animals on a low (5%) fat diet.[135] High fat diets have also been reported to cause an increase in the levels of fecal bile acids. These compounds have the ability to promote the tumor-initiating activity of chemical mutagens.[136]

It is reasonable to assume that constituents within the lumen of the large bowel might have tumor initiation and promoting activities that could stem in part from dietary habits. During the last 20 years efforts have been made to compare the composition of the intestinal microflora isolated from high and low risk populations, levels of fecal metabolites such as bile acids and cholesterol, and the metabolic activities of the intestinal microflora. The results of these studies suggest that the major kinds of bacteria in the large bowel do not change markedly with changes in dietary habits.[137] However, it has been reported by several laboratories that there is an increase in the levels of fecal bile acids and microbially generated secondary bile acids (deoxycholic and lithocholic acids) in individuals on high risk diets.

Hill et al.[138] originally proposed that the intestinal microflora might convert bile acids into carcinogens or cocarcinogens (promotors) by a series of unsaturation reactions on the steroid nucleus generating compounds similar in structure to polycyclic aromatic hydrocarbon carcinogens. However, these proposed compounds have not been detected in fecal extracts. The major fecal bile acids in man are deoxycholic and lithocholic acid which are generated from cholic acid and chenodeoxycholic acid, respectively, by the intestinal microflora. These secondary bile acids are not mutagenic in the Ames test (a quantitative bacterial assay for detecting DNA damage). However, lithocholic acid has unique biological properties, including production of bile duct hyperplasia and enhancement of tumorigenicity in animal model studies. Kelsey and Pienta[139] have shown that this compound can transform Syrian hamster embryo cells in culture. Deoxycholic acid has been shown to promote the action of primary chemical mutagens in animal model studies. Therefore, the secondary bile acids may be weak carcinogens and potential promoters of chemical mutagens.

The development of the Ames test[140] has allowed researchers to detect and quantitate chemical mutagens. Mutagens are almost always carcinogens. The presence of an ether-extractable mutagen in human feces was first reported by Bruce and co-workers[141] and these results have been confirmed by others.[142] Most of the mutagenic activity (as assayed by the Ames test) in feces appears to be due to a single type compound that is unstable in air and strongly UV absorbing. The formation of this compound is greatly enhanced by incubating human feces under anaerobic conditions in the presence of bile acids.[143] Moreover, intestinal *Bacteroides* sp. can biologically synthesize this mutagen if incubated with bile acids and a precursor extracted from feces of an individual producing large quantities of this mutagen. The elucidation of the structure of this mutagen has been reported by Wilkins and co-workers[144] and the results independently confirmed by Bruce et al.[145] The mutagen has been identified as 1-(1-glycero) tetradeca-1,3,5,9-pentaene (fecapentaene).

Preliminary data indicate that individuals in high risk populations secrete larger amounts of mutagenic compounds as compared to low risk populations. However, it is not presently known how diet effects the formation of these fecal mutagens nor if changes in the levels of colonic bile acids might alter this process.

H. Bile Acid Metabolism in Cystic Fibrosis

Numerous studies of bile acid metabolism in cystic fibrosis demonstrate that (1) fecal bile acid loss in these patients is increased, but returns to normal after pancreatic enzyme replacement,[146] (2) the bile acid pool is reduced, but enzymatic supplementation normalizes it,[147] (3) fasting plasma levels of bile acids are highly elevated in a subgroup of cystic fibrosis patients with steatorrhea,[148] and (4) in cystic fibrosis patients with liver disease serum levels of bile acids are elevated.[149] Bile acid loss in cystic fibrosis patients may equal that in regional ileitis.[150] The reason for significant fecal loss of bile acids in this disease is uncertain unless bile acids, rather than being absorbed, are bound to maldigested protein or lipids and excreted in feces. Pancreatic enzyme replacement improves digestion of protein and lipids and results in a more efficient preservation of bile acids.[147]

I. Bile Acid Metabolism in Infancy

In infants, bile acid uptake, synthesis, excretion, and ileal reabsorption are all decreased when compared to adults.[151] Specifically, the bile acid biosynthetic pathway differs from that of the adult. It appears that a significant proportion of bile acid synthesis in infants proceeds via the mitochondrial 26-hydroxylation pathway with preferential synthesis of chenodeoxycholic acid and an intermediate formation of lithocholic and 3β-hydroxy-5-cholenoic acids.[152,153] These two monohydroxy intermediates, which are known to be potentially cholestatic, make up 40% of the bile acids in the amniotic fluid at 36 weeks[154] and are present in full-term meconium and gallbladder bile of premature infants.[155] Moreover, these two intermediates were detected in the urine of infants with neonatal hepatitis and biliary atresia,[63] and in the serum of infants with cholestatic liver disease.[156] Their role in the pathogenesis of these disorders is uncertain.

The bile acid pool size and turnover kinetics have been determined in the full-term infant.[157] Cholic and chenodeoxycholic acid synthesis rates are approximately 50% that of the adult and even less in the premature infant. Consequently, there is a reduction in the amount of individual bile acids as well as the total pool size. The bile acid conjugation pattern is also altered in infants; there is preferential conjugation with taurine rather than glycine. Hepatic uptake of bile acid in infants is deficient and, when coupled with inadequate bile secretion, results in an elevation of plasma bile acids. Plasma bile acids may be elevated throughout the first year of infancy. All of these alterations may in part contribute to ''physiologic cholestasis'' of infancy.[156] Other aspects of bile acid metabolism during development and, in particular, the role of cholesterol 7α-hydroxylase are reviewed in Chapter 10.

In full-term infants, the intraluminal bile acid concentration is about 2 to 2.5 m*M* and, in premature infants, it is even lower.[151,155] This is well below the minimal concentration of 4 m*M* which is required for adequate fat absorption after a meal.[155,158] As a result, fat malabsorption is common in infants. The decreased luminal concentrations of bile acids is a result of a combination of factors such as diminution of bile acid pool size, reduction of bile acid synthesis and secretion, and/or altered distribution of bile acids between the EHC and plasma.

Several cholestatic syndromes have been described in infancy. Cerebrohepatorenal syndrome of Zelweger is a fatal autosomal recessive disease involving liver, brain, and kidney.[159] In these patients, the basic abnormality appears to be an absence of peroxisomes and defective mitochondrial functions. A marked increase of intermediates of cholic and chenodeoxycholic acids are found in urine of these patients, which suggest a block in the mitochondrial side chain oxidation.[160] Other rare diseases of cholestasis of infancy associated with abnormal bile acid metabolism are coprostanic acidemia[131,132] and Byler's disease.[161]

REFERENCES

1. **Mosbach, E. H. and Salen, G.,** Bile acid biosynthesis. Pathways and regulation, *Am. J. Dig. Dis.*, 19, 920, 1975.
2. **Danielsson, H. and Sjövall, J.,** Bile acid metabolism, *Annu. Rev. Biochem.*, 44, 233, 1975.
3. **Vlahcevic, Z. R., Schwartz, C. C., Gustafsson, J., Halloran, L. G., Danielsson, H., and Swell, L.,** Biosynthesis of bile acids in man. Multiple pathway to cholic acid and chenodeoxycholic acid, *J. Biol. Chem.*, 255, 2925, 1980.
4. **Anderson, K. E., Kok, E., and Javitt, N. B.,** Bile acid synthesis in man, Metabolism of 7α-hydroxy-cholesterol-^{14}C and 26-hydroxycholesterol-^{3}H, *J. Clin. Invest.*, 56, 226, 1975.
5. **Salen, G., Shefer, S., Setoguchi, T., and Mosbach, E. H.,** Bile alcohol metabolism in man. Conversion of 5β-cholestane-3α, 7α, 12α, 25-tetrol to cholic acid, *J. Clin. Invest.*, 56, 226, 1975.
6. **Lack, C. and Weiner, I. M.,** Intestinal bile salt transport, structure-activity relationship, *Am. J. Physiol.*, 207, 1, 1964.
7. **Reichen, J. and Paumgartner, G.,** Kinetics of taurocholate uptake by the perfused rat liver, *Gastroenterology*, 68, 132, 1974.
8. **Accatino, L. and Simon, F. R.,** Identification and characterization of bile acid receptor in isolated liver surface membranes, *J. Clin. Invest.*, 57, 496, 1976.
9. **Stiehl, A., Earnest, D. L., and Admirand, W. H.,** Sulfation and renal excretion of bile salts in patients with cirrhosis of the liver, *Gastroenterology*, 68, 534, 1978.
10. **Makino, I., Hashimoto, H., Shinozaki, Z. Q., et al.,** Sulfated and nonsulfated bile acids in urine, serum and bile of patients with hepatobiliary diseases, *Gastroenterology*, 68, 545, 1975.
11. **Swell, L., Gregory, D. H., and Vlahcevic, Z. R.,** Current concepts of the pathogenesis of cholesterol gallstones, *Med. Clin. N. Am.*, 58, 1449, 1974.
12. **Cowen, A. E., Korman, M. G., Hofmann, A. F., Cass, O. W., and Coffin, S. B.,** Metabolism of lithocholate in healthy man. II. Enterohepatic circulation, *Gastroenterology*, 61, 85, 1971.
13. **Vlahcevic, Z. R., Bell, C. C., Jr., Buhac, I., Farrar, J. T., and Swell, L.,** Diminished bile acid pool size in patients with gallstones, *Gastroenterology*, 59, 165, 1970.
14. **Vlahcevic, Z. R., Miller, J. R., Farrar, J. T., and Swell, L.,** Kinetics and pool size of primary bile acids in man, *Gastroenterology*, 61, 85, 1971.
15. **Shefer, S., Hauser, S., Bekerski, I., et al.,** Biochemical site of regulation of bile acid biosynthesis in the rat, *J. Lipid Res.*, 11, 404, 1970.
16. **Bergstrom, S. and Danielsson, H.,** On the regulation of bile acid formation in the rat liver, *Acta Phys. Scand.*, 43, 1, 1958.
17. **Eriksson, S.,** Biliary excretion of bile acids and cholesterol in bile fistula rats, *Proc. Soc. Exp. Biol. (N.Y.)*, 94, 578, 1957.
18. **Key, P. H., Bonorris, G. G., Marks, J. W., Chung, A., and Schoenfield, L. J.,** Biliary lipid synthesis and secretion in gallstone patients before and during treatment with chenodeoxycholic acid, *J. Lab. Clin. Med.*, 95, 816, 1980.
19. **Danzinger, R. G., Hofmann, A. F., Thistle, J. L., and Schoenfield, L. J.,** Effect of oral chenodeoxycholic acid on bile acid kinetics and biliary lipid composition in women with cholelithiasis, *J. Clin. Invest.*, 52, 2809, 1973.
20. **Davis, R. A., Highsmith, W. E., Malone-McNeal, M., et al.,** Bile acid synthesis by cultured hepatocytes: inhibition by mevinolin but not by bile acids, *J. Biol. Chem.*, 258, 4079, 1983.
21. **Admirand, W. H. and Small, D. M.,** The physicochemical basis of cholesterol gallstone formation in man, *J. Clin. Invest.*, 47, 1043, 1968.
22. **Holzbach, R. T., Marsh, M. E., Obzeweski, M., and Holan, K.,** Cholesterol solubility in bile; evidence that supersaturated bile is frequent in healthy man, *J. Clin. Invest.*, 52, 1467, 1973.
23. **Nervi, F. O. and Dietschy, J. M.,** The kinetic characteristics of inhibitions of hepatic cholesterogenesis by lipoproteins of intestinal origins, *J. Biol. Chem.*, 253, 1859, 1975.

24. **Shefer, S., Hauser, S., Lapar, V., and Mosbach, E. H.,** Regulatory effects of sterols and bile acids on hepatic 3-hydroxy-3-methyl-glutaryl-CoA reductase and cholesterol 7α-hydroxylase in the rat, *J. Lipid Res.*, 14, 573, 1973.
25. **Weis, H. J. and Dietschy, J. M.,** The interaction of various control mechanisms in determining the rate of hepatic cholesterogenesis in the rat, *Biochim. Biophys. Acta*, 398, 315, 1975.
26. **Schwartz, C. C., Berman, M., Vlahcevic, Z. R., Halloran, L. G., Gregory, D. H., and Swell, L.,** Multicompartmental analysis of cholesterol metabolism in man, *J. Clin. Invest.*, 61, 408, 1978.
27. **Brown, M. S. and Goldstein, J. L.,** Receptor-mediated control of cholesterol metabolism, *Science*, 191, 150, 1976.
28. **Brown, M. S., Kovanen, P. T., and Goldstein, J. L.,** Regulation of plasma cholesterol by lipoprotein receptors, *Science*, 212, 628, 1981.
29. **Goldstein, J. L. and Brown, M. S.,** The low density lipoprotein pathway and its relationship to atherosclerosis, *Annu. Rev. Biochem.*, 46, 897, 1977.
30. **Stein, O. and Stein, Y.,** The removal of cholesterol from Landschutz ascites cells by high-density apolipoprotein, *Biochim. Biophys. Acta*, 326, 232, 1973.
31. **Schwartz, C. C., Vlahcevic, Z. R., Halloran, L. G., and Swell, L.,** An *in vivo* evaluation in man of the transfer of esterified cholesterol between lipoproteins and into the liver and bile, *Biochim. Biophys. Acta*, 663, 143, 1981.
32. **Schwartz, C. C., Halloran, L. G., Vlahcevic, Z. R., Gregory, D. H., and Swell, L.,** Preferential utilization of free cholesterol from high density lipoproteins for biliary cholesterol secretion in man, *Science*, 200, 62, 1981.
33. **Halloran, L. G., Schwartz, C. C., Vlahcevic, Z. R., Gregory, D. H., and Swell, L.,** Effect of chenodeoxcholic acid administration on bile acid and cholesterol synthesis in man, *Surg. Forum*, 26, 439, 1975.
34. **Hui, D. Y., Innerarity, T. L., and Mahley, R. W.,** Lipoprotein binding to canine hepatic membranes, *J. Biol. Chem.*, 256, 5646, 1981.
35. **Mahley, R. W., Hui, D. Y., Innerarity, T. L., and Weisgraber, K. H.,** Two independent lipoprotein receptors on hepatic membranes of dog, swine, and man, *J. Clin. Invest.*, 68, 1197, 1981.
36. **Maley, R. W., Innerarity, T. L., Weisgraber, K. H., and Oh, S. Y.,** Altered metabolism *(in vivo* and *in vitro)* of plasma lipoproteins after selective chemical modification of lysine residues of the apoproteins, *J. Clin. Invest.*, 64, 743, 1979.
37. **Kovanen, P. T., Brown, M. S., and Goldstein, J. L.,** Increased binding of low density lipoprotein to liver membranes from rats treated with 17α-ethinyl estradiol, *J. Biol. Chem.*, 254, 11367, 1979.
38. **Chao, Y., Windler, E. E., Chen, G. C., and Havel, R. J.,** Hepatic catabolism of rat and human lipoproteins in rats treated with 17α-ethinyl estradiol, *J. Biol. Chem.*, 254, 11360, 1979.
39. **Windler, E. E., Kovanen, P. T., Chao, Y., Brown, M. S., Havel, R. J., and Goldstein, J. L.,** The estradiol-stimulated lipoprotein receptor of rat liver, *J. Biol. Chem.*, 255, 10464, 1980.
40. **Kovanen, P. T., Bilheimer, D. W., Goldstein, J. L., Jaramillo, J. J., and Brown, M. S.,** Regulatory role for hepatic low density lipoprotein receptors *in vivo* in the dog, *Proc. Natl. Acad. Sci. U.S.A.*, 78, 1194, 1981.
41. **Chao, Y. S., Jones, A. L., Hradek, G. T., Windler, E. E. T., and Havel, R. J.,** Autoradiographic localization of the sites of uptake, cellular transport, and catabolism of low density lipoproteins in the liver of normal and estrogen-treated rats, *Proc. Natl. Acad. Sci. U.S.A.*, 78, 597, 1981.
42. **Windler, E. E. T., Kovanen, P. T., Chao, Y. S., Brown, M. S., Havel, R. J., and Goldstein, J. L.,** The estradiol-stimulated lipoprotein receptor of rat liver, *J. Biol. Chem.*, 255, 10464, 1980.
43. **Theodor, E., Spritz, N., and Sleisenger, M. H.,** Metabolism of intravenously injected isotopic cholic acid in viral hepatitis, *Gastroenterology*, 55, 183, 1968.
44. **Korman, M. G., Hofmann, A. F., and Summerskill, W. H. J.,** Assessment of activity in chronic active liver disease; serum bile acids compared with conventional tests and histology, *N. Engl. J. Med.*, 292, 1209, 1974.
45. **Kuroda, T. and Okuda, K.,** Bile acids in hepatic bile in liver disease, *Acta Hepato. Gastroenterol.*, 21, 1209, 1974.
46. **Kaplowitz, N., Koke, Z. Q., and Javitt, N. B.,** Postprandial serum bile acids for the detection of hepatobiliary disease, *JAMA*, 225, 293, 1973.
47. **LaRusso, N. F., Hoffman, N. E., Hofmann, A. F., and Korman, M. G.,** Validity and sensitivity of an intravenous bile acid tolerance test in patients with liver disease, *N. Engl. J. Med.*, 292, 1209, 1976.
48. **Vlahcevic, Z. R., Buhac, I., Farrar, J. T., Bell, C. C., Jr., and Swell, L.,** Bile acid metabolism in patients with cirrhosis. I. Kinetic aspects of cholic acid metabolism, *Gastroenterology*, 60, 491, 1971.
49. **Vlahcevic, Z. R., Juttijudata, P., Bell, C. C., Jr., and Swell, L.,** Bile acid metabolism in patients with cirrhosis. II. Cholic and chenodeoxycholic acid metabolism, *Gastroenterology*, 62, 1174, 1972.
50. **Schoenfield, L. G. and Sjövall, J.,** Bile acids in the skin of patients with pruritic hepatobiliary disease, *Nature (London)*, 213, 93, 1967.

51. **Angelin, B., Einarsson, K., and Hellstrom, K.,** Effect of cholestyramine on bile acid kinetics in patients with portal cirrhosis of the liver, Evidence for selective defect in the formation of cholic acid, *Am. J. Dig. Dis.*, 23, 1115, 1978.
52. **Goldman, M., Vlahcevic, Z. R., Schwartz, C. C., Gustafsson, J., and Swell, L.,** Bile acid metabolism in cirrhosis. VIII. Quantitative evaluation of bile acid synthesis from [7β-^{3}H] 7α-hydroxycholesterol and [6-^{3}H] 26-hydroxycholesterol, *Hepatology*, 2, 59, 1982.
53. **Vlahcevic, Z. R., Goldman, M., Schwartz, C. C., Gustafsson, J., and Swell, L.,** Bile acid metabolism in cirrhosis. VII. Evidence for defective feedback control of bile acid synthesis, *Hepatology*, 1, 146, 1981.
54. **Vlahcevic, Z. R., Buhac, I., Bell, C. C., Jr., and Swell, L.,** Abnormal metabolism of secondary bile acids in patients with cirrhosis, *Gut*, 11, 420, 1970.
55. **Yoshida, T., McCormick, C., Swell, L., and Vlahcevic, Z. R.,** Bile acid metabolism in cirrhosis. IV. Characterization of the abnormality in deoxycholic acid metabolism, *Gastroenterology*, 68, 335, 1975.
56. **McCormick, W. C., Bell, C. C., Jr., Swell, L., and Vlahcevic, Z. R.,** Cholic acid synthesis as an index of the severity of liver disease in man, *Gut*, 14, 895, 1973.
57. **Bouchier, I. A. D.,** Postmortem study of the frequency of gallstones in patients with cirrhosis of the liver, *Gut*, 10, 705, 1969.
58. **Nicholas, P., Rinaudo, P. A., and Conn, H. O.,** Increased incidence of cholelithiasis in Laennec's cirrhosis, *Gastroenterology*, 63, 112, 1972.
59. **Vlahcevic, Z. R., Yoshida, T., Juttijudata, P., Bell, C. C., Jr., and Swell, L.,** Bile acid metabolism in cirrhosis. III. Biliary lipid secretion in patients with cirrhosis and its relevance to gallstone formation, *Gastroenterology*, 64, 298, 1973.
60. **Linsheer, W. G., Patterson, I. F., Moore, E. W., et al.,** Medium and long chain fat absorption in cirrhosis, *J. Clin. Invest.*, 45, 1317, 1966.
61. **Sun, D. C. H., Albacete, R. H., and Chen, T. K.,** Malabsorption studies in cirrhosis of the liver, *Arch. Int. Med.*, 119, 567, 1967.
62. **Back, P.,** Identification and quantitative determination of urinary bile acids excreted in cholestasis, *Clin. Chim. Acta*, 4, 199, 1973.
63. **Makino, I., Sjövall, J., Norman, A., and Strandvik, B.,** Excretion of 3β-hydroxy-5-cholenoic- and 3α-hydroxy-5α-cholanoic acids in urine of infants with biliary atresia, *FEBS Lett.*, 15, 161, 1971.
64. **Back, P.,** Evidence for bile salt glucuronides in cholestasis, in *Advances in Bile Acid Research*, Matern, S., Hackenschmidt, J., Back, P., and Gerok, W., Eds., F. K. Schattauer Verlag, New York, 1974.
65. **Murphy, G. M., Janse, F. H., and Billing, B. H.,** Unsaturated monohydroxy bile acids in primary cirrhosis, *Biochem. J.*, 129, 491, 1972.
66. **Stiehl, A.,** Disturbances of bile acid metabolism in cholestasis, *Clin. Gastroenterol.*, 6(1), 45, 1977.
67. **Vlahcevic, Z. R., Bell, C. C., Jr., Gregory, D. H., Buker, G., Juttijudatta, P., and Swell, L.,** Relationship of bile acid pool size to the formation of lithogenic bile in female Indians of the Southwest, *Gastroenterology*, 62, 73, 1972.
68. **Bell, C. C., Jr., McCormick, W. C., Gregory, D. H., Law, D. H., Vlahcevic, Z. R., and Swell, L.,** Relationship of bile acid pool size to the formation of lithogenic bile in the male Indians of the Southwest, *Surg. Gynecol. Obstet.*, 134, 473, 1972.
69. **Grundy, S. M., Metzger, A. L., and Adler, R. D.,** Mechanisms of lithogenic bile formation in American Indian women with cholesterol gallstones, *J. Clin. Invest.*, 51, 3026, 1972.
70. **Grundy, S. M., Duane, W. C., Adler, R. D., Aron, J. M., and Metzger, A. L.,** Biliary lipid outputs in young women with cholesterol gallstones, *Metabolism*, 23, 67, 1974.
71. **Van Besten, L., Connor, W. E., and Bell, S.,** The effect of dietary cholesterol on the composition of human bile, *Surgery*, 73, 266, 1973.
72. **Kern, F., Jr., Everson, G. T., DeMark, B., et al.,** Biliary lipids, bile acids and gallbladder function in the human female; effects of pregnancy and ovulatory cycle, *J. Clin. Invest.*, 68, 1229, 1981.
73. **Coyne, M. J., Bonorris, G. G., Goldstein, L. T., and Schoenfield, L. J.,** Effect of chenodeoxycholic acid and phenobarbital on the rate limiting enzymes of hepatic cholesterol and bile acid synthesis in patients with gallstones, *J. Lab. Clin. Med.*, 87, 281, 1976.
74. **Schwartz, C. C., Berman, M., Vlahcevic, Z. R., Halloran, L. G., Gregory, D. H., and Swell, L.,** Multicompartmental analysis of cholesterol metabolism in man. Characterization of the hepatic bile acid and biliary cholesterol precursor sites, *J. Clin. Invest.*, 61, 408, 1978.
75. **Naunyn, B.,** A Treatise on Cholelithiasis, *New Sydenham Society*, 1896, p. 22.
76. **Thistle, J. L. and Hofmann, A. F.,** Efficacy and specificity of chenodeoxycholic acid for dissolving gallstones, *N. Engl. J. Med.*, 289, 655, 1973.
77. **Iser, J. H., Dowling, R. H., Mik, Y. I. H., and Bell, D. G.,** Chenodeoxycholic acid treatment of gallstones, *N. Engl. J. Med.*, 293, 378, 1975.
78. **Coyne, M. J., Bonorris, G. C., Chung, A., Goldstein, L. I., Lahuna, D., and Schoenfield, L. J.,** Treatment of gallstones with chenodeoxycholic acid and phenobarbital, *N. Engl. J. Med.*, 292, 604, 1975.

79. **Makino, I., Shinozake, G., Yoshino, K., and Nakagawa, S.,** Dissolution of cholesterol gallstones by ursodeoxycholic acid, *Jpn. J. Gastroenterol.*, 72, 690, 1975.
80. **Thistle, J. L., Turcott, J., Ott, B. Y., Carlson, G. L., LaRusso, N. F., and Hofmann, A. F.,** Ursodeoxycholic acid unsaturates bile at a lower dose than chenodeoxycholic acid, *Gastroenterology*, 74, 103, 1978.
81. **Thistle, J. L., Hofmann, A. F., Ott, B. J., and Stephens, D. H.,** Chenotherapy for gallstone dissolution. I. Efficacy and safety, *JAMA*, 239, 1041, 1978.
82. **Schoenfield, L. J. and Lachin, J. M.,** The steering committee, The National Cooperative Gallstone Study Group. Chenodiol (chenodeoxycholic acid) for dissolution of gallstones: The National Cooperative Gallstone Study: a controlled trial of efficacy and safety, *Ann. Intern. Med.*, 95, 257, 1981.
83. **Erlinger, S., Le Go, A., Husson, J., and Fevery, J.,** Franco-Belgiam Cooperative Study of Ursodeoxycholic Acid in the Medical Dissolution of Gallstone: A Double-Blind, Randomized, Dose-Response Study, and Comparison with Chenodeoxycholic Acid, *Hepatology*, 4, 308, 1984.
84. Tokyo Cooperative Gallstone Study Group, Efficacy and indications of ursodeoxycholic acid treatment for dissolving gallstones. A multicenter double blind trial, *Gastroenterology*, 78, 542, 1980.
85. **Angelin, B., Ewerth, S., and Einarsson, K.,** Ursodeoxycholic acid treatment in cholesterol gallstone disease; effects on hepatic 3-hydroxy-3-methylglutaryl coenzyme A reductase activity, biliary lipid composition, and plasma lipid levels, *J. Lipid Res.*, 24, 461, 1983.
86. **LaRusso, N. F. and Thistle, J. L.,** Effect of litholytic bile acids on cholesterol absorption in gallstone patients, *Gastroenterology*, 84, 265, 1983.
87. **Gallo-Torres, H. E., Miller, O. N., and Hamilton, J. G.,** A comparison of the effect of bile salts on the absorption of cholesterol from the intestine of the rat, *Biochim. Biophys. Acta*, 176, 605, 1969.
88. **Raicht, R. F., Cohen, B. I., and Mosbach, E. H.,** Effects of sodium taurochenodeoxycholate and sodium taurocholate on cholesterol absorption in the rat, *Gastroenterology*, 67, 1155, 1974.
89. **Tangedahl, T. N., Thistle, J. L., Hofmann, A. F., and Matseshe, J. W.,** Effect of the beta-sitosterol alone or in combination with chenic acid on cholesterol saturation of bile and cholesterol absorption in gallstone patients, *Gastroenterology*, 76, 1341, 1979.
90. **Einarsson, K. and Grundy, S. M.,** Effects of feeding cholic acid and chenodeoxycholic acid on cholesterol absorption and hepatic secretion of biliary lipids in man, *J. Lipid Res.*, 21, 23, 1980.
91. **Mok, H. Y. I. and Grundy, S. M.,** Cholesterol and bile acid absorption during bile acid therapy in obese subjects undergoing weight reduction, *Gastroenterology*, 78, 62, 1980.
92. **Ponz de Leon, M., Carulli, N., Loria, P., Iori, R., and Zironi, F.,** The effect of chenodeoxycholic acid (CDCA) on cholesterol absorption, *Gastroenterology*, 77, 223, 1979.
93. **Ponz de Leon, M., Carulli, N., Loria, P., Iori, R., and Zironi, F.,** Cholesterol absorption during bile acid feeding. Effect of ursodeoxycholic acid (UDCA) administration, *Gastroenterology*, 78, 214, 1980.
94. **Ahlberg, J., Angelin, B., and Einarsson, K.,** Hepatic 3-hydroxy-3-methyl-glutaryl coenzyme A reductase activity and biliary lipid composition in man: Relation to cholesterol gallstone disease and effects of cholic acid and chenodeoxycholic acid treatment, *J. Lipid Res.*, 22, 410, 1981.
95. **Maton, P. N., Ellis, H. J., Higgins, M. J. P., and Dowling, R. H.,** Hepatic HMG-CoA reductase in human cholelithiasis: effects of chenodeoxycholic and ursodeoxycholic acids, *Eur. J. Clin. Invest.*, 10, 325, 1980.
96. **Carulli, N., Ponz de Leon, M., Zironi, F., Pinetti, A., Smerieri, R., Iori, R., and Loria, P.,** Hepatic cholesterol and bile acid metabolism in subjects with gallstones: comparative effects of short term feeding of chenodeoxycholic and ursodeoxycholic acid, *J. Lipid Res.*, 21, 35, 1980.
97. **Salen, G., Colalillo, A., Verga, D., Bagan, E., Tint, G. S., and Shefer, S.,** Effect of high and low doses of ursodeoxycholic acid on gallstone dissolution in humans, *Gastroenterology*, 78, 1412, 1980.
98. **Angelin, B., Backman, L., Einarsson, K., Eriksson, L., and Ewerth, S.,** Hepatic cholesterol metabolism in obesity: activity of microsomal 3-hydroxy-3-methylglutaryl coenzyme A reductase, *J. Lipid Res.*, 23, 770, 1982.
99. **Einarsson, K., Hellstrom, K., and Kallner, M.,** Bile acid kinetics in relation to sex, serum lipids, body weights and gallbladder disease in patients with various types of hyperlipoproteinemia, *J. Clin. Invest.*, 54, 1301, 1974.
100. **Ament, M. E., Shimoda, S. S., Saunders, D. R., and Rubin, E.,** Pathogenesis of diarrhea in three cases of small intestinal stasis syndrome, *Gastroenterology*, 63, 728, 1972.
101. **Gracey, M., Papadimitrio, J., and Bower, G.,** Ultrastructural changes in the small intestines of rats with self-filling blind loops, *Gastroenterology*, 67, 646, 1974.
102. **Northfield, T. C.,** Intraluminal precipitation of bile acids in stagnant loop syndrome, *Gut*, 14, 341, 1973.
103. **Heaton, K. W.,** Disturbances of bile acid metabolism in intestinal diseases, *Clin. Gastroenterol.*, 6(1), 69, 1977.
104. **Sherr, H. P., Sasaki, Y., Newman, A., et al.,** Detection of bacterial deconjugation of bile salts by convenient breath analysis technique, *N. Engl. J. Med.*, 285, 656, 1971.

105. **Fromm, H., Thomas, P. J., and Hormann, A. F.,** Sensitivity and specificity in tests of distal ileal function: prospective comparison of bile acid and vitamin B_{12} absorption in ileal resection patients, *Gastroenterology*, 64, 1077, 1973.
106. **King, C. E., Toskes, P. P., Guilarte, T. R., et al.,** Comparison of the One-gram d-[^{14}C]-xylose breath test in patients with small-intestine bacterial overgrowth, *Dig. Dis. Sci.*, 25, 53, 1980.
107. **Meihoff, W. E. and Kern, F.,** Bile salt malabsorption in regional ileitis, ileal resection and mannitol induced diarrhea, *J. Clin. Invest.*, 47, 261, 1968.
108. **Mekhjian, H. S., Phillips, S. F., and Hofmann, A. R.,** Colonic secretion of water and electrolytes induced by bile acids; perfusion studies in man, *J. Clin. Invest.*, 50, 1569, 1971.
109. **Binder, H. J. and Rawlings, C. L.,** Effect of conjugated dihydroxy bile salts on electrolyte transport in rat colon, *J. Clin. Invest.*, 52, 1460, 1973.
110. **Binder, H. J., Filburne and Volpe, B. T.,** Bile salt alteration of colonic electrolytic transport: role of cyclic adenosine monophosphate, *Gastroenterology*, 68, 503, 1975.
111. **Hofmann, A. F. and Poley, J. R.,** Cholestyramine treatment of diarrhea associated with ileal resection, *N. Engl. J. Med.*, 281, 397, 1969.
112. **Hofmann, A. F. and Poley, J. R.,** Role of bile acid malabsorption in pathogenesis of diarrhea and steatorrhea in patients with ileal resection, *Gastroenterology*, 62, 918, 1972.
113. **McJunkin, B., Fromm, H., Sarva, R. P., and Amin, P.,** Factors in the mechanism of diarrhea in bile acid malabsorption; fecal pH — a key determinant, *Gastroenterology*, 80, 1454, 1981.
114. **Low-Beer, T. S., Heaton, K. W., Pomare, E. W., and Read, A. E.,** The effect of coeliac disease upon bile salts, *Gut*, 14, 204, 1971.
115. **Banwell, J. G. and Gorbach, S. L.,** Tropical sprue, *Gut*, 10, 328, 1969.
116. **Cassels, J. S., Banwell, J. G., Gorbach, S. L., Mitra, R., and Mazumdep, D. N. G.,** Tropical sprue and malnutrition in West Benjal. IV. Bile salt deconjugation in tropical sprue, *Am. J. Clin. Nutr.*, 23, 1579, 1970.
117. **Thaysen, E. H. and Pedersen, L.,** Idiopathic bile acid catharsis, *Gut*, 17, 965, 1976.
118. **Pedersen, L., Arnfred, T., and Thaysen, H. E.,** Rapid screening of increased bile acid deconjugation and bile acid malabsorption by means of the glycine 1-(^{14}C) cholylglycine assay, *Scand. J. Gastroenterol.*, 8, 665, 1973.
119. **Backer, O., Gudmand-Hoyer, E., Andersen, B., et al.,** Randomized trial of jejunoileal bypass versus medical treatment in morbid obesity, *Lancet*, 2, 1255, 1979.
120. **Fromm, H., Sarva, R. P., Ravitch, M. M., et al.,** Effects of jejunoileal bypass on the enteroheaptic circulation of bile acids, bacterial flora in the upper small intestine, and absorption of vitamin B_{12}, *Metabolism*, 32, 1133, 1983.
121. **Guth, P. H.,** Pathogenesis of gastric mucosal injury, *Annu. Rev. Med.*, 33, 183, 1982.
122. **Miranda, M., Defilippi, C., and Valenzueca, J. E.,** Abnormalities of interdigestive complex and increased duodenogastric reflux in gastric ulcer patients, *Dig. Dis. Sci.*, 30, 16, 1981.
123. **Rhodes, J.,** Etiology of gastric ulcer, *Gastroenterology*, 63, 171, 1972.
124. **Fisher, R. and Cohen, S.,** Pyloric sphincter dysfunction in patients with gastric ulcer, *N. Engl. J. Med.*, 288, 273, 1973.
125. **Henderson, R. D., Mugashe, F. L., Jeejeebohy, K. N., et al.,** Synergism of acid and bile salts in the production of experimental esophagitis, *Can. J. Surg.*, 16, 12, 1973.
126. **Salen, G.,** Cholesterol deposition in cerebrotendinous xanthomatosis. A possible mechanism, *Ann. Intern. Med.*, 75, 843, 1971.
127. **Setoguchi, T., Salen, G., Tint, G. S., and Mosbach, E. H.,** A biochemical abnormality in cerebrotendinous xanthomatosis: impairment of bile acid biosynthesis associated with incomplete degradation of the cholesterol side chain, *J. Clin. Invest.*, 53, 1393, 1974.
128. **Salen, G. and Grundy, S. M.,** The metabolism of cholesterol and bile acids in cerebrotendinous xanthomatosis, *J. Clin. Invest.*, 52, 2822, 1973.
129. **Oftebro, H., Björkhem, I., Skrede, S., Schreiner, A., and Pedersen, J. I.,** Cerebrotendinous xanthomatosis. A defect in mitochondria 26-hydroxylation required for normal biosynthesis of cholic acid, *J. Clin. Invest.*, 65, 1418, 1980.
130. **Berginer, V. M., Salen, G., and Shefer, S.,** Long-term treatment of cerebrotendinous xanthomatosis with chenodeoxycholic acid, *N. Engl. J. Med.*, 311, 1649, 1984.
131. **Eyssen, H., Parmentier, G., Compernolle, F., Boon, J., and Eggermont, E.,** Trihydroxycoprostanoic acid in the duodenal fluid of two children, *Biochem. Biophys. Acta*, 273, 212, 1972.
132. **Hanson, R. F., Isenberg, J. N., Williams, G. C., Hachey, D., Szezysanik, P., Klein, P. D., and Sharp, H. L.,** The metabolism of 3α, 7α, 12α-trihydroxy-5β-cholestane-26-oic acid in two siblings with cholestasis due to intrahepatic bile duct anomalies, *J. Clin. Invest.*, 56, 577, 1975.
133. **Doll, R.,** The geographical distribution of cancer, *Br. J. Cancer*, 23, 1, 1969.
134. **Burkitt, D. P.,** Epidemiology of cancer of the colon and rectum, *Cancer (Philadelphia)*, 28, 3, 1971.

135. **Reddy, B. S., Wantanabe, K., and Weisburger, J. H.,** Effect of high fat diet on colon carcinogenesis is in 344 rats treated with 1,2-dimethylhydrazine, methylazoxymethanol acetate or methylnitrosoura, *Cancer Res.,* 37, 4156, 1977.
136. **Reddy, B. S., Narisawa, T., Weisburger, J. H., and Wynder, E. L.,** Promoting effect of sodium deoxycholate on adenocarcinomas in germ free rats, *J. Natl. Cancer Inst.,* 56, 411, 1976.
137. **Moore, W. E. C. and Holdman, L. V.,** Human fecal flora: the normal flora of 20 Japanese-Hawaiians, *Appl. Microbiol.,* 29, 961, 1974.
138. **Hill, M. J., Drasar, B. S., Aries, J., et al.,** Bacteria and etiology of cancer of the large bowel, *Lancet,* 1, 95, 1971.
139. **Kelsey, M. I. and Pienta, R. J.,** Transformation of hamster embryo cells by neutral sterols and bile acids, *Toxicol. Lett.,* 9, 177, 1981.
140. **Ames, B. N., McCann, J., and Yamasaki, E.,** Methods for detecting carcinogens and mutagens with the salmonella mammalian microsome mutagenicity test, *Mutation Res.,* 31, 347, 1975.
141. **Bruce, W. R., Varghese, A. J., Funer, R., and Land, P. C.,** A mutagen in the feces of normal humans, in *Origins of Human Cancer,* Book C, Risk Assessment Cold Spring Harbor Conferences on Cell Proliferation, Vol. 4, Hiat, H. H., Watson, J. D., and Winsten, J. A., Eds., Cold Spring Harbor Laboratory, N.Y., 1977, 1641.
142. **Venitt, S.,** Mutagens in human feces: are they relevant to cancer of the large bowel, *Mutation Res.,* 98, 265, 1982.
143. **VanTassel, R. L., Macdonald, D. K., and Wilkins, T. D.,** Stimulation of mutagen production in human feces by bile and bile acids, *Mutation Res.,* 103, 233, 1982.
144. **Hirani, N., Kingston, D. G. I., VanTassel, R. L., and Wilkins, T. D.,** Structure elucidation of a potent mutagen from human feces, *J. Am. Chem. Soc.,* 104, 6149, 1982.
145. **Gapta, I., Baptista, J., Bruce, W. R., et al.,** Structures of fecapentaenes, the mutagens of bacterial origin isolated from human feces, *Biochemistry,* 22, 241, 1983.
146. **Goodchild, M. C., Murphy, G. M., Howell, A. M., et al.,** Aspects of bile acid metabolism in cystic fibrosis, *Arch. Dis. Child,* 50, 769, 1975.
147. **Watkins, J. B., Tercyak, A., and Szczepanik, P.,** Bile salt kinetics in cystic fibrosis: influence of pancreatic enzyme replacement, *Gastroenterology,* 73, 1023, 1977.
148. **Lloyd-Still, J. D. and Demers, L. M.,** Serum glycine conjugated bile acids in periatric hepatobiliary disorders, *Am. J. Clin. Pathol.,* 71, 444, 1979.
149. **Arborgh, B., Edlund, A., Norman, A., and Strandvik, V.,** Urinary bile acid excretion in correlation to liver histopathology in cystic fibrosis, *Scand. J. Gastroenterol.,* 15, 73, 1980.
150. **Weber, A. M., Roy, C. C., Chartrand, C., et al.,** Relationship between bile acid malabsorption and pancreatic insufficiency in cystic fibrosis, *Gut,* 17, 295, 1976.
151. **Freese, D.,** Intracellular cholestatic syndromes of infancy, *Semin. Liver Dis.,* 2, 255, 1982.
152. **Back, P. and Ross, K.,** Identification of 3β-hydroxy-5-cholenoic acid in human meconium, *Hoppe-Seyler's Z. Physiol. Chem.,* 354, 83, 1973.
153. **Mitropoulous, K. and Myant, N.,** The formation of lithocholic acid, chenodeoxycholic acid and α and β muricholic acid from cholesterol incubated with rat liver mitochondria, *Biochem. J.,* 103, 472, 1967.
154. **Deleze, G., Paumgartner, G., Karlaganig, G., et al.,** Bile acid pattern in human amniotic fluid; an index of fetal liver maturity, *Eur. J. Clin. Invest.,* 8, 41, 1978.
155. **Mittono, H.,** Studies on the toxicity of bile acids in infants, 2. Bile acid metabolism and individual differences in premature and newborn infants, *Acta Pediatr. Jpn.,* 21, 25, 1979.
156. **Balistreri, W., Suchy, F., Farrel, M., and Heubi, J.,** Pathologic versus physiologic cholestasis: elevated serum concentration of secondary bile acid in the presence of hepatobiliary disease, *J. Pediatr.,* 98, 399, 1951.
157. **Watkins, J., Ingail, D., Szczepanik, P., et al.,** Bile salt metabolism in the newborn. Measurement of pool size by a stable isotope technique, *N. Engl. J. Med.,* 288, 431, 1973.
158. **Hofmann, A.,** The function of bile salts in fat absorption, *Biochem. J.,* 89, 57, 1963.
159. **Danks, D., Tippet, P., Adams, C., and Campbell, D.,** Cerebrohepatorenal syndrome of Zelweger, *J. Pediatr.,* 86, 382, 1975.
160. **Hanson, R., Szczepanik-Van Leeuven, P., Williams, G., et al.,** Defects in bile acid synthesis in Zelweger syndrome, *Science,* 203, 1107, 1979.
161. **Linarelli, L., Williams, C., and Phillips, M.,** Byler's disease. Fatal intrahepatic cholestasis, *J. Pediatr.,* 81, 484, 1972.

Index

INDEX

A

B

C

D

E

F

G

H

I

J

L

M

N

O

P

R

S

T

U

V

W

X

Z